genetics

genetics

VOLUME 2
E–I

Richard Robinson

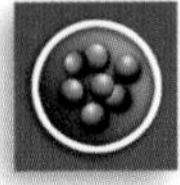

MACMILLAN
REFERENCE
USA™

THOMSON
GALE™

New York • Detroit • San Diego • San Francisco • Cleveland • New Haven, Conn. • Waterville, Maine • London • Munich

Genetics
Richard Robinson

For more information, contact
Macmillan Reference USA
300 Park Avenue South, 9th Floor
New York, NY 10010
Or you can visit our Internet site at
http://www.gale.com

Volume ISBN Numbers
0-02-865607-5 (Volume 1)
0-02-865608-3 (Volume 2)
0-02-865609-1 (Volume 3)
0-02-865610-5 (Volume 4)

LIBRARY OF CONGRESS CATALOGING- IN-PUBLICATION DATA

Genetics / Richard Robinson, editor in chief.
p. ; cm.
Includes bibliographical references and index.
ISBN 0-02-865606-7 (set : hd.)
1. Genetics—Encyclopedias.
[DNLM: 1. Genetics—Encyclopedias—English. 2. Genetic Diseases, Inborn—Encyclopedias—English. 3. Genetic Techniques—Encyclopedias—English. 4. Molecular Biology—Encyclopedias—English. QH 427 G328 2003] I. Robinson, Richard, 1956–
QH427 .G46 2003
576'.03—dc21
2002003560

Printed in Canada
10 9 8 7 6 5 4 3 2 1

For Your Reference

The following section provides a group of diagrams and illustrations applicable to many entries in this encyclopedia. The molecular structures of DNA and RNA are provided in detail in several different formats, to help the student understand the structures and visualize how these molecules combine and interact. The full set of human chromosomes are presented diagrammatically, each of which is shown with a representative few of the hundreds or thousands of genes it carries.

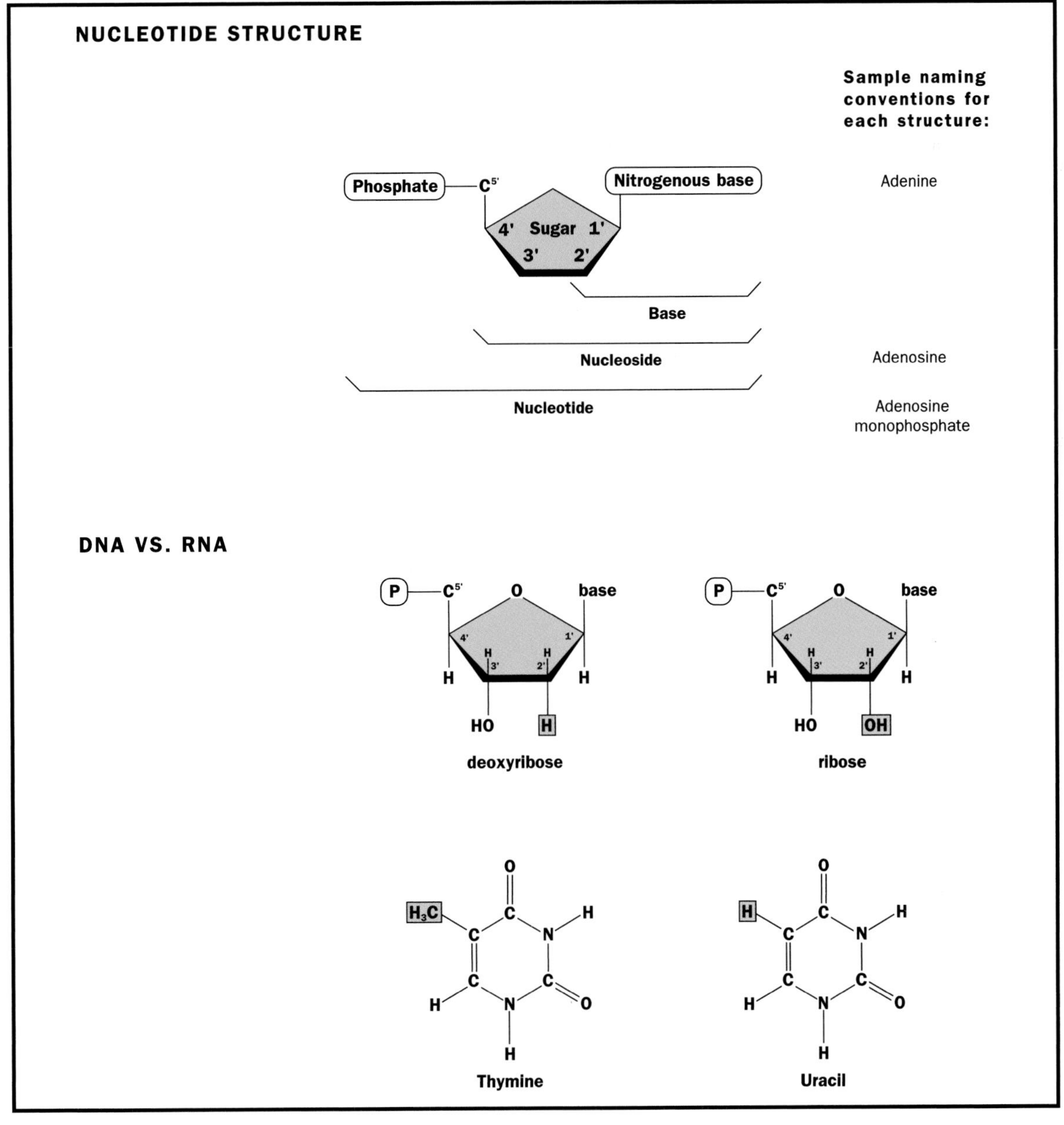

NUCLEOTIDE STRUCTURES

Purine-containing DNA nucleotides

Adenine

Guanine

Pyrimidine-containing DNA nucleotides

Thymine

Cytosine

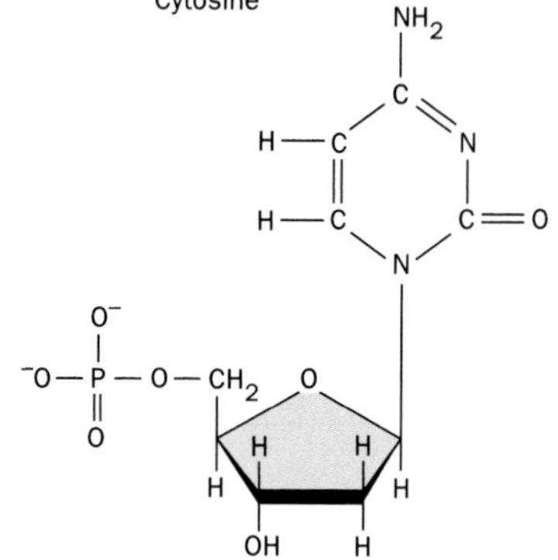

CANONICAL B-DNA DOUBLE HELIX

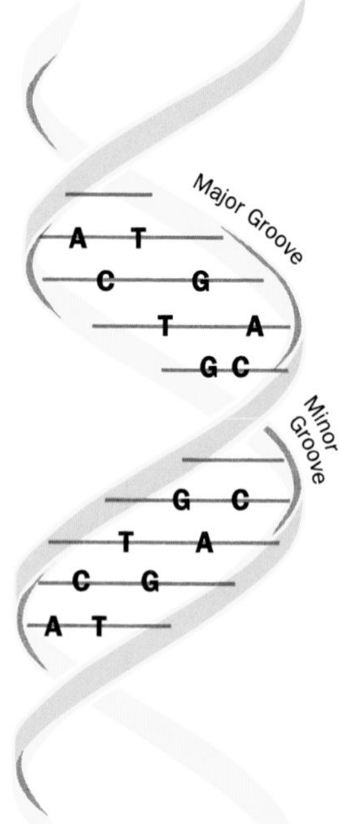

Ribbon model

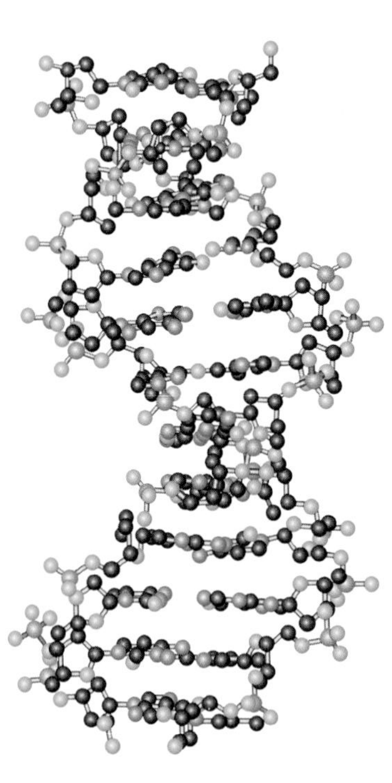
Ball-and-stick model

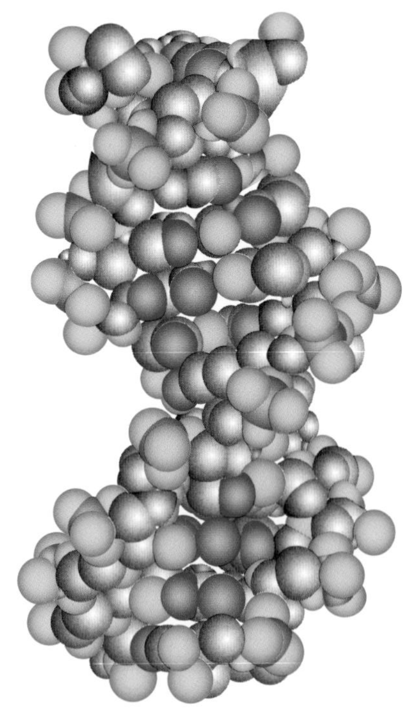
Space-filling model

DNA NUCLEOTIDES PAIR UP ACROSS THE DOUBLE HELIX; THE TWO STRANDS RUN ANTI-PARALLEL

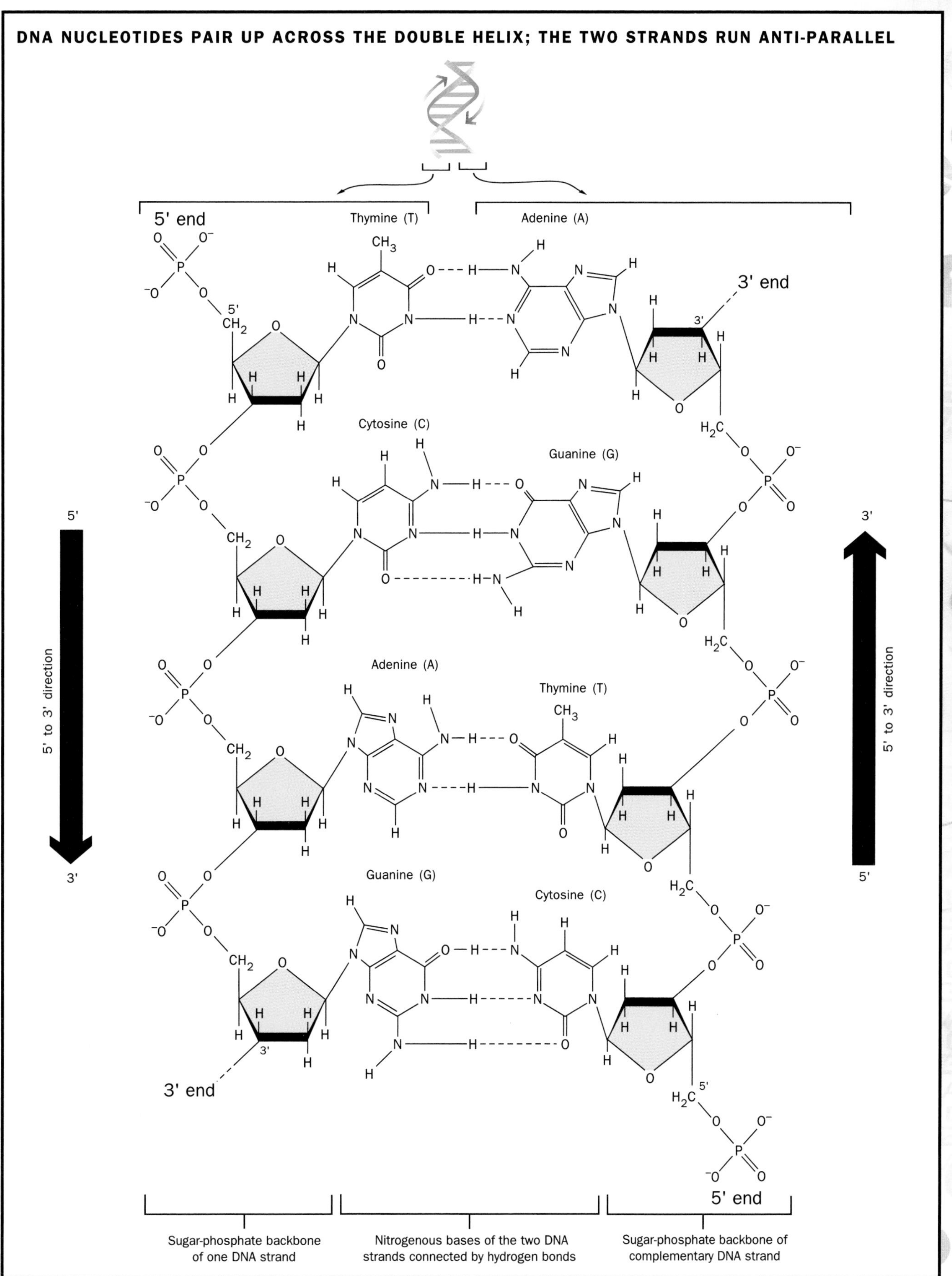

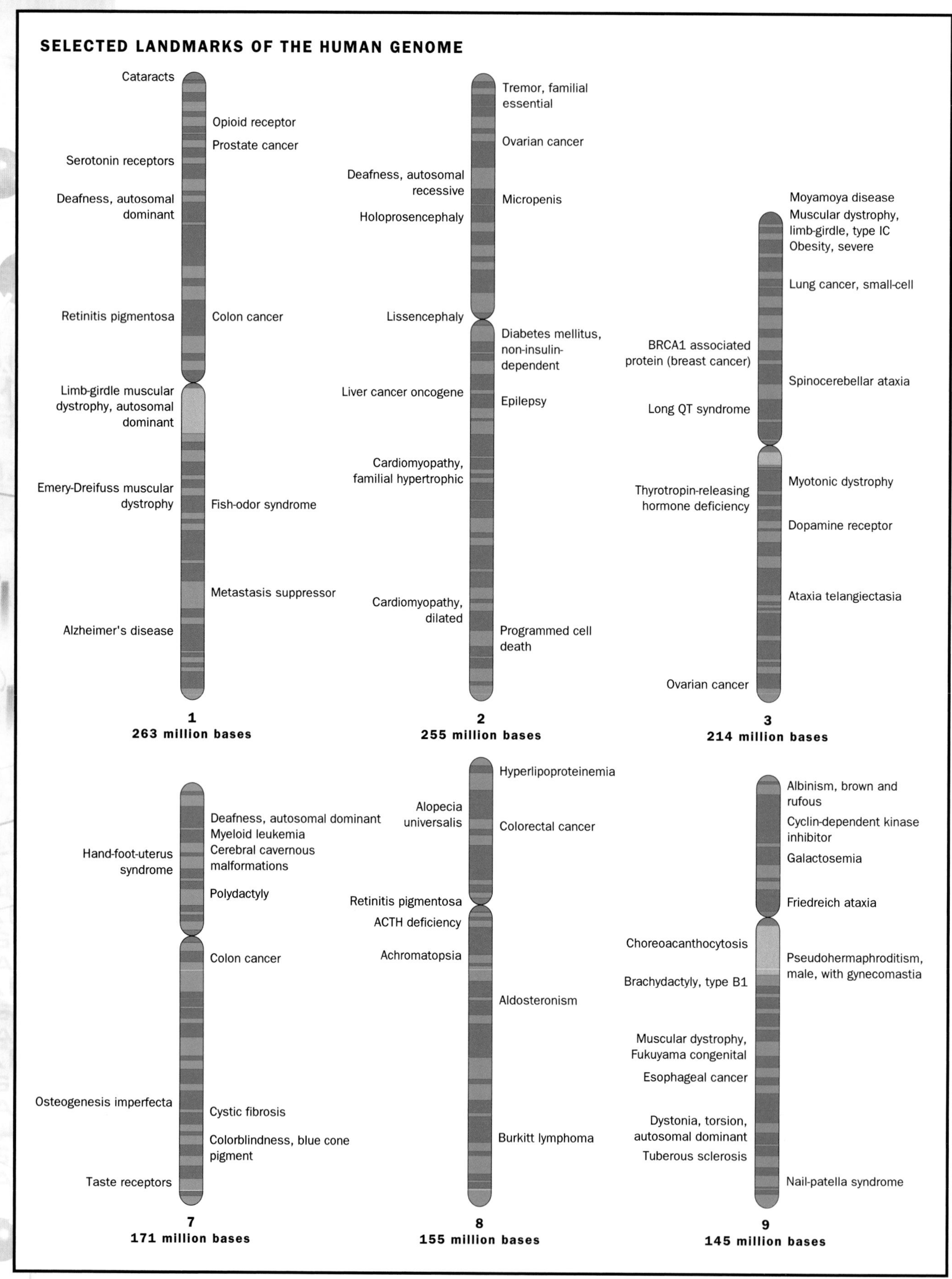
SELECTED LANDMARKS OF THE HUMAN GENOME
Cataracts
Opioid receptor
Prostate cancer
Serotonin receptors
Deafness, autosomal dominant
Retinitis pigmentosa
Colon cancer
Limb-girdle muscular dystrophy, autosomal dominant
Emery-Dreifuss muscular dystrophy
Fish-odor syndrome
Metastasis suppressor
Alzheimer's disease
1
263 million bases
Tremor, familial essential
Ovarian cancer
Deafness, autosomal recessive
Micropenis
Holoprosencephaly
Lissencephaly
Diabetes mellitus, non-insulin-dependent
Liver cancer oncogene
Epilepsy
Cardiomyopathy, familial hypertrophic
Cardiomyopathy, dilated
Programmed cell death
2
255 million bases
Moyamoya disease
Muscular dystrophy, limb-girdle, type IC
Obesity, severe
Lung cancer, small-cell
BRCA1 associated protein (breast cancer)
Spinocerebellar ataxia
Long QT syndrome
Thyrotropin-releasing hormone deficiency
Myotonic dystrophy
Dopamine receptor
Ataxia telangiectasia
Ovarian cancer
3
214 million bases
Deafness, autosomal dominant
Myeloid leukemia
Cerebral cavernous malformations
Hand-foot-uterus syndrome
Polydactyly
Colon cancer
Osteogenesis imperfecta
Cystic fibrosis
Colorblindness, blue cone pigment
Taste receptors
7
171 million bases
Hyperlipoproteinemia
Alopecia universalis
Colorectal cancer
Retinitis pigmentosa
ACTH deficiency
Achromatopsia
Aldosteronism
Burkitt lymphoma
8
155 million bases
Albinism, brown and rufous
Cyclin-dependent kinase inhibitor
Galactosemia
Friedreich ataxia
Choreoacanthocytosis
Pseudohermaphroditism, male, with gynecomastia
Brachydactyly, type B1
Muscular dystrophy, Fukuyama congenital
Esophageal cancer
Dystonia, torsion, autosomal dominant
Tuberous sclerosis
Nail-patella syndrome
9
145 million bases

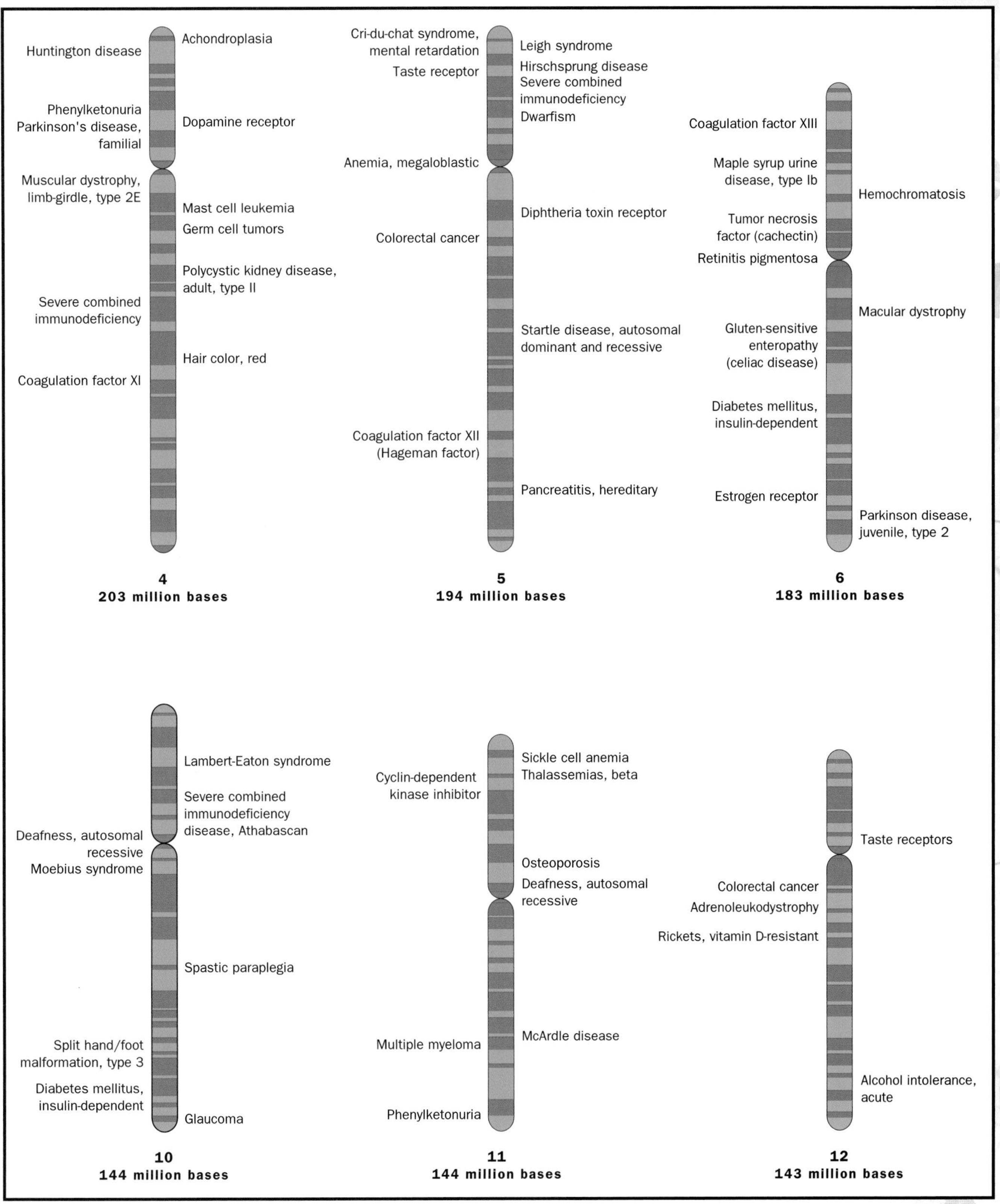

Huntington disease
Achondroplasia
Phenylketonuria
Parkinson's disease, familial
Dopamine receptor
Muscular dystrophy, limb-girdle, type 2E
Mast cell leukemia
Germ cell tumors
Polycystic kidney disease, adult, type II
Severe combined immunodeficiency
Hair color, red
Coagulation factor XI
4
203 million bases
Cri-du-chat syndrome, mental retardation
Taste receptor
Leigh syndrome
Hirschsprung disease
Severe combined immunodeficiency
Dwarfism
Anemia, megaloblastic
Diphtheria toxin receptor
Colorectal cancer
Startle disease, autosomal dominant and recessive
Coagulation factor XII (Hageman factor)
Pancreatitis, hereditary
5
194 million bases
Coagulation factor XIII
Maple syrup urine disease, type Ib
Hemochromatosis
Tumor necrosis factor (cachectin)
Retinitis pigmentosa
Macular dystrophy
Gluten-sensitive enteropathy (celiac disease)
Diabetes mellitus, insulin-dependent
Estrogen receptor
Parkinson disease, juvenile, type 2
6
183 million bases
Lambert-Eaton syndrome
Severe combined immunodeficiency disease, Athabascan
Deafness, autosomal recessive
Moebius syndrome
Spastic paraplegia
Split hand/foot malformation, type 3
Diabetes mellitus, insulin-dependent
Glaucoma
10
144 million bases
Sickle cell anemia
Thalassemias, beta
Cyclin-dependent kinase inhibitor
Osteoporosis
Deafness, autosomal recessive
McArdle disease
Multiple myeloma
Phenylketonuria
11
144 million bases
Taste receptors
Colorectal cancer
Adrenoleukodystrophy
Rickets, vitamin D-resistant
Alcohol intolerance, acute
12
143 million bases

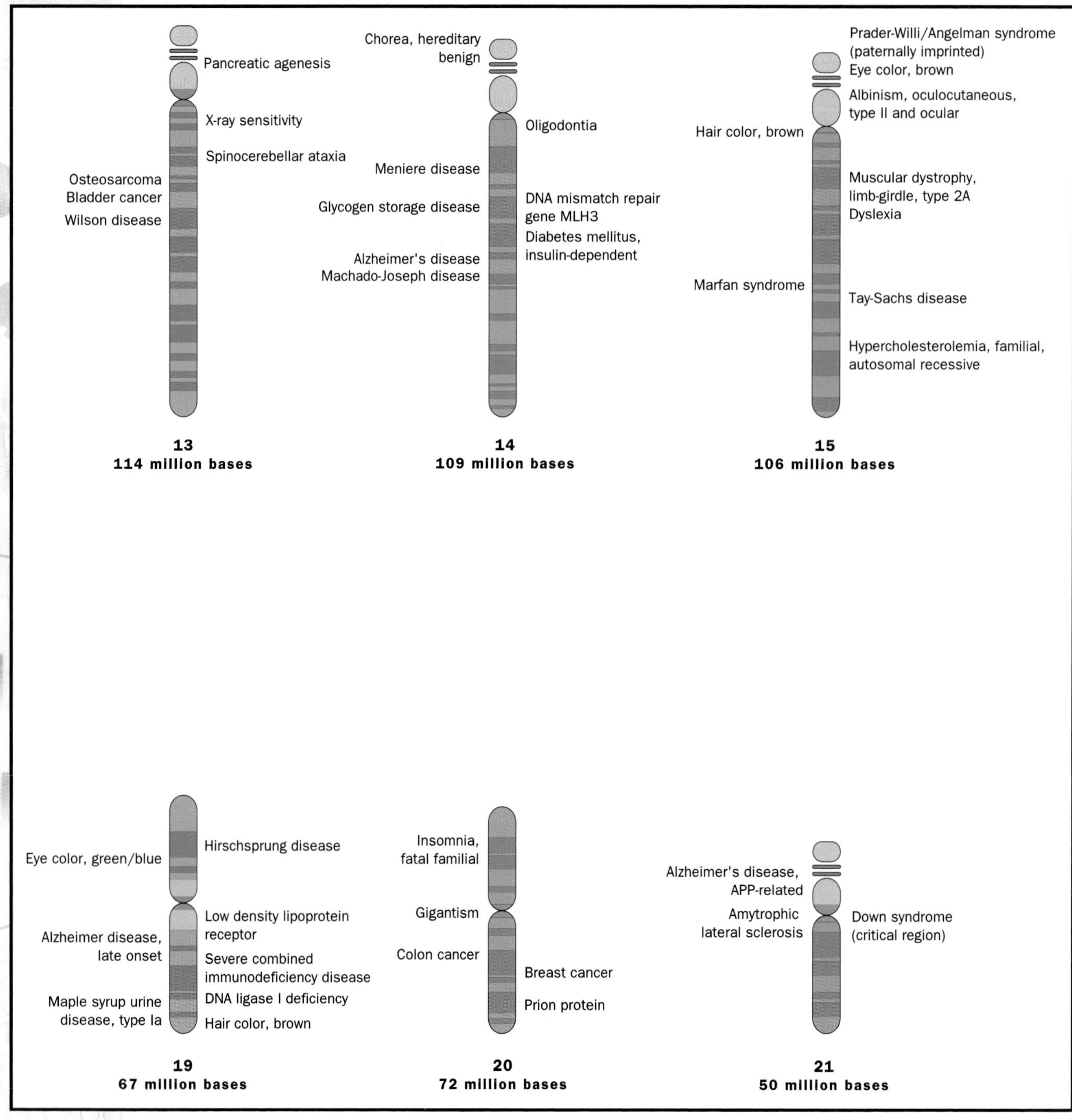
Pancreatic agenesis
X-ray sensitivity
Spinocerebellar ataxia
Osteosarcoma
Bladder cancer
Wilson disease
13
114 million bases
Chorea, hereditary benign
Oligodontia
Meniere disease
Glycogen storage disease
DNA mismatch repair gene MLH3
Diabetes mellitus, insulin-dependent
Alzheimer's disease
Machado-Joseph disease
14
109 million bases
Prader-Willi/Angelman syndrome (paternally imprinted)
Eye color, brown
Albinism, oculocutaneous, type II and ocular
Hair color, brown
Muscular dystrophy, limb-girdle, type 2A
Dyslexia
Marfan syndrome
Tay-Sachs disease
Hypercholesterolemia, familial, autosomal recessive
15
106 million bases
Eye color, green/blue
Hirschsprung disease
Low density lipoprotein receptor
Alzheimer disease, late onset
Severe combined immunodeficiency disease
DNA ligase I deficiency
Maple syrup urine disease, type Ia
Hair color, brown
19
67 million bases
Insomnia, fatal familial
Gigantism
Colon cancer
Breast cancer
Prion protein
20
72 million bases
Alzheimer's disease, APP-related
Amytrophic lateral sclerosis
Down syndrome (critical region)
21
50 million bases

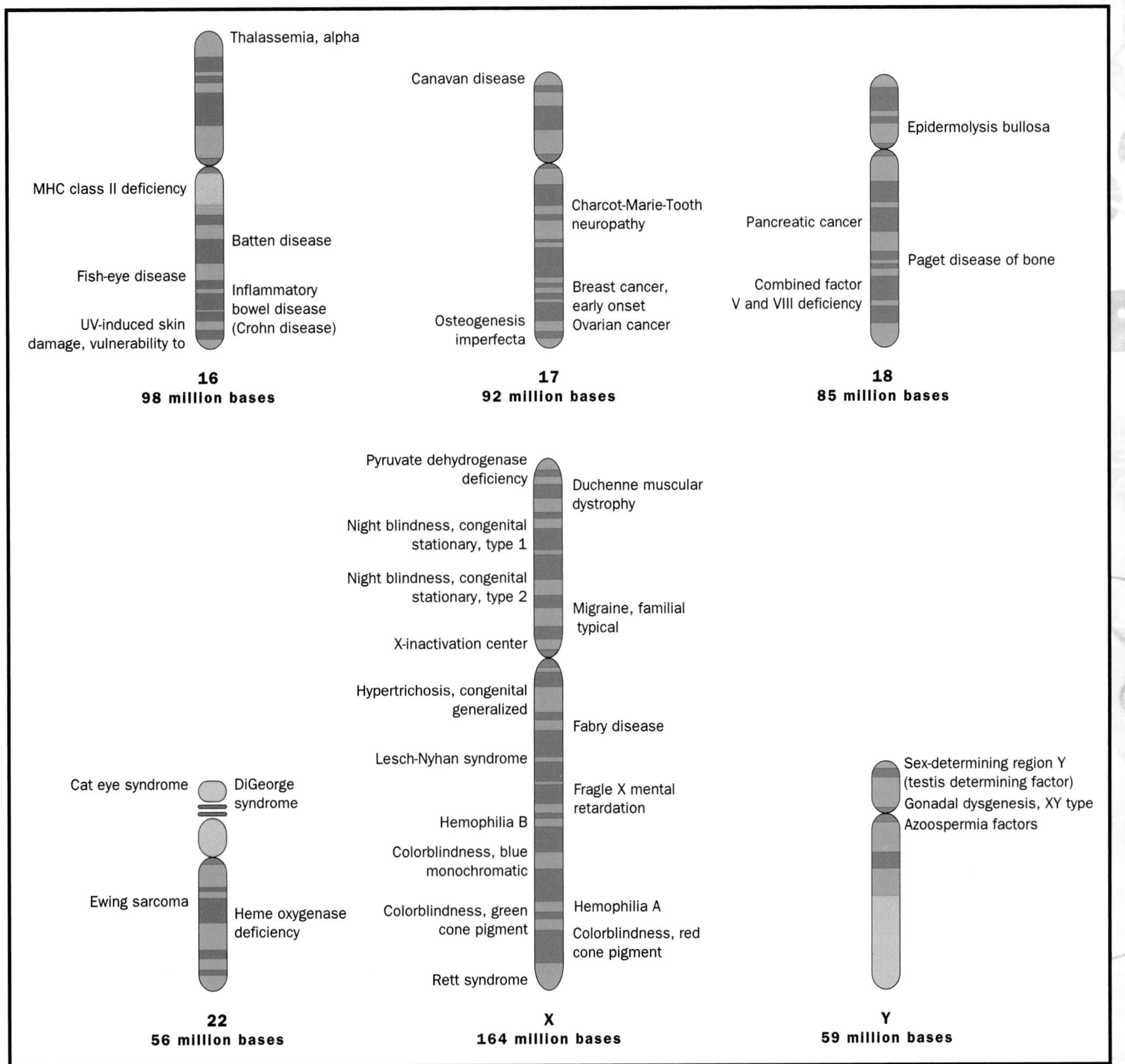
Thalassemia, alpha
MHC class II deficiency
Batten disease
Fish-eye disease
Inflammatory bowel disease (Crohn disease)
UV-induced skin damage, vulnerability to
16
98 million bases
Canavan disease
Charcot-Marie-Tooth neuropathy
Breast cancer, early onset
Ovarian cancer
Osteogenesis imperfecta
17
92 million bases
Epidermolysis bullosa
Pancreatic cancer
Paget disease of bone
Combined factor V and VIII deficiency
18
85 million bases
Cat eye syndrome
DiGeorge syndrome
Ewing sarcoma
Heme oxygenase deficiency
22
56 million bases
Pyruvate dehydrogenase deficiency
Duchenne muscular dystrophy
Night blindness, congenital stationary, type 1
Night blindness, congenital stationary, type 2
Migraine, familial typical
X-inactivation center
Hypertrichosis, congenital generalized
Fabry disease
Lesch-Nyhan syndrome
Fragle X mental retardation
Hemophilia B
Colorblindness, blue monochromatic
Hemophilia A
Colorblindness, green cone pigment
Colorblindness, red cone pigment
Rett syndrome
X
164 million bases
Sex-determining region Y (testis determining factor)
Gonadal dysgenesis, XY type
Azoospermia factors
Y
59 million bases

Contributors

Eric Aamodt
Louisiana State University Health Sciences Center, Shreveport
Gene Expression: Overview of Control

Maria Cristina Abilock
Applied Biosystems
Automated Sequencer
Cycle Sequencing
Protein Sequencing
Sequencing DNA

Ruth Abramson
University of South Carolina School of Medicine
Intelligence
Psychiatric Disorders
Sexual Orientation

Stanley Ambrose
University of Illinois
Population Bottleneck

Allison Ashley-Koch
Duke Center for Human Genetics
Disease, Genetics of
Fragile X Syndrome
Geneticist

David T. Auble
University of Virginia Health System
Transcription

Bruce Barshop
University of California, San Diego
Metabolic Disease

Mark A. Batzer
Louisiana State University
Pseudogenes
Repetitive DNA Elements
Transposable Genetic Elements

Robert C. Baumiller
Xavier University
Reproductive Technology
Reproductive Technology: Ethical Issues

Mary Beckman
Idaho Falls, Idaho
DNA Profiling
HIV

Samuel E. Bennett
Oregon State University Department of Genetics
DNA Repair
Laboratory Technician
Molecular Biologist

Andrea Bernasconi
Cambridge University, U.K.
Multiple Alleles
Nondisjunction

C. William Birky, Jr.
University of Arizona
Inheritance, Extranuclear

Joanna Bloom
New York University Medical Center
Cell Cycle

Deborah Blum
University of Wisconsin, Madison
Science Writer

Bruce Blumberg
University of California, Irvine
Hormonal Regulation

Suzanne Bradshaw
University of Cincinnati
Transgenic Animals
Yeast

Carolyn J. Brown
University of British Columbia
Mosaicism

Michael J. Bumbulis
Baldwin-Wallace College
Blotting

Michael Buratovich
Spring Arbor College
Operon

Elof Carlson
The State Universtiy of New York, Stony Brook
Chromosomal Theory of Inheritance, History
Gene
Muller, Hermann
Polyploidy
Selection

Regina Carney
Duke University
College Professor

Shu G. Chen
Case Western Reserve University
Prion

Gwen V. Childs
University of Arkansas for Medical Sciences
In situ Hybridization

Cindy T. Christen
Iowa State University
Technical Writer

Patricia L. Clark
University of Notre Dame
Chaperones

Steven S. Clark
University of Wisconsin
Oncogenes

Nathaniel Comfort
George Washington University
McClintock, Barbara

P. Michael Conneally
Indiana University School of Medicine
Blood Type
Epistasis
Heterozygote Advantage

Howard Cooke
Western General Hospital: MRC Human Genetics Unit
Chromosomes, Artificial

Denise E. Costich
Boyce Thompson Institute
Maize

Terri Creeden
March of Dimes
Birth Defects

Kenneth W. Culver
Novartis Pharmaceuticals Corporation
Genomics
Genomics Industry
Pharmaceutical Scientist

Mary B. Daly
Fox Chase Cancer Center
Breast Cancer

Pieter de Haseth
Case Western Reserve University
Transcription

Rob DeSalle
American Museum of Natural History
Conservation Geneticist
Conservation Biology: Genetic Approaches

Elizabeth A. De Stasio
Lawerence University
Cloning Organisms

Danielle M. Dick
Indiana University
Behavior

Michael Dietrich
Dartmouth College
Nature of the Gene, History

Christine M. Disteche
University of Washington
X Chromosome

Gregory Michael Dorr
University of Alabama
Eugenics

Jennie Dusheck
Santa Cruz, California
Population Genetics

Susanne D. Dyby
U.S. Department of Agriculture: Center for Medical, Agricultural, and Veterinary Entomology
Classical Hybrid Genetics
Mendelian Genetics
Pleiotropy

Barbara Emberson Soots
Folsom, California
Agricultural Biotechnology

Susan E. Estabrooks
Duke Center for Human Genetics
Fertilization
Genetic Counselor
Genetic Testing

Stephen V. Faraone
Harvard Medical School
Attention Deficit Hyperactivity Disorder

Gerald L. Feldman
Wayne State University Center for Molecular Medicine and Genetics
Down Syndrome

Linnea Fletcher
Bio-Link South Central Regional Coordinater, Austin Community College
Educator
Gel Electrophoresis
Marker Systems
Plasmid

Michael Fossel
Executive Director, American Aging Association
Accelerated Aging: Progeria

Carol L. Freund
National Institute of Health: Warren G. Magnuson Clinical Center
Genetic Testing: Ethical Issues

Joseph G. Gall
Carnegie Institution
Centromere

Darrell R. Galloway
The Ohio State University
DNA Vaccines

Pierluigi Gambetti
Case Western Reserve University
Prion

Robert F. Garry
Tulane University School of Medicine
Retrovirus
Virus

Perry Craig Gaskell, Jr.
Duke Center for Human Genetics
Alzheimer's Disease

Theresa Geiman
National Institute of Health: Laboratory of Receptor Biology and Gene Expression
Methylation

Seth G. N. Grant
University of Edinburgh
Embryonic Stem Cells
Gene Targeting
Rodent Models

Roy A. Gravel
University of Calgary
Tay-Sachs Disease

Nancy S. Green
March of Dimes
Birth Defects

Wayne W. Grody
UCLA School of Medicine
Cystic Fibrosis

Charles J. Grossman
Xavier University
Reproductive Technology
Reproductive Technology: Ethical Issues

Cynthia Guidi
University of Massachusetts Medical School
Chromosome, Eukaryotic

Patrick G. Guilfoile
Bemidji State University
DNA Footprinting
Microbiologist
Recombinant DNA
Restriction Enzymes

Richard Haas
University of California Medical Center
Mitochondrial Diseases

William J. Hagan
College of St. Rose
Evolution, Molecular

Jonathan L. Haines
Vanderbilt University Medical Center
Complex Traits
Human Disease Genes, Identification of Mapping
McKusick, Victor

Michael A. Hauser
Duke Center for Human Genetics
DNA Microarrays
Gene Therapy

Leonard Hayflick
University of California
Telomere

Shaun Heaphy
University of Leicester, U.K.
Viroids and Virusoids

John Heddle
York University
Mutagenesis
Mutation
Mutation Rate

William Horton
Shriners Hospital for Children
Growth Disorders

Brian Hoyle
Square Rainbow Limited
Overlapping Genes

Anthony N. Imbalzano
University of Massachusetts Medical School
Chromosome, Eukaryotic

Nandita Jha
University of California, Los Angeles
Triplet Repeat Disease

John R. Jungck
Beloit College
Gene Families

Richard Karp
Department of Biological Sciences, University of Cincinnati
Transplantation

David H. Kass
Eastern Michigan University
Pseudogenes
Transposable Genetic Elements

Michael L. Kochman
University of Pennsylvania Cancer Center
Colon Cancer

Bill Kraus
Duke University Medical Center
Cardiovascular Disease

Steven Krawiec
Lehigh University
Genome

Mark A. Labow
Novartis Pharmaceuticals Corporation
Genomics
Genomics Industry
Pharmaceutical Scientist

Ricki Lewis
McGraw-Hill Higher Education; The Scientist
Bioremediation
Biotechnology: Ethical Issues
Cloning: Ethical Issues

Genetically Modified Foods
Plant Genetic Engineer
Prenatal Diagnosis
Transgenic Organisms: Ethical Issues

Lasse Lindahl
University of Maryland, Baltimore
Ribozyme
RNA

David E. Loren
University of Pennsylvania School of Medicine
Colon Cancer

Dennis N. Luck
Oberlin College
Biotechnology

Jeanne M. Lusher
Wayne State University School of Medicine; Children's Hospital of Michigan
Hemophilia

Kamrin T. MacKnight
Medlen, Carroll, LLP: Patent, Trademark and Copyright Attorneys
Attorney
Legal Issues
Patenting Genes
Privacy

Jarema Malicki
Harvard Medical School
Zebrafish

Eden R. Martin
Duke Center for Human Genetics
Founder Effect
Inbreeding

William Mattox
University of Texas/Anderson Cancer Center
Sex Determination

Brent McCown
University of Wisconsin
Transgenic Plants

Elizabeth C. Melvin
Duke Center for Human Genetics
Gene Therapy: Ethical Issues
Pedigree

Ralph R. Meyer
University of Cincinnati
Biotechnology and Genetic Engineering, History of
Chromosome, Eukaryotic
Genetic Code
Human Genome Project

Kenneth V. Mills
College of the Holy Cross
Post-translational Control

Jason H. Moore
Vanderbilt University Medical School
Quantitative Traits
Statistical Geneticist
Statistics

Dale Mosbaugh
Oregon State University: Center for Gene Research and Biotechnology
DNA Repair
Laboratory Technician
Molecular Biologist

Paul J. Muhlrad
University of Arizona
Alternative Splicing
Apoptosis
Arabidopsis thaliana
Cloning Genes
Combinatorial Chemistry
Fruit Fly: *Drosophila*
Internet
Model Organisms
Pharmacogenetics and Pharmacogenomics
Polymerase Chain Reaction

Cynthia A. Needham
Boston University School of Medicine
Archaea
Conjugation
Transgenic Microorganisms

R. John Nelson
University of Victoria
Balanced Polymorphism
Gene Flow
Genetic Drift
Polymorphisms
Speciation

Carol S. Newlon
University of Medicine and Dentistry of New Jersey
Replication

Sophia A. Oliveria
Duke University Center for Human Genetics
Gene Discovery

Richard A. Padgett
Lerner Research Institute
RNA Processing

Michele Pagano
New York University Medical Center
Cell Cycle

Rebecca Pearlman
Johns Hopkins University
Probability

Fred W. Perrino
Wake Forest University School of Medicine
DNA Polymerases
Nucleases
Nucleotide

David Pimentel
Cornell University: College of Agriculture and Life Sciences
Biopesticides

Toni I. Pollin
University of Maryland School of Medicine
Diabetes

Sandra G. Porter
Geospiza, Inc.
Homology

Eric A. Postel
Duke University Medical Center
Color Vision
Eye Color

Prema Rapuri
Creighton University
HPLC: High-Performance Liquid Chromatography

Anthony J. Recupero
Gene Logic
Bioinformatics
Biotechnology Entrepreneur
Proteomics

Diane C. Rein
BioComm Consultants
Clinical Geneticist
Nucleus
Roundworm: *Caenorhabditis elegans*
Severe Combined Immune Deficiency

Jacqueline Bebout Rimmler
Duke Center for Human Genetics
Chromosomal Aberrations

Keith Robertson
Epigenetic Gene Regulation and Cancer Institute
Methylation

Richard Robinson
Tucson, Arizona
Androgen Insensitivity Syndrome
Antisense Nucleotides
Cell, Eukaryotic
Crick, Francis
Delbrück, Max
Development, Genetic Control of
DNA Structure and Function, History
Eubacteria
Evolution of Genes
Hardy-Weinberg Equilibrium
High-Throughput Screening
Immune System Genetics
Imprinting
Inheritance Patterns
Mass Spectrometry
Mendel, Gregor
Molecular Anthropology
Morgan, Thomas Hunt
Mutagen
Purification of DNA
RNA Interferance
RNA Polymerases
Transcription Factors
Twins
Watson, James

Richard J. Rose
Indiana University
Behavior

Howard C. Rosenbaum
Science Resource Center, Wildlife Conservation Society
Conservation Geneticist
Conservation Biology: Genetic Approaches

Astrid M. Roy-Engel
Tulane University Health Sciences Center
Repetitive DNA Elements

Joellen M. Schildkraut
Duke University Medical Center
Public Health, Genetic Techniques in

Silke Schmidt
Duke Center for Human Genetics
Meiosis
Mitosis

David A. Scicchitano
New York University
Ames Test
Carcinogens

William K. Scott
Duke Center for Human Genetics
Aging and Life Span
Epidemiologist
Gene and Environment

Gerry Shaw
MacKnight Brain Institute of the University of Flordia
Signal Transduction

Alan R. Shuldiner
University of Maryland School of Medicine
Diabetes

Richard R. Sinden
Institute for Biosciences and Technology: Center for Genome Research
DNA

Paul K. Small
Eureka College
Antibiotic Resistance
Proteins
Reading Frame

Marcy C. Speer
Duke Center for Human Genetics
Crossing Over
Founder Effect
Inbreeding
Individual Genetic Variation
Linkage and Recombination

Jeffrey M. Stajich
Duke Center for Human Genetics
Muscular Dystrophy

Judith E. Stenger
Duke Center for Human Genetics
Computational Biologist
Information Systems Manager

Frank H. Stephenson
Applied Biosystems
Automated Sequencer
Cycle Sequencing
Protein Sequencing
Sequencing DNA

Gregory Stewart
State University of West Georgia
Transduction
Transformation

Douglas J. C. Strathdee
University of Edinburgh
Embryonic Stem Cells
Gene Targeting
Rodent Models

Jeremy Sugarman
Duke University Department of Medicine
Genetic Testing: Ethical Issues

Caroline M. Tanner
Parkinson's Institute
Twins

Alice Telesnitsky
University of Michigan
Reverse Transcriptase

Daniel J. Tomso
National Institute of Environmental Health Sciences
DNA Libraries
Escherichia coli
Genetics

Angela Trepanier
Wayne State University Genetic Counseling Graduate Program
Down Syndrome

Peter A. Underhill
Stanford University
Y Chromosome

Joelle van der Walt
Duke University Center for Human Genetics
Genotype and Phenotype

Jeffery M. Vance
Duke University Center for Human Genetics
Gene Discovery
Genomic Medicine
Genotype and Phenotype
Sanger, Fred

Gail Vance
Indiana University
Chromosomal Banding

Jeffrey T. Villinski
University of Texas/MD Anderson Cancer Center
Sex Determination

Sue Wallace
Santa Rosa, California
Hemoglobinopathies

Giles Watts
Children's Hospital Boston
Cancer
Tumor Suppressor Genes

Kirk Wilhelmsen
Ernest Gallo Clinic & Research Center
Addiction

Michelle P. Winn
Duke University Medical Center
Physician Scientist

Chantelle Wolpert
Duke University Center for Human Genetics
Genetic Counseling
Genetic Discrimination
Nomenclature
Population Screening

Harry H. Wright
University of South Carolina School of Medicine
Intelligence
Psychiatric Disorders
Sexual Orientation

Janice Zengel
University of Maryland, Baltimore
Ribosome
Translation

Stephan Zweifel
Carleton College
Mitochondrial Genome

Table of Contents

VOLUME 3

VOLUME 4

genetics

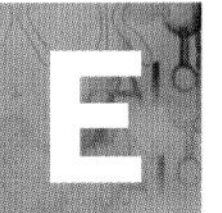

Educator

An educator is a person who systematically works to improve another's understanding of a topic. The role of educator encompasses both those who teach in classrooms and the more informal educators who, for example, work in zoos, museums, and recreational areas. The work of educators varies depending on the institution that employs them and the age or grade level of the people he or she teaches.

K-12 Educator

Educators who teach children from kindergarten to 12th grade have many responsibilities. They must prepare lesson plans for all of their classes, tailoring their plans to the instructional backgrounds and abilities of the children in each class. Class sizes may range from twenty to thirty-five students, and a teacher may be assigned as many as six different classes. Thus, in addition to teaching, educators at this level must be skilled in classroom management, for they must often deal with behavioral problems and even act as surrogate parents for younger children.

Class preparation includes drafting lesson plans, writing and grading tests, evaluating homework, and sometimes purchasing classroom supplies with their own money. Teachers are held accountable for the success of their students, and are required to participate in professional development programs (courses or seminars) throughout their careers. Finally, they must maintain communication with the parents, participate in numerous educational and administrative activities, and sponsor student organizations such as the student council, sports teams, and school clubs.

Two-Year College Faculty

Teachers at two-year colleges are responsible for preparing lessons for two to four different classes, and they usually teach between four and five courses each semester. They, too, have mandatory professional development. In addition to their teaching duties they must serve on school committees dealing with such issues as hiring or technology. They are responsible to their department heads, deans, and other higher administrators, but they generally do not have to deal with parents.

Class sizes vary from twelve to thirty students, depending upon the subject and type of course being taught. Lecture courses, for example, tend to have higher enrollments than laboratory courses. For lab courses, educators sometimes may have to do their own set-up, if the department does not employ a laboratory technician. Some two-year schools hire their faculty with the possibility of tenure, whereas others offer only short-term contracts, typically for one to three years. Higher positions are in the administration.

Four-Year College Faculty

Educators at a four-year college must conduct publishable research, direct a laboratory, and mentor graduate students, in addition to teaching one to four undergraduate courses each semester or year. On first joining a faculty, educators typically receive the rank of assistant professor. On contract renewal, promotion to associate professor is the norm, with the rank of full professor being accorded upon achieving tenure.

Graduate student teaching assistants (TAs) are often available to assist the educator. TAs usually do all of the grading and supervise the lab classes. Four-year college educators also usually have the assistance of a laboratory technician.

To receive tenure, educators at this level must pass a review by a committee of their peers in their department, in which their success in research, publications, and teaching is evaluated. Most departments require their faculty to serve on one or more internal committees, and service on college- or university-wide committees may also be required, particularly of junior (non-tenured) faculty.

Educational Requirements

Educators at the K-12 level must have, minimally, a bachelor's degree in education. Often, however, they earn a degree in an academic subject and take master's level courses in education. Public school educators must pass a state-administered test, after which they receive certification allowing them to teach. Private schools may not require certification.

Educators at two- and four-year colleges usually have degrees in their area of specialization, but state certification is not required. Some two-year institutions may require master's or doctoral training in education in addition to a degree in an academic subject, but educators at the college level usually gain their teaching experience as teaching assistants. In some cases, an academic degree is not required to teach in a two-program, if the teacher has had enough experience working in the field for which students are to be trained. Four-year colleges expect their educators to hold a graduate degree, often a doctorate, and usually require a record of published research.

Compensation and Other Benefits

Pay scales for educators vary depending upon the grade level taught and the institution at which they teach. At the K-12 level, teachers can earn from \$21,000 to \$80,000, depending on years of experience and the school district in which they teach. At two-year colleges, starting salaries average about \$35,000, but the upper salary limit is about the same as for K-12. Four-year colleges and universities typically offer starting salaries of about \$40,000, and can rise to more than $100,000 for full professors whose research and publications have earned them renown in their fields.

Educators typically are free of teaching duties during the summer months, but those who work at the college level are expected to conduct research and publish even when classes are not in session. College-level teachers are also often eligible for sabbaticals, during which teaching duties are not required, but they are expected to use this time for a research project. A love of teaching and learning is needed for educators at all levels, but each student age group makes different demands on the teacher. SEE ALSO COLLEGE PROFESSOR.

Linnea Fletcher

Embryonic Stem Cells

Mammalian embryonic stem (ES) cells have the special property of being able to differentiate into virtually every cell type. Because ES cells can be genetically manipulated **in vitro** and can be transplanted into embryos and adults, they are a powerful tool in biological experiments and hold promise for future medical therapies.

in vitro "in glass"; in lab apparatus, rather than within a living organism

The ability to differentiate into all cell types, a property known as pluripotency, arises from the fact that ES cells are isolated from in vitro outgrowths of early stage embryos (in the mouse, at three and one-half days, at the **blastocyst** stage). These outgrowths are cultured in specialized conditions—often in the presence of support cells, called feeder cells, which do not proliferate, and specific growth factors. The ES cells proliferate rapidly in culture, and clonal (identical) populations can readily be initiated from single cells.

blastocyst early stage of embryonic development

Until 1998 the only mammalian embryonic stem cells isolated were those from the mouse. The first human embryonic stem cells were isolated in 1998 from embryos created through in vitro fertilization. In the United States, government funding for research on human ES cells was restricted in 2002 to a small set of existing human ES cell lines. Only privately funded research remained able to use other sources of cells.

In the following years, most publicized research on human ES cells will concentrate on manipulating cells in culture, attempting to understand the signals that cells require to proliferate and differentiate into various adult cell types, such as nerve or muscle cells.

In 2001 there was a highly publicized report of human cloning using human ES cells, but the resulting embryos did not progress past the twelve-cell stage, and there was no reliable evidence that the ES cells ever contributed to the development of the embryos. Nonetheless, most researchers thought obstacles would likely be overcome, and that the use of ES cells for therapy or cloning would become technically possible.

Mouse ES cell research is much further advanced. Research in the mouse has included injection of ES cells into blastocysts, an early stage in embryo development. Once injected, ES cells fully participate in embryonic development and form part of every tissue in the embryo. The resulting mouse is called a chimera, since it usually contains a mixture of cells derived from

Figure 1. Mouse embryonic stem cells in culture.

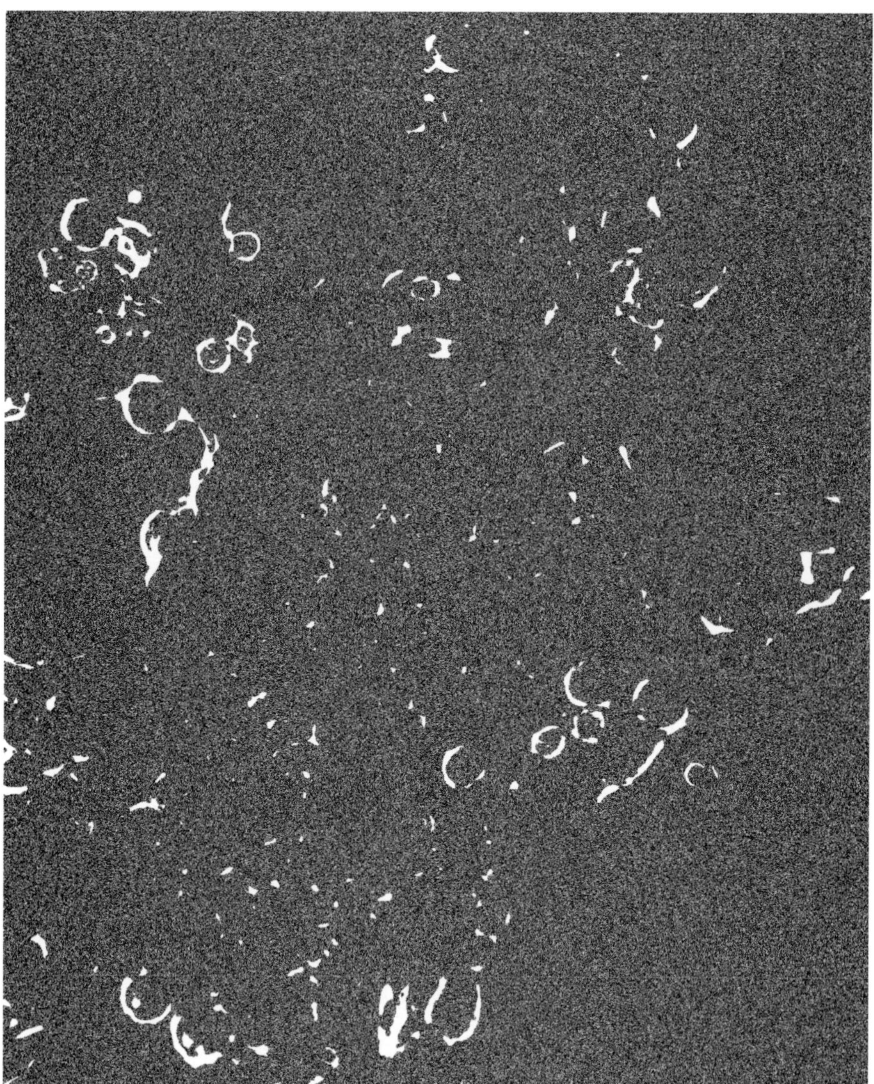

both the ES cells and the host blastocyst. An important feature of a chimera is that its ES cells will contribute to forming the germ cells, which eventually form the gametes of the animals. Consequently, mating of the chimera with recipient females allows the (ES cell derived) sperm to deliver their genetic material into an egg, and subsequent offspring will carry genes from the ES cell. Thus genetic manipulations made in ES cells in culture can be transmitted through chimeras to intact animals.

As with other cell lines grown in vitro, it is possible to add new genes, termed "transgenes," or to modify existing genes in ES cells using a technique known as gene targeting. Gene targeting involves using specialized DNA **vectors** that share substantial DNA sequence similarity with the gene that needs to be modified. When introduced into cultured ES cells, the gene-targeting vector undergoes "homologous recombination," a process in which the vector is integrated into the existing chromosomal gene and leaves the ES cells genetically modified.

vectors carriers

Removing a gene product is often referred to as creating a "knockout," since it prevents a protein from being expressed. Modifying a specific gene

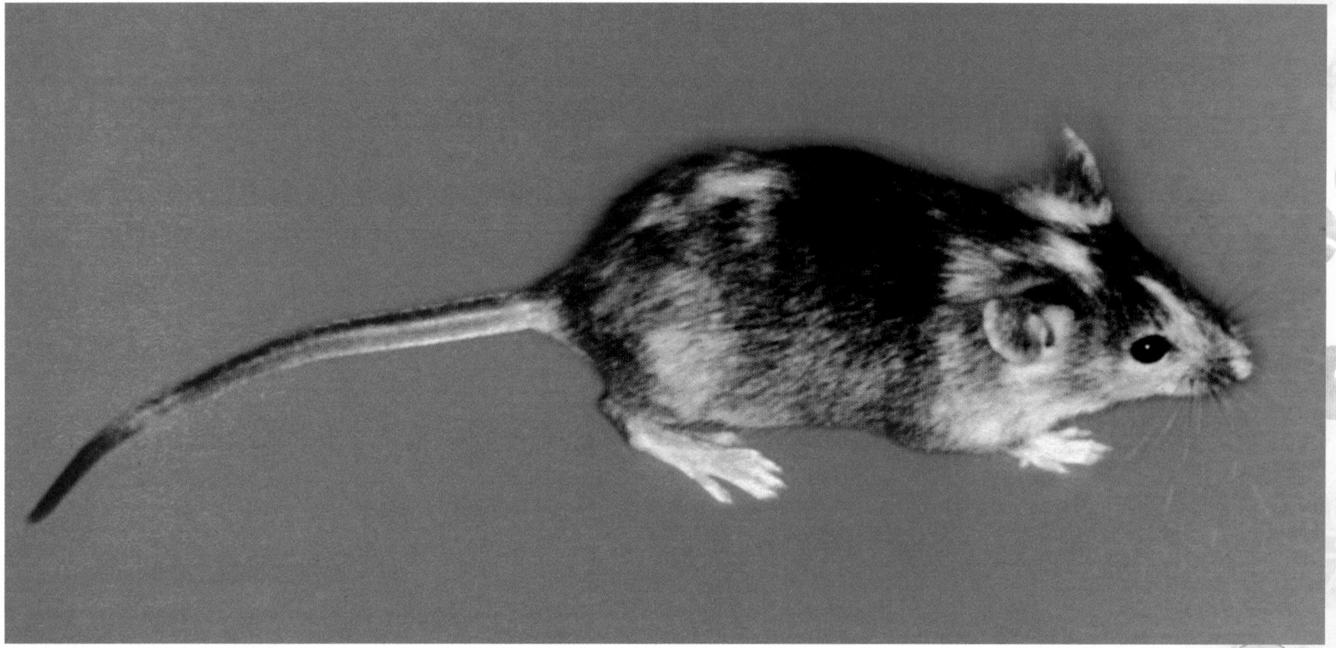

in the genome to include a single DNA nucleotide change (mutation) or incorporating a genetic marker (such as a gene that produces a colored protein that can be seen with a microscope) is accomplished by a similar method and is often referred to as creating a "knock-in"—since a piece of DNA is inserted into a specific part of the genome.

Figure 2. A typical chimera. ES cell-derived tissue is identified as light patches in the mouse coat.

The genetically modified ES cells can be used to create mouse chimeras, whose offspring will carry the genetic modifications into the ES cells. These mice can be identified by analysis of DNA made from a small tissue sample, usually taken from the mouse's tail. The genomic DNA can be analyzed by polymerase chain reaction (**PCR**) or **Southern blot** to identify the mice carrying the genetic change. These offspring will be **heterozygous** for the introduced mutation and can be interbred to generate a second generation, some of which will be **homozygous** for the introduced mutation. The homozygous mice will lack both copies of the normal gene—any differences to normal development displayed by the mice as a result of losing the gene of interest can be analyzed. This, in turn, can provide valuable clues to the function of the gene and its normal role in mouse physiology.

PCR polymerase chain reaction, used to amplify DNA

Southern blot a technique for separating DNA fragments by electrophoresis and then identifying a target fragment with a DNA probe

heterozygous characterized by possession of two different forms (alleles) of a particular gene

homozygous containing two identical copies of a particular gene

This experimental approach to genetic manipulation of mice, which was pioneered in the 1980s, has been used to modify thousands of mouse genes and has played a significant role in understanding many physiological and pathological processes. For example, mutations in the p53 gene show its involvement in cancer, and mutations in the psd–95 gene show involvement in learning and memory. Variations on these methods are also used to model human disease mutations in mice and to test drug therapies.

ES cells also have potential for treating human diseases through "cell-based" therapy. In this context, the ES cells (or their differentiated derivative cells) are transplanted into patients whose own cells are defective or degenerated. This strategy could have applications for diseases such as Parkinson's disease (which affects brain cells), diabetes (pancreatic cells) and heart disease (which affects heart muscle cells). An important clinical issue

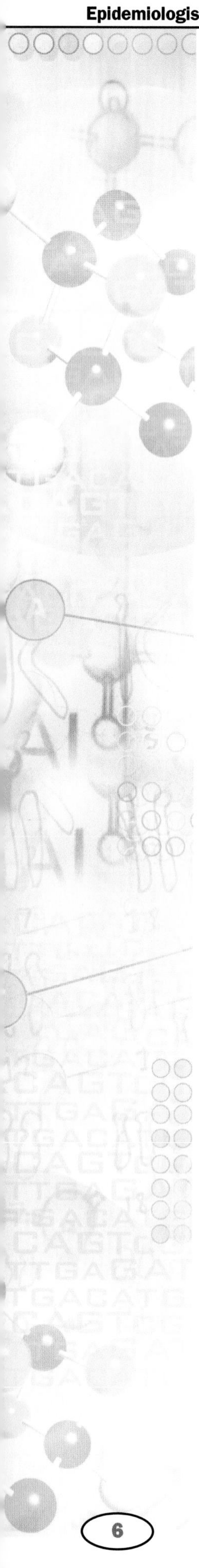

at this point will be whether ES cells not derived from the patient will be rejected by the patient's immune system. If so, one strategy for dealing with this problem would be to use a patient's own cells to create an embryo by nuclear transfer, from which ES cells compatible with that patient could then be derived.

The use of human blastocysts in both research and therapy remains controversial because it is necessary to destroy the blastocyst to generate an ES cell line from it. As a result, work with human embryos is governed by strict regulations in many countries. SEE ALSO GENE TARGETING; MARKER SYSTEMS; REPRODUCTIVE TECHNOLOGY; RODENT MODELS; TRANSGENIC ANIMALS.

Seth G. N. Grant and Douglas J. C. Strathdee

Bibliography

Donehower, L. A., et al. "Mice Deficient for p53 Are Developmentally Normal but Susceptible to Spontaneous Tumors." *Nature* 356 (1992): 215–221.

Holland, Suzanne, Karen Lebacqz, and Laurie Zoloth, eds. *The Human Embryonic Stem Cell Debate: Science, Ethics, and Public Policy.* New York: MIT Press, 2001.

Juengst, Eric, and Michael Fossel. "The Ethics of Embryonic Stem Cells—Now and Forever, Cells without End." *Journal of the American Medical Association* 284 (2000): 3180–3184.

Migaud, M., "Enhanced Long-Term Potentiation and Impaired Learning in Mice with Mutant Postsynaptic Density-95 Protein." *Nature* 396 (1998): 433–439.

Thomson, James, et al. "Embryonic Stem Cell Lines Derived from Human Blastocysts." *Science* 282 (1998): 1145–1147.

Turksen, Kursad, ed. *Embryonic Stem Cells: Methods and Protocols.* Totowa, NJ: Humana Press, 2002.

Internet Resources

Embryonic Stem Cells. University of Wisconsin-Madison. <http://www.news.wisc.edu/packages/stemcells/>.

Stem Cell Primer. National Institutes of Health. <http://www.nih.gov/news/stemcell/primer.htm>.

Epidemiologist

Epidemiologists are scientists that study the factors influencing the health status of populations. These populations may be defined by geography (such as the residents of a particular city), occupation (such as members of the armed forces), or any other common trait (such as age, race, or sex). Epidemiologists look for trends in measures of the health of the population, such as the average life span, the leading causes of death, and the number of cases of a disease that are found in the population. To determine what causes certain trends or health problems, epidemiologists collect large amounts of data about individuals in the population. They analyze these data to determine who is sick, when they got sick, and what factors the sick people have in common. The process is similar to that which investigators use to search for clues to solve a crime. Thus, epidemiologists are often described as "disease detectives."

Career Requirements, Employment, and Compensation

A career in epidemiology generally requires a master's or doctoral degree in public health. Epidemiologists work in many different settings: universi-

ties, industry, government, and nongovernmental research or health organizations. Those with master's degrees generally start as project officers or staff members, coordinating data collection and analysis for health studies and as they grow in experience they will advance within their organizations. The doctoral degree often leads to academic careers on university faculty and leadership positions in other research organizations and industry. Physicians who earn master's or doctoral degrees in public health also often fill these positions. Epidemiologists who receive additional training (usually as doctoral or postdoctoral students) in human and statistical genetics are often called "genetic epidemiologists," in recognition of their specialty within epidemiology. Compensation varies widely, depending on the level of education, employment setting, and experience. In 2001 the starting salary for a new graduate with a master's degree and no previous work experience might be in the $30,000 to $40,000 range; a Ph.D. or M.D./M.P.H. with several years' experience might earn over $100,000 in industry.

Many Roles, Many Rewards

Epidemiologists interested in the influence of genetic factors on health may play several types of roles on a research project, including assisting in the overall design of the study, developing instruments to collect nongenetic risk-factor data, and using that data to investigate possible interactions between genetic and nongenetic (environmental) factors that influence health. Genetic epidemiologists perform a similar role, using study designs and statistical approaches developed specifically for the analysis of human genetics data.

The professional rewards of a career in epidemiology are the excitement of discovery and the knowledge that epidemiologic studies can be used to help people improve or maintain their health over time. Epidemiologic research has significantly improved the public's health over the past century. Research results have been used to identify new medicines to treat disease, to educate the public about the health effects of cigarette smoking and inactive lifestyles, and to improve sanitation and water treatment, significantly reducing the burden of infectious disease in heavily populated areas. SEE ALSO GENE AND ENVIRONMENT; POPULATION SCREENING; PUBLIC HEALTH, GENETIC TECHNIQUES IN; STATISTICAL GENETICIST.

William K. Scott

Bibliography

Jaret, Peter. "The Disease Detectives." *National Geographic* (January 1991): 114–140.

Epistasis

Epistasis, first defined by the English geneticist William Bateson in 1907, is the masking of the expression of a gene at one position in a chromosome, or **locus**, at one or more genes at other positions. Epistasis should not be confused with dominance, which refers to the interaction of genes at the same locus. The human genome contains from 30,000 to 70,000 gene loci. Some of them are involved in numerous interactions, making it difficult to identify their role in development and metabolism. As we learn more about

locus site on a chromosome (plural, loci)

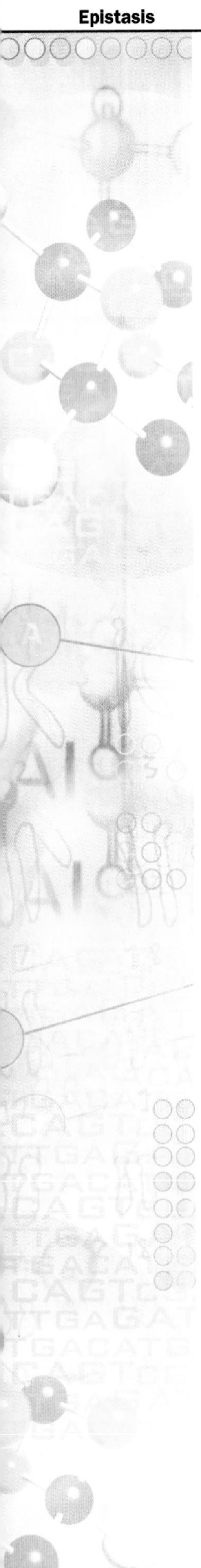

The "A" locus is epistatic to the "B" locus. The "B" locus can influence coat color only if there is at least one dominant "A" allele at the "A" locus.

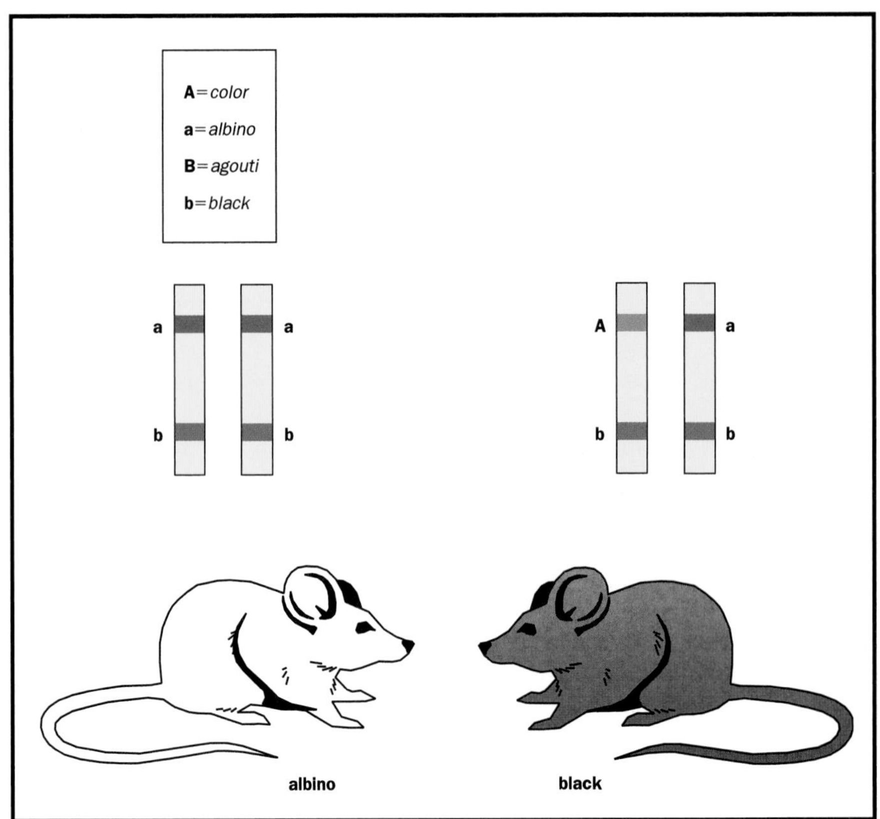

the human and other genomes, it becomes clear that the borrowed phrase "no gene is an island" is an appropriate expression to describe the interplay among gene loci.

Puzzling Inheritance Patterns Explained

phenotype observable characteristics of an organism

antigen a foreign substance that provokes an immune response

dominant controlling the phenotype when one allele is present

homozygous containing two identical copies of a particular gene

There are many examples of epistasis. One of the first to be described in humans is the Bombay **phenotype**, involving the ABO blood group system. Individuals with this phenotype lack a protein called the H **antigen** (genotype hh), which is used to form A and B antigens. Even though such individuals may have A or B genes, they appear to be blood group O because they lack the H antigen.

Another well-known example is coat color in mice. Two coat-color loci are involved. At locus A, color is **dominant** over albino (lack of pigment). At locus B, the coat color agouti is dominant over black. A mouse that is **homozygous** for the albino gene will show no pigment regardless of its genotype at the other locus. Thus the A and B loci are epistatic.

It is likely that the phenomenon of lack of penetrance, in which a dominant gene fails to be expressed, is often due to epistasis. There are many cases where dominant disorders, such as polydactyly (in which individuals have extra fingers or toes), appear to "skip generations." The nonexpression of the dominant gene is likely due to the alleles the individual has at an independent locus that is epistatic to the polydactyly locus. Lack of penetrance may also be accompanied by variable expressivity, where a gene is only partially expressed. As the molecular basis of these disorders becomes known, the reason for nonpenetrance will be easier to determine.

Such interactions between loci probably occur in the genetic etiology of complex traits such as the psychiatric disorders schizophrenia and manic depression. David Lykken, a genetic psychologist at the University of Minnesota, coined the term "emergenesis" to describe multiple gene interactions involved in a specific complex trait. After comparing EEG (electroencephalogram, or "brain wave") data from identical and fraternal twins, Lykken concluded that multiple-level interactions of independent or partly independent genes must be involved.

Epistatic interactions make it difficult to identify loci conferring risk for complex disorders, and they may be a major reason that researchers have made only slow progress in mapping susceptibility genes for complex disorders. To locate interacting loci involved in the genetic origins of complex diseases requires collecting DNA samples from a large number of families where two or more individuals have the disorder. Such large-scale studies are usually difficult to conduct.

Interactions among Proteins

As the Bombay phenotype demonstrates, it is actually proteins, not the genes, that interact. After identifying interacting loci, the next step is determining the proteins that the genes at those loci encode, and the properties of those proteins.

The emerging field that involves the study of proteins and protein interactions is called proteomics. New techniques are now available to locate proteins that interact with one another. In the yeast two-hybrid system, one such technique, one protein is used as bait, and a pool of unknown proteins, referred to as prey proteins, are tested to see if any of them bind to the bait. Binding, if it occurs, triggers a reaction that causes yeast cells to turn blue. In one experiment testing a protein's interactions with more than 1,000 other proteins, 950 interactions were found. Not all of these interactions are likely to occur or be important in the organism, but such results indicate how common, and complex, protein interactions are in living organisms. SEE ALSO BLOOD TYPE; COMPLEX TRAITS; INHERITANCE PATTERNS; PROTEOMICS; PSYCHIATRIC DISORDERS.

P. Michael Conneally

Bibliography

Blum, Kenneth, and Ernest P. Noble, eds. *Handbook of Psychiatric Genetics.* New York: CRC Press, 1996.

Ezell, Carol. "Beyond the Human Genome." *Scientific American* 283 (2000): 64–69.

Mange, Arthur P., and Elaine J. Mange. *Genetics: Human Aspects.* Sunderland, MA: Sinauer Associates, 1990.

Race, Robert R., and Ruth Sanger. *Blood Groups in Man*, 6th ed. Oxford: Blackwell Scientific Publications, 1975.

Escherichia coli

Escherichia coli (*E. coli*) is a very common bacterium that normally inhabits the digestive tract of animals, including humans. It is widespread in the natural world and can also be found in soil and water. It is a member of the

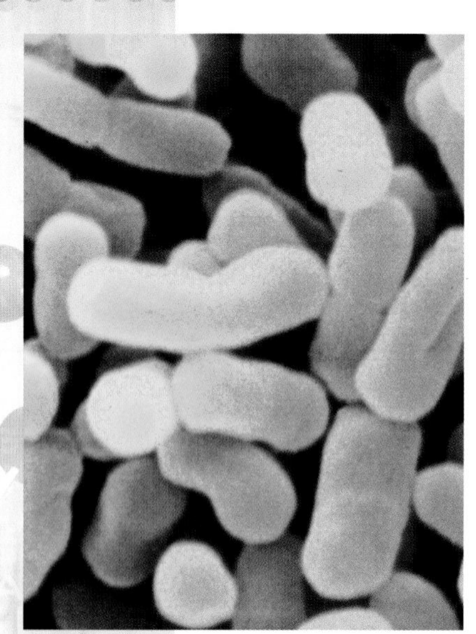

Escherichia coli

bacterial family Enterobacteriaciae, which also includes the bacteria *Shigella*, *Salmonella*, and *Yersinia*, among others. Some of these organisms, including *E. coli*, can cause serious diseases under certain conditions.

Attributes of *E. coli*

E. coli is important to human health because it is a source of vitamins B12 and K, which it manufactures from undigested food in the large intestine. Unlike many other intestinal bacteria, *E. coli* can survive and grow in the presence of oxygen (although it can also grow without oxygen), which makes it a useful experimental model organism in the laboratory.

Even though *E. coli* is a single species of bacteria, many different varieties (called strains) of the species exist. Each has different characteristics, and while some are safe model organisms, others can cause potentially deadly disease. This is the case with *E. coli* 0157:H7, which is considered a dangerous **pathogen** which can infect humans. This strain is significantly different from the commonly used laboratory strains, which do not cause disease.

pathogen disease-causing organism

Importance in Laboratory Studies

E. coli is the most well-understood bacterium in the world, and is an extremely important model organism in many fields of research, particularly molecular biology, genetics, and biochemistry. It is easy to grow under laboratory conditions, and research strains are very safe to work with. As with many bacteria, *E. coli* grows quickly, which allows many generations to be studied in a short time. In fact, under ideal conditions, *E. coli* cells can double in number after only 20 minutes.

Furthermore, a very large number of *E. coli* bacteria can be grown in a small space—many millions in a drop of broth, for example. These are important characteristics in genetic experiments, which often involve selecting a single bacterial cell from among millions of candidates, then allowing it to reproduce into high numbers again to perform additional experiments.

Many vital techniques, such as molecular cloning and overexpression of cloned genes, were initially developed in *E. coli* and are still simpler and more effective in the bacterium. Crucial experiments that illuminated the details of fundamental biological processes such as DNA replication, **transcription**, and **translation** were performed for the first time or with greatest success in *E. coli*. The bacterium is still a primary resource in many modern laboratories. Even research efforts that focus on other organisms, including humans or crop plants, often use *E. coli* extensively as a tool to facilitate cloning and DNA sequencing.

transcription messenger RNA formation from a DNA sequence

translation synthesis of protein using mRNA code

Discoveries Made in *E. coli*

Some of the discoveries made in *E. coli* have provided an invaluable framework for understanding biological processes in more complex organisms. As mentioned above, many fundamental processes that are shared by all living things are most easily studied in this simple bacterial model. Furthermore, *E. coli* has served as a model for understanding the biology of other bacteria.

The ways in which *E. coli* interacts with the human body are in many cases very similar to the ways that other disease-causing organisms act. Therefore, this model organism has been important in the study of human health, and has allowed researchers to ask questions about bacteria in general (for example, how antibiotics stop infections, or how the immune system fights off disease).

Genome Sequenced Early

Sequencing of the *E. coli* K-12 strain **genome** (a popular model strain) was completed in 1997; subsequently, at least two collections of the pathogenic 0157:H7 strain have been completely sequenced. The bacterium has a genome of approximately 4.3 million base pairs of DNA, and carries about 4,400 genes. Interestingly, only about 50 percent of the predicted genes have been described and characterized, a surprisingly low percentage for such a well-understood organism. For this and other reasons, *E. coli* remains one of the most significant model organisms used today. SEE ALSO CHROMOSOME, PROKARYOTIC; EUBACTERIA; GENOME; HUMAN GENOME PROJECT; MODEL ORGANISMS; PLASMID.

genome the total genetic material in a cell or organism

Daniel J. Tomso

Bibliography

Madigan, Michael T., John M. Martinko, and Jack Parker. *Brock Biology of Microorganisms*, 9th ed. Upper Saddle River, NJ: Prentice Hall, 2000.

Eubacteria

The Eubacteria, also called just "bacteria," are one of the three main domains of life, along with the Archaea and the Eukarya. Eubacteria are prokaryotic, meaning their cells do not have defined, membrane-limited nuclei. As a group they display an impressive range of biochemical diversity, and their numerous members are found in every habitat on Earth. Eubacteria are responsible for many human diseases, but also help maintain health and form vital parts of all of Earth's ecosystems.

Structure

Like archeans, eubacteria are prokaryotes, meaning their cells do not have nuclei in which their DNA is stored. This distinguishes both groups from the eukaryotes, whose DNA is contained in a nucleus. Despite this structural resemblance, the Eubacteria are not closely related to the Archaea, as shown by analysis of their RNA (see below).

Eubacteria are enclosed by a cell wall. The wall is made of cross-linked chains of peptidoglycan, a **polymer** that combines both amino acid and sugar chains. The network structure gives the wall the strength it needs to maintain its size and shape in the face of changing chemical and **osmotic** differences outside the cell. Penicillin and related antibiotics prevent bacterial cell growth by inactivating an enzyme that builds the cell wall. Penicillin-resistant bacteria contain an enzyme that chemically modifies penicillin, making it ineffective.

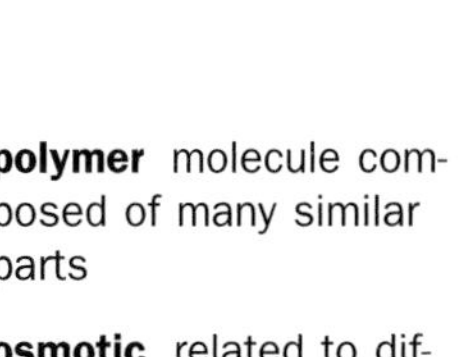

polymer molecule composed of many similar parts

osmotic related to differences in concentrations of dissolved substances across a permeable membrane

Some types of bacteria have an additional layer outside the cell wall. This layer is made from lipopolysaccharide (LPS), a combination of **lipids** and sugars. There are several consequences to possessing this outer layer. Of least import to the bacteria but significant for researchers, this layer prevents

lipids fat or waxlike molecules, insoluble in water

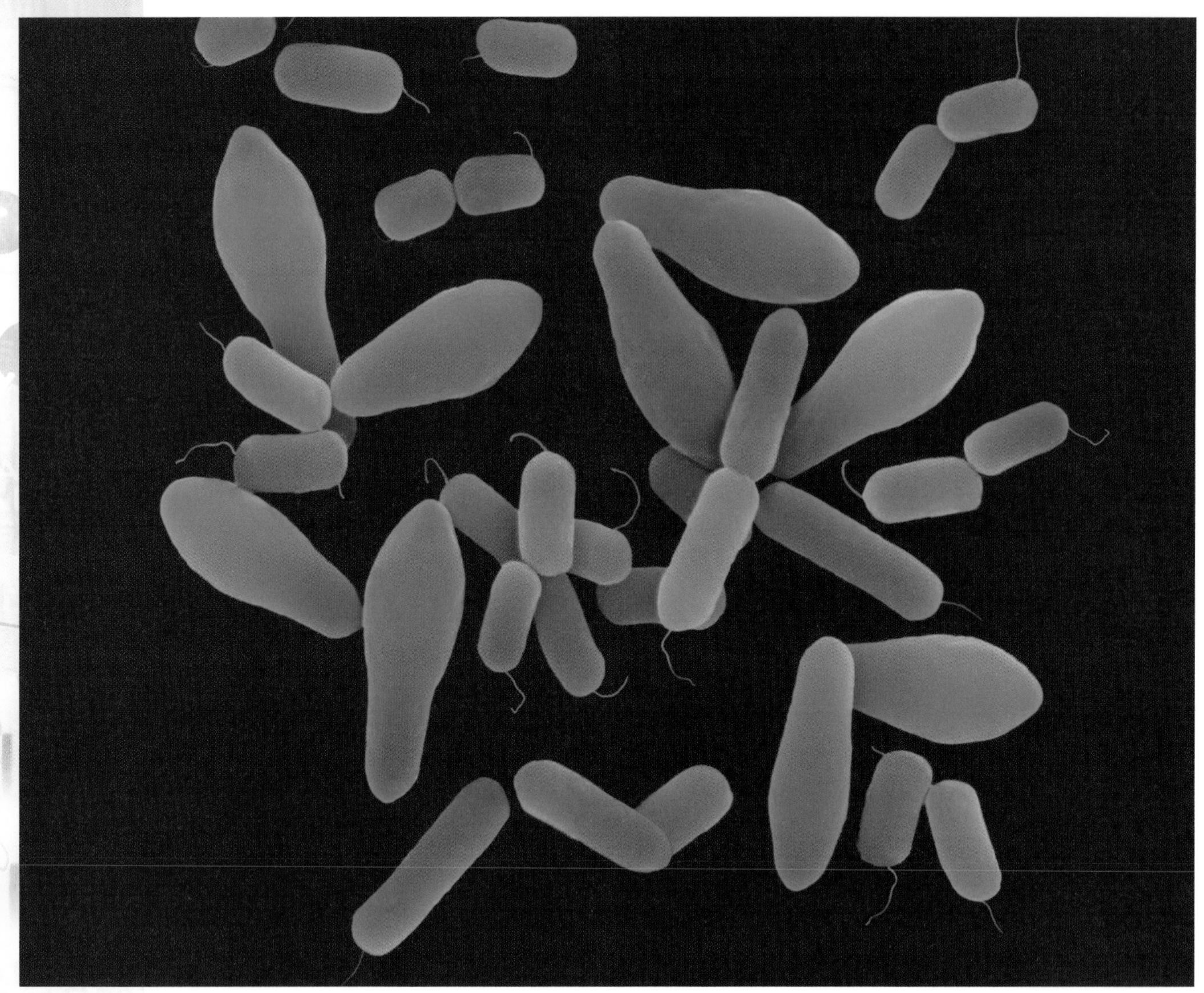

Magnified 1,600 times, a scanning electron micrograph of *Clostridium perfringens* captures a Gram-positive, rod-shaped bacterium forming endospores. This bacterium causes food poisoning, wound infections, and gas gangrene.

them from retaining a particular dye (called Gram stain) that is used to classify bacteria. Bacteria that have this LPS layer are called Gram-negative, in contrast to Gram-positive bacteria, which do not have an outer LPS layer and which do retain the stain. Of more importance to both the bacteria and the organisms they infect is that one portion of the LPS layer, called endotoxin, is particularly toxic to humans and other mammals. Endotoxin is partly to blame for the damage done by infection from *Salmonella* and other Gram-negative species.

Within the cell wall is the plasma membrane, which, like the eukaryotic plasma membrane, is a phospholipid bilayer studded with proteins. Embedded in the membrane and extending to the outside may be flagella, which are whiplike protein filaments. Powered by molecular motors at their base, these spin rapidly, propelling the bacterium through its environment.

Within the plasma membrane is the bacterial cytoplasm. Unlike eukaryotes, bacteria do not have any membrane-bound organelles, such as mitochondria or chloroplasts. In fact, these two organelles are believed to have evolved from eubacteria that took up residence inside an ancestral eukaryote.

Bacterial cells take on one of several common shapes, which until recently were used as a basis of classification. Bacilli are rod shaped; cocci are spherical; and spirilli are spiral or wavyshaped. After division, bacterial cells may remain linked, and these form a variety of other shapes, from clusters to filaments to tight coils.

Metabolism

Despite the lack of internal compartmentalization, bacterial metabolism is complex, and is far more diverse than eukaryotic metabolism. Within the Eubacteria there are species that perform virtually every biochemical reaction known (and much bacterial chemistry remains to be discovered). Most of the vitamins humans require in our diet can be synthesized by bacteria, including the vitamin K humans absorb from the *Escherichia coli* (**E. coli**) bacteria in our large intestines.

E. coli common bacterium of the human gut, used in research as a model organism

The broadest and most significant metabolic distinction among the Eubacteria is based on the source of energy they use to power their metabolism. Like humans, many bacteria are heterotrophs, consuming organic (carbon-containing) high-energy compounds made by other organisms. Other bacteria are chemolithotrophs, which use inorganic high-energy compounds, such as hydrogen gas, ammonia, or hydrogen sulfide. Still others are phototrophs, using sunlight to turn simple low-energy compounds into high-energy ones, which they then consume internally.

For all organisms, extraction of energy from high-energy compounds requires a chemical reaction in which electrons move from atoms that bind them loosely to atoms that bind them tightly. The difference in binding energy is the profit available for powering other cell processes. In almost all eukaryotes, the ultimate electron acceptor is oxygen, and water and carbon dioxide are the final waste products. Some bacteria use oxygen for this purpose as well. Others use sulfur (forming hydrogen sulfide, which has a strong odor), carbon (forming flammable methane, common in swamps), and a variety of other compounds.

Bacteria that use oxygen are called aerobes. Those that do not are called anaerobes. This distinction is not absolute, however, since many organisms can switch between the two modes of metabolism, and others can tolerate the presence of oxygen even if they do not use it. Some bacteria die in oxygen, however, including members of the **Gram positive** *Clostridium* genus. *Clostridium botulinum* produces botulinum toxin, the deadliest substance known. *C. tetani* produces tetanus toxin, responsible for tetanus and "lockjaw," while other *Clostridium* species cause gangrene.

Gram positive able to take up Gram stain, used to classify bacteria

Life Cycle

When provided with adequate nutrients at a suitable temperature and pH, *E. coli* bacteria can double in number within 20 minutes. This is faster than most species grow, and faster than *E. coli* grows under natural conditions. Regardless of the rate, the growth of a bacterium involves synthesizing double the quantity of all its parts, including membrane, proteins, **ribosomes**, and DNA. Separation of daughter cells, called binary fission, is accomplished by creating a wall between the two halves. The new cells may eventually separate, or may remain joined.

ribosomes protein-RNA complexes at which protein synthesis occurs

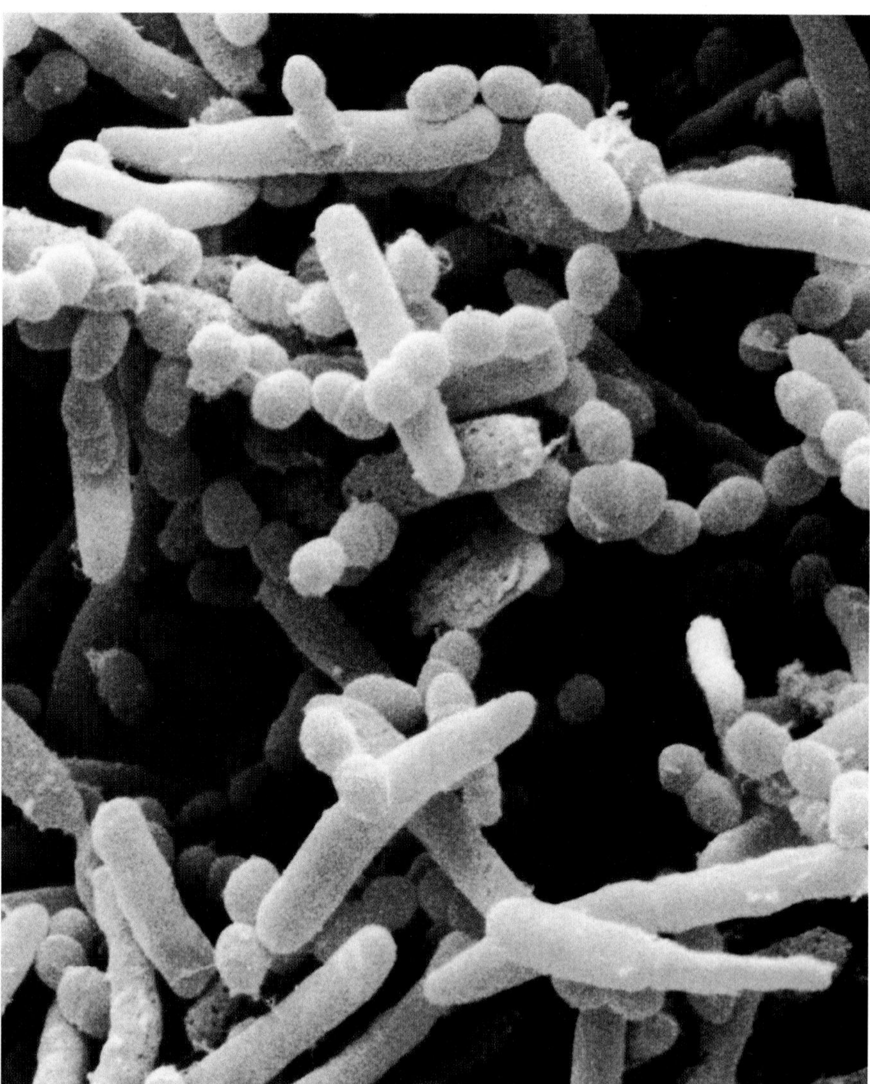

Scanning electron micrograph of bacteria in rod and cocci form.

When environmental conditions are harsh, some species (including members of the genus *Clostridium*) can form a special resistive structure within themselves called an endospore. The endospore contains DNA, ribosomes, and other structures needed for life, but is metabolically inactive. It has a protective outer coat and very low water content, which help it survive heating, freezing, radiation, and chemical attack. Endospores are known to have survived for several thousand years, and may be capable of surviving for much longer, possibly millions of years. When exposed to the right conditions (presence of warmth and nutrients), the endospore quickly undergoes conversion back into an active bacterial cell.

DNA

Most eubacteria have DNA that is present in a single large circular chromosome. In addition, there may be numerous much smaller circles, called **plasmids**. Plasmids usually carry one or a few genes. These often are for specialized functions, such as metabolism of a particular nutrient or antibiotic.

plasmids small rings of DNA found in many bacteria

nucleoid region of the bacterial cell in which DNA is located

Despite the absence of a nucleus, the chromosome is usually confined to a small region of the cell, called the **nucleoid**, and is attached to the inner membrane. The bacterial genome is smaller than that of a eukaryote. For example, *E. coli* has only 4.6 million base pairs of DNA, versus three billion in humans. As in eukaryotes, the DNA is tightly coiled to fit it into the cell. Unlike eukaryotes, however, the DNA is not attached to histone proteins.

Much of what we know about DNA replication has come from study of bacteria, particularly *E. coli*, and the details of this process are discussed elsewhere in this encyclopedia. Unlike eukaryotic replication, prokaryotic replication begins at a single point, and proceeds around the circle in both directions. The result is two circular chromosomes, which are separated during cell division. Plasmids replicate by a similar process.

Gene Transfer

While bacteria do not have sex like multicellular organisms, there are several processes by which they obtain new genes: **conjugation**, transformation, and transduction. Conjugation can occur between two appropriate bacterial strains when one (or both) extends hairlike projections called pili to contact the other. The chromosome, or part of one, may be transferred from one bacterium to the other. In addition, plasmids can be exchanged through these pili. Some bacteria can take up DNA from the environment, a process called transformation. The DNA can then be incorporated into the host chromosome.

conjugation a type of DNA exchange between bacteria

Some bacterial viruses, called phages, can carry out transduction. With some phages, the virus temporarily integrates into the host chromosome. When it releases itself, it may carry some part of the host DNA with it. When it goes on to infect another cell, this extra DNA may be left behind in the next round of integration and release. Other phages, called generalized transducers, package fragments of the chromosome into the phage instead of their own **genome**. When the transducing phage infects a new cell, they inject bacterial DNA. These phages lack their own genome and are unable to replicate in the new cell. The inserted bacterial DNA may recombine (join in with) the host bacterial chromosome.

genome the total genetic material in a cell or organism

Gene Regulation and Protein Synthesis

Gene expression in many bacteria is regulated through the existence of operons. An operon is a cluster of genes whose protein products have related functions. For instance, the *lac* operon includes one gene that transports lactose sugar into the cell and another that breaks it into two parts. These genes are under the control of the same **promoter**, and so are transcribed and translated into protein at the same time. RNA polymerase can only reach the promoter if a repressor is not blocking it; the *lac* repressor is dislodged by lactose. In this way, the bacterium uses its resources to make lactose-digesting **enzymes** only when lactose is available.

promoter DNA sequence to which RNA polymerase binds to begin transcription

enzymes proteins that control a reaction in a cell

Other genes are expressed constantly at low levels; their protein products are required for "housekeeping" functions such as membrane synthesis and DNA repair. One such enzyme is DNA gyrase, which relieves strain in the double helix during replication and repair. DNA gyrase is the target for the antibiotic ciproflaxin (sold under the name Cipro), effective against *Bacillus anthracis*, the cause of anthrax. Since eukaryotes do not have this type of DNA gyrase, they are not harmed by the action of this antibiotic.

As in eukaryotes, translation (protein synthesis) occurs on the ribosome. Without a nucleus to exclude it, the ribosome can attach to the messenger RNA even while the RNA is still attached to the DNA. Multiple ribosomes can attach to the same mRNA, making multiple copies of the same protein.

The ribosomes of eubacteria are similar in structure to those in eukaryotes and archaea, but differ in molecular detail. This has two important consequences. First, sequencing ribosomal RNA molecules is a useful tool for understanding the evolutionary diversification of the Eubacteria. Organisms with more similar sequences are presumed to be more closely related. The same tool has been used to show that Archaea and Eubacteria are not closely related, despite their outward similarities. Indeed, Archaea are more closely related to eukaryotes (including humans) than they are to Eubacteria.

Second, the differences between bacterial and eukaryotic ribosomes can be exploited in designing antibacterial therapies. Various unique parts of the bacterial ribosome are the targets for numerous antibiotics, including streptomycin, tetracycline, and erythromycin. SEE ALSO ARCHAEA; CELL, EUKARYOTIC; CHROMOSOME, PROKARYOTIC; CONJUGATION; *ESCHERICHIA COLI*; OPERON; PLASMID; TRANSDUCTION; TRANSFORMATION.

Richard Robinson

Bibliography

Madigan, Michael T., John M. Martinko, and Jack Parker. *Brock Biology of Microorganisms*, 9th ed. Upper Saddle River, NJ: Prentice Hall, 2000.

Margulis, Lynn, and Karlene Schwartz. *Five Kingdoms*, 3rd ed. New York: W. H. Freeman, 1998.

Eugenics

While the idea of improving humans through selective breeding is at least as old as the ancient Greeks, it gained widespread prominence after 1869. In 1883, Sir Francis Galton coined the word "eugenics," from the Greek word *eugenes*, meaning "well-born" or "hereditarily endowed with noble qualities," to describe this new science of directed human evolution. Galton's work, and the subsequent rediscovery of Gregor Mendel's genetic studies, convinced many scientists and social reformers that eugenic control over heredity could improve human life.

Galton's ideas swept America during the Progressive Era of the early twentieth century. At that time, many scientists and laypeople believed that eugenics could facilitate social progress by eradicating problems ranging from alcoholism and prostitution to poverty and disease. What better way to prevent such misfortunes, eugenicists asked, than to prevent the birth of people genetically susceptible to them? Eugenics seemed to offer an efficient and humane solution to society's ills. Unfounded hope in this imperfect science, however, ultimately contributed to repressive social policies, including marriage and immigration restriction, forced sterilization, segregation, and, in the case of Nazi Germany, euthanasia ("mercy killing") and genocide, all in the name of human betterment.

British Origins

Charles Darwin's theories of evolution by natural selection rocked the scientific world in 1859, and prompted his cousin, Galton, to study human evolution. Galton's first book, *Hereditary Genius* (1869), analyzed famous European families and concluded that "genius," which he defined as the ability to succeed in life, tended to run in families. Galton believed that individuals inherited the traits that destined them to either success or failure. Thus, success resulted from biology, not from the wealth or poverty of a person's background, and controlled breeding might permanently improve the human race.

Galton hoped to speed and direct human evolution. Writing in *Inquiries into the Human Faculty and Its Development* (1883), Galton defined eugenics as "the science of improving stock . . . to give the more suitable races or strains of blood a better chance of prevailing speedily over the less suitable than they otherwise would have had." Familiar with farmers' achievements in breeding more-valuable plants and animals, Galton believed that such methods were "equally applicable to men, brutes, and plants."

Galton identified those fit folk who should have children and stigmatized those he deemed unfit for parenthood. He also believed then-accepted notions of "racial" superiority and inferiority, had more to do with class and cultural prejudice than with biological difference. Galton assumed that wealthy people like himself were fit, whereas poor folk were unfit. Northern European "white" people stood atop the evolutionary scale of fitness, followed by "whites" from southeast Europe, Asians, Native Americans, Africans, and Australian Aborigines.

Positive and Negative Eugenics

Galton identified positive and negative eugenics as the two basic methods to improve humanity. Positive eugenics used education, tax incentives, and childbirth stipends to encourage **procreation** among fit people. Education would convince fit parents to have more children, out of a desire to increase the common good. Lower taxes on larger families and the provision of a small birth payment for each "eugenic" child would provide further inducements. Conversely, eugenically educated but unfit people would selflessly forgo procreation, to prevent the propagation of their hereditary "taint." Believing that neither altruism nor self-interest would be enough to control the unfit, however, many eugenicists also advocated negative eugenics.

procreation reproduction

Negative eugenics sought to limit procreation through marriage restriction, segregation, sexual sterilization, and, in its most extreme form, euthanasia. In an attempt to decrease procreation among the "unfit," laws prohibited marriage to people with diseases, or other conditions believed to be hereditary. Similar restrictions banned marriage between people of different races, in order to prevent **miscegenation**. Popular in the United States, antimiscegenation laws sought to use science to legitimize racial prejudice. Since marriage restriction failed to stop extramarital procreation, eugenicists argued for more intrusive interventions.

miscegenation racial mixing

Many of these more intrusive interventions relied upon segregation. For example, individuals judged unfit might be segregated in institutions such as insane asylums, tuberculosis sanatoriums, and homes for the so-called

feebleminded or mentally retarded. Isolated from "normal" society, these people were also segregated by sex within the institution to prevent procreation. Segregation through incarceration, however, proved too costly to be applied to all but the most severely handicapped.

Compulsory sexual sterilization of those individuals deemed "feebleminded" or "slow" promised eugenic and economic benefits for society. Once sterilized, such individuals posed no eugenic risk; sexual intercourse would never result in pregnancy. Sterilized individuals could therefore return to society and work, rather than remaining an economic "burden" in an institution. Many social reformers argued that compulsory sterilization was more humane than locking people away during their childbearing years.

In the case of individuals afflicted with gross physical or mental abnormalities, the most radical eugenic intervention was proposed: euthanasia. While many eugenicists theorized about euthanasia, very few seriously considered it as a real possibility. This would change with the advent of Nazi eugenics in Germany.

Mendelian Inheritance, Intelligence Testing, and American Eugenics

Galton's eugenic ideas found fertile ground in America after 1900, when scientists rediscovered Mendel's findings regarding the inheritance of physical traits in pea plants. Mendel's notions of "dominant" and "recessive" genetic traits, easily identified in "lower" organisms such as plants and animals, convinced people that human eugenic improvement was possible. Scientists assumed that even complex human traits such as intelligence and behavior behaved as simple genetic "unit characters," such as height or color in peas.

The advent of intelligence testing in the 1900s provided a new way to quantify Galton's notion of genius. American eugenicists assessed an individual's eugenic worth by combining his intelligence quotient (IQ) with a Galtonian study of the family pedigree. Psychologist Henry Herbert Goddard published one famous study, *The Kallikak Family*, in 1912. Goddard traced two family lines that originated with a common male ancestor, whom he called Martin Kallikak (from the Greek words for beautiful [*kalos*] and bad [*kakos*]). One branch appeared healthy and eugenic, descended from Martin's marriage to a "respectable" woman. The second branch was composed of "Defective degenerates" (alcoholics, criminals, prostitutes, and particularly the mentally "feebleminded") born of Martin's dalliances with a "feebleminded" tavern mistress. Goddard thus "proved" the inheritance of feeblemindedness, and its social cost.

Convinced that "feeblemindedness" and other complex antisocial behaviors behaved like simple Mendelian traits, eugenicists lobbied for compulsory sterilization laws. Between 1907 and the mid-1930s, such laws were adopted by thirty-two American states. The U.S. Supreme Court upheld these laws in 1927, when Justice Oliver Wendell Holmes ruled that, "the principle that sustains compulsory vaccination of schoolchildren is broad enough to cover the cutting of the fallopian tubes. . . . Three generations of imbeciles are enough." "Feebleminded" individuals were prominent

People gather for the first Fitter Family contest, sponsored by the American Eugenics Society, at the Kansas State Free Fair in 1920. At the height of the Eugenics movement's popularity such exhibits were well attended—people participated in testing meant to evaluate their "eugenic fitness." Winners of such contests were white, with Northern and/or Western European heritage.

among the more than 60,000 individuals sterilized in the United States under eugenic sterilization laws between 1927 and 1979.

Nazi Eugenics

Eugenics is commonly associated with the Nazi racial hygiene program that began in 1933 and ended in May 1945, with Germany's defeat near the end of World War II. Although the German eugenics movement existed long before the Nazis came to power, scholars have shown that Nazi eugenicists were inspired by American eugenic studies and sterilization, as well as their antimiscegenation and immigration restriction laws.

Goddard's *Kallikak* study was well respected among German eugenicists who, like American eugenicists, emphasized genetics as the basis of human differences. German racial hygienists also praised the American eugenicist Madison Grant's racist book, *The Passing of the Great Race*, which Adolf Hitler referred to as his "Bible." The 1933 Nazi "Law for the Prevention of Genetically Diseased Progeny," relied on American examples, especially the model law drafted by American eugenicist Harry Hamilton Laughlin. In 1936 the University of Heidelberg awarded Laughlin an honorary degree for his

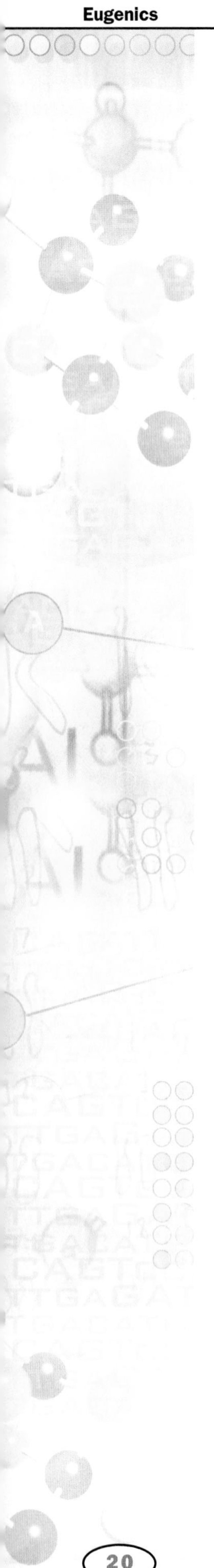

contributions to "racial hygiene." Laughlin's degree was ordered and signed by Hitler.

The Nazis instituted state-supported positive eugenics programs that encouraged "racially fit" women to reproduce, as well as a massive negative eugenics program that included euthanasia. Ultimately, the Nazis sterilized about 400,000 people and euthanized another 70,000 individuals that were judged to be feebleminded or otherwise unfit. The euthanasia program foreshadowed the extermination of six million Jewish victims, along with millions of others, notably Gypsies and homosexuals, in the Holocaust.

Demise of Eugenics

Early on, some scientists objected to the eugenicists' insistence that heredity overwhelmed environmental influences in shaping human life. Others objected to the eugenicists' methodologies, noting that family studies often relied on hearsay evidence and biased observation rather than direct, quantifiable, empirical measurements. Others challenged the eugenicists' reliance on **phenotypic** traits such as body form to diagnose presumed underlying genetic causes. Developments in genetics increasingly undermined this simplistic reasoning from phenotype to genotype.

phenotypic related to the observable characteristics of an organism

Instead, genetic studies increasingly revealed the complex nature of most human phenotypic traits. Human traits rarely result from the action of single gene pairs, and expression depends on complex environmental influences. Moreover, many genes induce pleiotropic effects: that is, a single gene may influence more than one phenotypic characteristic. If multiple genes cause single traits, or if single genes are involved in many effects, then any attempt to "breed out" traits becomes virtually impossible. Moreover, if, as most eugenicists believed, negative traits are recessive factors in single-gene disorders, then most "bad" genes are harbored in apparently normal, **heterozygous** carriers. The Hardy-Weinberg theorem, formulated in 1908, made it clear that eugenic selection directed solely against affected individuals would barely reduce the incidence of a trait in the larger population. To decrease such defects by half would require forty generations (1,000 years) of perfect negative eugenics.

heterozygous characterized by possession of two different forms (alleles) of a particular gene

The Hardy-Weinberg theorem alone, unfortunately, did not dissuade most geneticists from eugenics. Many continued to believe, as geneticist Herbert Spencer Jennings wrote in 1930, that preventing the "propagation of even one congenitally defective individual puts a period to at least one line of operation of this devil. To fail to do at least so much would be a crime." Nevertheless, the most bigoted aspects of eugenics dwindled after 1946, as scientists recoiled from the horrors of Nazi atrocities.

The dream of improving human life through genetic intervention remains with us today. While genetic knowledge and technology have changed since the Holocaust, the cultural and political context surrounding the pursuit of genetic improvement has undergone even greater transformations. The goal of present genetic intervention is not group improvement, but individual therapy. Modern conceptions of individual, patient, and human rights reduce the risk of abuses committed in the name of eugenics. While negative eugenics has been largely eliminated as neither possible nor socially acceptable, positive eugenics are still considered desirable among some people who propose genetic engineering for the development of children with superior traits. The ethical issues surrounding genetic engineering and cloning are still debated in light of the history of the eugenics movement. SEE ALSO CLONING: ETHICAL ISSUES; GENE THERAPY: ETHICAL ISSUES; HARDY-WEINBERG EQUILIBRIUM; INHERITANCE PATTERNS; REPRODUCTIVE TECHNOLOGY: ETHICAL ISSUES.

Gregory Michael Dorr

Bibliography

Gould, Stephen J. *The Mismeasure of Man,* revised and expanded ed. New York: W. W. Norton and Company, 1996.

Kevles, Daniel J. *In the Name of Eugenics: Genetics and the Uses of Human Heredity.* New York: Alfred A. Knopf, 1985.

Kuhl, Stefan. *The Nazi Connection: Eugenics, American Racism, and German National Socialism.* New York: Oxford University Press, 1994.

Paul, Diane B. *Controlling Human Heredity, 1865 to the Present.* Atlantic Highlands, NJ: Humanities Press, 1995.

———. *The Politics of Heredity: Essays on Eugenics, Biomedicine, and the Nature-Nurture Debate.* Albany: State University of New York Press, 1998.

Pernick, Martin. *The Black Stork: Eugenics and the Death of Defective Babies in American Medicine and Motion Pictures since 1915.* New York: Oxford University Press, 1996.

Proctor, Robert. *Racial Hygiene: Medicine under the Nazis.* Cambridge, MA: Harvard University Press, 1988.

Reilly, Philip R. *The Surgical Solution: A History of Involuntary Sterilization in the United States.* Baltimore, MD: Johns Hopkins University Press, 1991.

Selden, Steven. *Inheriting Shame: The Story of Eugenics and Racism in America.* New York: Teachers College Press, 1999.

Internet Resource

"Image Archive on the American Eugenics Movement." <http://www.eugenicsarchive.org/eugenics>.

Evolution, Molecular

All life on Earth is cellular and uses DNA to store genetic information. However, evidence suggests that, on ancient Earth, much complex chemical activity preceded cellular development, and it was probably not DNA-based at the start. What was the nature of this activity, and how did it lead to life? "Molecular evolution" is a term used to describe the stages that preceded the origin of life on Earth. The term implies that information-containing molecules were subject to the process of natural selection, whereby genetic structures were capable of both replication (the copying of specific **nucleic acid** sequences) and **mutation**. In addition, it is theorized that certain chemical reactions may have taken place on early Earth, before true evolution began, and some of these reactions may have helped to form these informational molecules that enabled **replication** and mutation.

nucleic acid DNA or RNA

mutation change in DNA sequence

replication duplication of DNA

The Antiquity of Life

The oldest known fossils date from about 3.5 billion years ago, and have been found in Western Australia. Called stromatolites, these domelike rocks

Stromatolites, like these in the shallow waters of the Bahamas, have existed for over three million years. These rocks contain fossils of various cyanobacteria such as colonial chroococcalean forms, and filamentous *Palaeolyngbya*.

are formed today by photosynthetic organisms (cyanobacteria) that live in shallow waters. Large communities of cyanobacteria form microbial mats that trap sediment, providing a sturdy foundation on which another layer of bacteria can grow.

The search for older fossils has been frustrated by the scarcity of pristine rocks that have not been altered by high temperatures and pressures. However, evidence from geological deposits in Greenland dated at 3.87 billion years ago may point to the presence of life at an even earlier time. Since our planet itself is approximately 4.6 billion years old, and several hundred million years were needed for the surface to cool to "hospitable" temperatures, the origin of life must have been quite rapid compared to the vast span of terrestrial history.

Building the Building Blocks: RNA Nucleotides

A productive hypothesis, which has stimulated the design of laboratory experiments that might mimic the formation of biochemical compounds on early Earth, has been the notion that ribonucleic acid (RNA) was the primordial genetic material. Many scientists believe that RNA arose *before* DNA, and that DNA later took over the information-storage role from RNA. Among the reasons for this view are the modern routes by which pieces of DNA are made. For example, the characteristic sugar that forms the backbone is generated by an **enzyme** that removes an oxygen (the "deoxy" in deoxyribonucleic acid) from the corresponding RNA sugar, prior to assembly of the chain. This enzyme probably evolved after the appearance of RNA components on early Earth.

enzyme a protein that controls a reaction in a cell

RNA chains are composed of repeating units called **nucleotides**. Each nucleotide consists of a phosphate and a ribose sugar, to which one of the four "bases" (uracil, adenine, cytosine, and guanine) is appended. While phosphate minerals were common on Earth early in its history, each of the other components has been the target of laboratory simulations to determine how they might have been formed.

nucleotides the building blocks of RNA or DNA

The bases appear to have been relatively easy to create. Hydrogen cyanide, a possible ingredient in the early oceans and lakes, reacts in the presence of ultraviolet light to give off adenine. Other cyanide derivatives (along with urea) can produce the other bases. A reasonable natural setting for this chemistry would have been a shallow lake or lagoon, as envisioned by Stanley Miller in his preparation of cytosine.

More challenging to those studying the origins of life has been determining how ribose may have been formed and to explore how it may have reacted with the bases (especially cytosine and uracil) and acquired the phosphate to form RNA. Formaldehyde (the same chemical used to preserve specimens) is a likely **prebiotic** molecule that reacts with itself to give a very complex mixture of products that includes traces of ribose. A more effective route, however, starts with formaldehyde and a derivative known as glycolaldehyde phosphate: these react under alkaline conditions to give mainly a ribose compound with two phosphates attached, and the reaction is promoted by certain minerals.

prebiotic before the origin of life

Heating a mixture of ribose with either adenine or guanine to dryness results in some bond formation between the base and the sugar, but this strategy has failed when cytosine or uracil was used in place of adenine and guanine. More research is still needed to establish plausible routes to these RNA precursors on early Earth, and some skeptics have even proposed that a simpler type of backbone may have preceded that found in nucleic acids today.

Linking Subunits into Chains

Mineral catalysts have also been useful in forming chains of RNA-like molecules from the activated precursors. James Ferris has focused on the ability of montmorillonite (a type of clay) to promote the assembly of the ribose-phosphate backbone, starting with a special, uracil-containing compound known as a phosphorimidazolide. Although phosphorimidazolides do not occur in nature today, they are highly reactive species and closely related to the modern building blocks of RNA. Most scientists do not maintain that these compounds were present in the **primordial soup**, but they are convenient substitutes for the natural precursors of RNA.

primordial soup hypothesized prebiotic environment rich in life's building blocks

When the adenine derivative of this compound binds to the mineral surface, the products include chains of up to ten units long, primarily with the "biologically correct" bonds between adjacent riboses. By repeated additions of the starting material, the process can extend the structure up to fifty units. The uracil derivative reacts in a similar fashion, although the chains are somewhat shorter. These data suggest that binding to mineral surfaces may have been important in controlling the proximity and orientation of molecules that could give rise to the first RNA-like fragments, and set the stage for subsequent replication.

RNA Replication without Enzymes

Once formed, how would the first RNA chains cause copies of themselves to be created? Replication is the process by which DNA or RNA makes copies of specific base sequences. However, the "copy" is not identical to the original, called a template. Rather, it is analogous to a photographic negative, with predicable differences from the original. In the case of RNA, the

new strand has a different pattern of bases, determined by the specific interactions (**hydrogen bonding**) between the old strand and the new one. Thus, adenine bonds to uracil (causing it to become part of the copy), and guanine "directs" the incorporation of cytosine in the same way. As an example, a **template** sequence abbreviated as AACCAA would be replicated as UUGGUU (the letters stand for each of the four bases). Of course, the faithful transfer of genetic information is a much more complex process, involving an array of complicated protein enzymes and requiring special "activated" precursors for each of the bases.

hydrogen bonding weak bonding between the H of one molecule or group and a nitrogen or oxygen of another

template a master copy

Leslie Orgel and his associates have carried out extensive studies of replication since 1980. Their goals were, first, to establish whether this process could occur without the help of **enzymes** (which would not have been present in the environment of early Earth); second, to analyze the accuracy of the copies; third, to explore the limits on the types of sequences that could be copied; and fourth, to determine if the copies could then serve as templates for self-perpetuating replication. These aspects have met with different degrees of success, as discussed below.

enzymes proteins that control a reaction in a cell

Orgel's first breakthrough came with the reaction of activated guanine derivatives (again using phosphorimidazolides) on a template consisting of repeating cytosines, which was thus very similar to a small piece of RNA, except that it had only one type of base. In the presence of a zinc catalyst, the guanine derivatives bound to the template and formed chains with more than thirty guanines linked to one another. (Without the template, the only products were those with two or three bases.) Equally striking was that the guanine chain contained mainly the same type of phosphate-ribose backbone as in native RNA, and the template preferentially bound the "correct" base more than 99 percent of the time. Even if the activated derivatives of uracil, adenine, and cytosine were present, they become incorporated into the product with less than 1 percent efficiency. These early experiments demonstrated that templates could accurately form long chains with the appropriate bond between the sugar and phosphate.

The copying of other sequences using this approach has been more difficult, however, partly because of the unreactive structures that the templates often form. For example, RNA chains containing only adenine mixed with other chains of uracil tend to form aggregates that hinder replication. Guanine is even more unusual, in that it organizes into arrays with four chains locked together, which also prevents replication, although this problem may be overcome by employing very dilute reaction conditions (more relevant to early Earth). The greatest success has been achieved with mixed, cytosine-rich templates that contain adenine, guanine, or uracil as isolated bases separated by at least three cytosines.

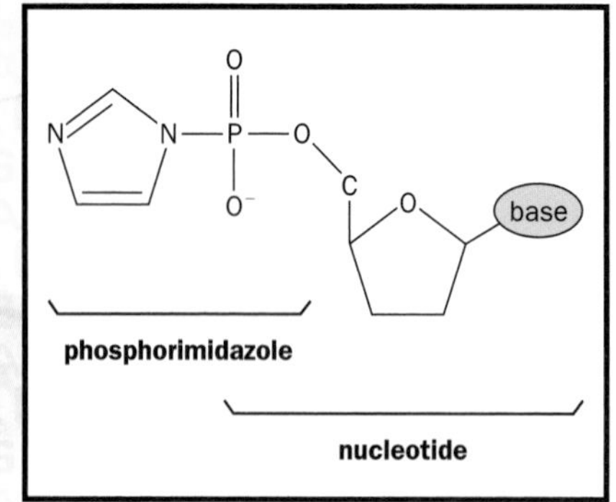

The phosphorimidazole activates the nucleotide, promoting the linking of multiple units. This synthetic compound may mimic compounds present in prebiotic environments.

A further difficulty is that none of the systems studied by Orgel is capable of replication beyond the first stage: the copies can never serve as templates themselves, because they remain tightly bound to the original template. However, Gunter von Kiedrowski made an important advance in 1994, when he showed that sets of three bases containing guanine and cytosine on a DNA backbone could assemble on a template six units long and then separate. For example, two fragments of GGC could link together on a CCGCCG framework, and then break apart to form a GGCGGC template, available for further replication. The reaction conditions were quite different from what might have existed on early Earth, especially the chemical used to form the bond between the GGC units, but it supports the concept that true replication of this type might be possible.

RNA Can Act as an Enzyme

In cells, replication is controlled by protein **catalysts** called enzymes. However, since proteins are thought not to have been present on early Earth, how could replication and its related reactions been catalyzed? A key discovery, which has affected how molecular biologists and biochemists view RNA, is that it can also act as a catalyst. In the early 1980s Thomas Cech and Sidney Altman independently discovered RNA (that is, non-protein-based) catalysts in a variety of cell types, whereas scientists had previously thought that only proteins have this power. The concept that RNA could both store information and accelerate biochemical processes made it a much more likely candidate for catalyzing reactions in early life.

catalysts substances that speed up a reaction without being consumed (e.g., enzyme)

The catalytic RNAs (known as **ribozymes**) found in nature today mainly promote reactions that involve removing certain sequences from an RNA chain before it can be used in protein formation. However, a variety of exotic tools in the molecular biologists' arsenal have now allowed the creation of new ribozymes by **in vitro** evolution (evolution in a test tube), a method developed by Jack Szostak. He and his associate, David Bartel, showed in 1993 that a novel ribozyme with the ability to link together two RNA chains could be evolved in the laboratory through successive rounds of selection and amplification (creating many copies of the most effective sequence from the previous round). After eight rounds, the best catalyst was faster by a factor of three million, compared to the uncatalyzed reaction.

ribozyme RNA-based catalyst

in vitro in glass"; in lab apparatus, rather than within a living organism

Bartel has applied this strategy toward the development of the first RNA that catalyzes replication: it binds a smaller RNA template, which then creates a copy that is up to 14 units long. When a set of these replicated products was analyzed, the average accuracy for incorporation of the correct base was 96.7 percent (lowest accuracy was achieved for adenine, greatest for guanine). Although the ribozyme itself was 189 units long and thus represents a highly complex structure that would probably not have formed spontaneously in the early ocean, the experiment demonstrates that RNA can promote the critical step of "peeling off" the copy to allow another cycle of product formation. Such evolutionary exercises thus provide a powerful method for exploring the relationship between ribozyme sequence and the catalytic activity.

Goals for Future Research

Theories of molecular evolution have been based on the paradigm that RNA (or a similar structure) served as the first genetic molecule. Many gaps remain in our understanding, from the assembly of the individual units, to the replication of RNA sequences. Further advances in the study of catalysis, especially with minerals and ribozymes, may provide new avenues for future researchers to explore. SEE ALSO EVOLUTION OF GENES; NUCLEOTIDE; RIBOZYME; RNA PROCESSING.

William J. Hagan

Bibliography

Ferris, James P. "Chemical Replication." *Nature* 369 (1994): 184–185.

Fry, Iris. *The Emergence of Life on Earth: A Historical and Scientific Overview.* New Brunswick, NJ: Rutgers University Press, 2000.

Hayes, John M. "The Earliest Memories of Life on Earth." *Nature* 384 (1996): 21–22.

Joyce, Gerald F. "RNA Evolution and the Origins of Life." *Nature* 338 (1989): 217–224.

———. "Directed Molecular Evolution." *Scientific American* 267 (Dec. 1992): 90–97.

Orgel, Leslie E. "Molecular Replication." *Nature* 358 (1992): 203–209.

———. "The Origin of Life on the Earth." *Scientific American* 271 (Oct. 1994): 77–83.

Von Kiedrowski, Gunter. "Origins of Life: Primordial Soup or Crepes?" *Nature* 381 (1996): 20–21.

Wills, Christopher, and Jeffrey Bada. *The Spark of Life: Darwin and the Primeval Soup.* Cambridge, MA: Perseus Publishing, 2000.

Evolution of Genes

genome the total genetic material in a cell or organism

Individual genes and whole **genomes** change over time. Indeed, evolution of genes ultimately accounts for the evolution of organisms that is seen in the fossil record: Humans evolved from earlier apes, and those creatures from their ancestors, by gene changes in the earlier creatures. Just as the fossil record can be examined to understand the patterns of organismic evolution, so too can genes be compared to understand genomic evolution.

Natural Selection

heritable genetic

Most changes that occur in genes are subject to natural selection, the process first outlined by Charles Darwin in 1859. In natural selection, a **heritable** change arises by chance. If the organism with that change is better able to survive and reproduce, it will leave more descendants in future generations. These descendants will also carry the new genetic change, and as they reproduce, the change will become more widespread in the population. On the other hand, if the change decreases an organism's survival rate, it will be lost from the population. It is also possible to have a neutral change, with no immediate effect on survival. Such "hidden" genetic variation within a population provides grist for evolution when it offers a selective advantage under new environmental conditions.

It is important to remember that a genome (all the DNA of an organism) is more than just its genes. The genome includes vast amounts of DNA outside of genes, and this too is subject to change over time. In fact, non-gene portions usually change at a faster rate than genes, because many of these changes have little or no effect on the organism's survival.

Evolution of genes and genomes includes sequence changes to existing genes, gene duplication, recombination of gene segments, and the varied actions of transposable elements as they move through the genome.

Point Mutations in Existing Genes

amino acids building blocks of protein

Genes are long strings of four nucleotides (abbreviated A, T, C, and G) whose order dictates the order of **amino acids** in proteins (or nucleotides in RNA). A point mutation is a change in a single nucleotide position at a particular point in a gene. Point mutations can be changes that convert one type of nucleotide to another (a C to a T, for instance), or cause the deletion or addition of a single nucleotide. While the rate of mutation is slow, over long periods (millions of years) most possible sequence changes will have occurred several times in a population, and so natural selection is likely to have acted on most genes in almost every modern population.

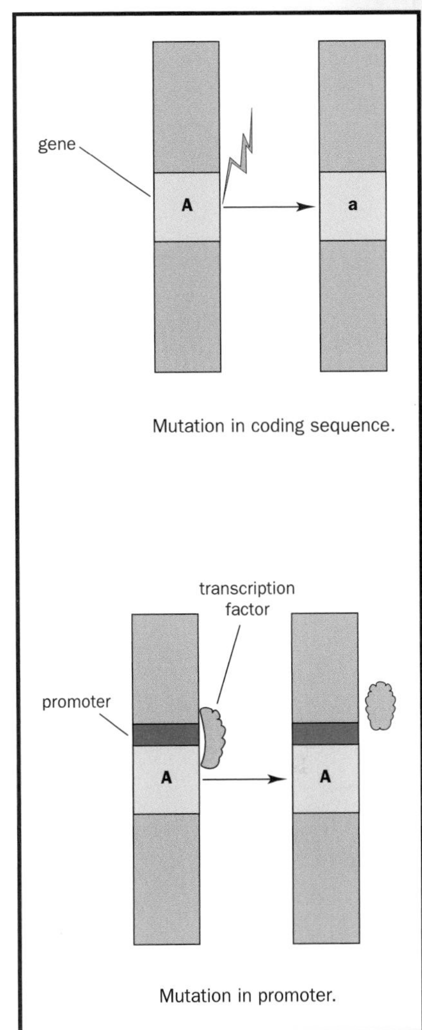

Mutations in the coding region may change the protein that results. Mutations in the promoter can affect how transcription factors bind, altering the level of gene expression.

Some point mutations can change the amino acid sequence of the resulting protein, altering its properties. For instance, a digestive enzyme's attraction for its substrate (the food molecule it breaks down) may be altered to allow the organism to digest new foods, or prevent it from digesting old ones. The sickle form of hemoglobin arose because of a point mutation that changed a single amino acid in hemoglobin. The result was a molecule that bound oxygen less tightly under certain conditions, conferring resistance to malaria, but also causing sickle cell disease, a type of hemoglobinopathy.

Other point mutations may leave the protein unchanged, but alter the conditions under which it is expressed. Genes are expressed (that is, are "read" to cause protein production) when **transcription factors** bind to a region of the gene known as the promoter. Promoters interact with other DNA regions, called enhancers, which are often a long distance from the gene along the chromosome, but are nonetheless close to it because of folding of the DNA. Mutations in either the promoter or enhancer region can have profound effects on the sensitivity of expression to hormones, temperature changes, and other regulatory influences. For instance, a human gene coding for one form of an enzyme called pancreatic elastase appears to have been silenced by an evolutionarily recent enhancer mutation.

Not all DNA sequence changes will lead to changes in the encoded protein or in the way that it is expressed. For many amino acids, there are several DNA "synonyms" that all code for the same amino acid. GGG, GGC, GGA, and GGT all code for the amino acid proline, for example. This is known as the degeneracy of the genetic code. Also, mutations that code for chemically similar amino acids may have no significant effect on protein function. For instance, leucine (AAT) and isoleucine (TAT) are both small, nonpolar amino acids often found on the interior of proteins. Mutations that exchange one for the other may have little effect on protein structure or function.

Genes in **eukaryotes** also contain noncoding regions, called introns. Mutations in these regions often have no effect, since their sequences do not code for part of the finished protein. Changes to the ends of introns and certain internal sequences may have an effect, however, since these control the removal of the intron sequence from the RNA copy after transcription. A mutation here can prevent intron removal, altering the finished protein. One form of the hemoglobin disease beta-thalassemia is due to an intron mutation.

transcription factors proteins that increase the rate of gene transcription

eukaryotes organisms with cells possessing a nucleus

Alleles

Humans and most other multicellular creatures are diploid, meaning they carry two sets of virtually identical chromosomes (one inherited from the mother, one from the father). The two members of a chromosome pair (called homologous chromosomes) carry identical sets of genes, so that each

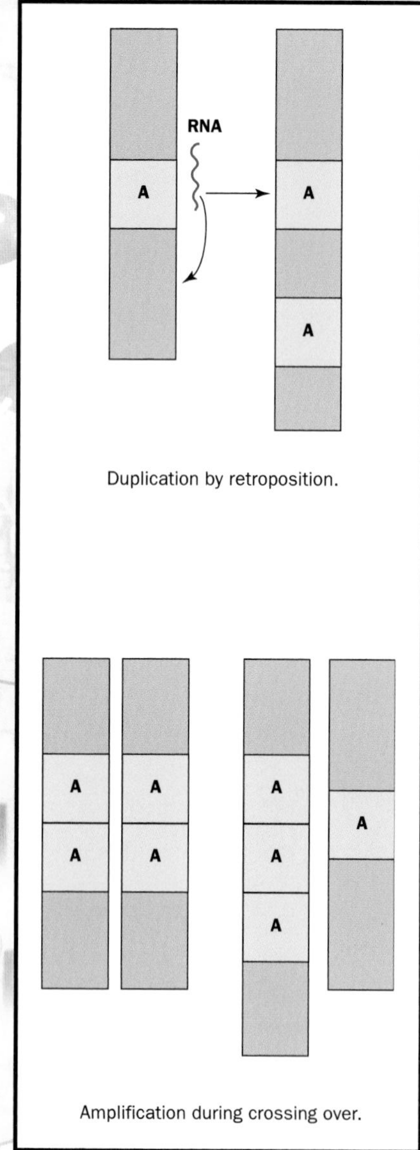

Two common mechanisms of gene duplication.

alleles particular forms of genes

organism has two copies of each gene. A point mutation in one of these creates a new form of the gene. Different forms of the same gene are called **alleles**. Creation of alleles is one of the most common forms of gene evolution. The existence of diploidy allows a greater tolerance of new alleles, since a mutation to one allele still leaves the other one functioning, and in many cases this may be sufficient for survival. The existence of alleles allows greater genetic diversity in a population, which may increase that population's ability to adapt to changing environmental conditions.

Gene Duplication

Occasionally a gene on a single chromosome will be duplicated to create a pair of identical genes. Duplication may occur for any one of several reasons. One type of duplication occurs when the RNA transcript of a gene is "reverse transcribed" back into a DNA sequence and reinserted elsewhere in the genome (a process called retroposition), leading to a new, possibly functional copy of the gene in a new location and subject to different regulatory systems. This appears to have occurred with the human gene for phosphoglycerate kinase 2, which is involved in energy use in the cell.

More likely, a retroposed gene will be functionless, since it will not have the promoters and enhancers it needs for expression. Typically, one copy of a duplicated gene is either nonfunctional or accumulates mutations that render it so. After a long period of evolutionary time, the duplicate gene may acquire so many mutations that it may be difficult to see its relationship to its parent gene. Nonfunctional copies of previously functional genes are called pseudogenes.

Analysis of genomes shows that many gene copies are found lying next to each other, linked head to tail in an arrangement called a "tandem repeat." This may occur because of errors of the normal recombination machinery that is responsible for DNA repair and crossing over during meiosis. Tandem repeats are susceptible to amplification, which is the further increase in the number of copies. This can occur during crossing over. Normal crossing over pairs up identical segments on homologous chromosomes, and then exchanges them. If the chromosomes each have a tandem repeat, the crossover machinery may line up incorrectly, leaving one homologue with three gene copies and one with only one. Repeating this process over ensuing generations can lead to dozens of extra gene copies.

Duplication of much larger portions of a genome is also possible, including whole chromosomes (called chromosomal aberrations) and even the entire genome (called polyploidy). In each case, the number of copies of a gene increases. Such copies are usually removed by natural selection, but it is sometimes advantageous to have several gene copies, particularly for those genes that code for ribosomal RNAs. These are present in dozens or even hundreds of copies, allowing rapid production of new ribosomes during cell growth.

While gene duplication is a rare event in the short term, it is frequent enough in the long term to have been a central feature in the evolution of the genome of eukaryotic organisms. As with alleles, gene duplication frees up a gene copy to accumulate mutations with less selective penalty. Over time, such a gene may change its function slightly, or even acquire a new

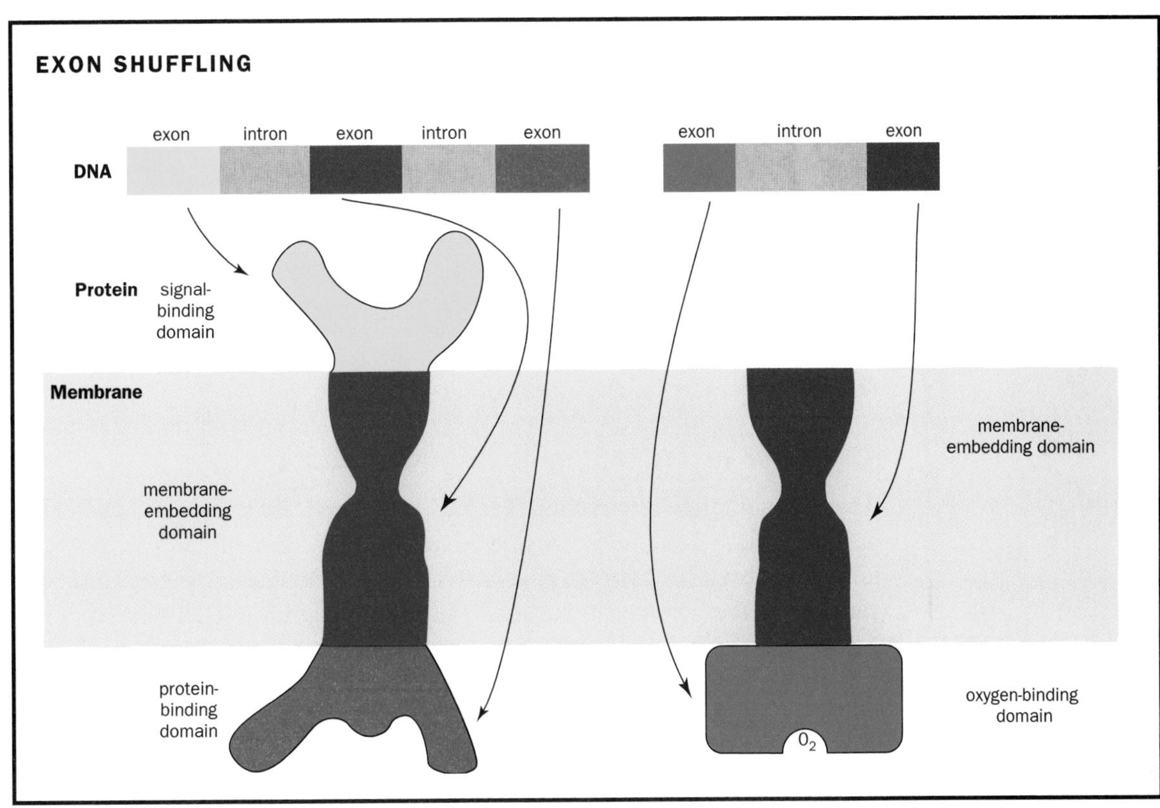

The "exon shuffling" model of protein evolution suggests that exons encode functional domains of protein, and can be shuffled to create proteins with novel functions.

function, that increases the capabilities of the organism. The modification of duplicated genes therefore provides the diversity that is acted on by natural selection. An increase in the rate of gene duplication is thought to be one factor contributing to the diversification of multicellular organisms during the so-called Cambrian explosion 600 million years ago, which gave rise to most of the basic animal forms that exist today.

Gene Families

The set of genes that evolve from a single ancestral gene comprise a gene family. In humans, gene families are found in many important groups of proteins, and more are being discovered as the human genome is explored. These include the globins, which carry oxygen in the blood (hemoglobin) and store it in muscle (myoglobin); the immunoglobulins (specifically, the heavy chain of the immunoglobulin), which form antibodies of the immune system; the actins, which move the cell cytoskeleton and muscle; the collagens, which form cartilage and other structural materials; and the homeotic genes, master controllers of embryonic development. There are many other examples as well.

A new member of a particular gene family is discovered by comparing its sequence to known members. This is usually done by computerized database search, and is one of the challenges of **bioinformatics**, a new specialty devoted to collecting and analyzing large amounts of biological information.

bioinformatics use of information technology to analyze biological data

Pseudogenes

Over time, some gene copies mutate to lose their function entirely. Such so-called pseudogenes may arise through accumulation of mutations that

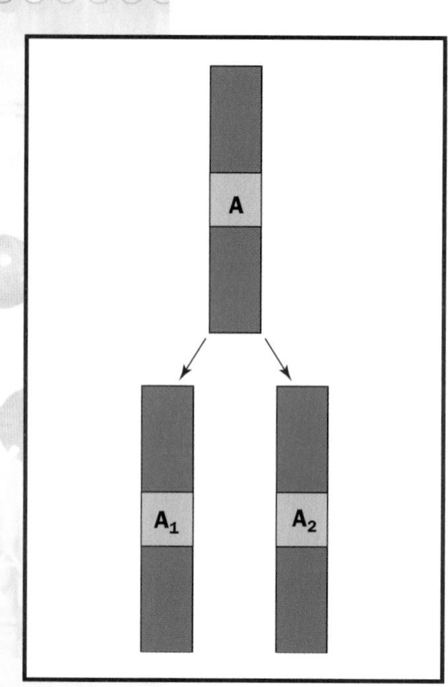

A gene family is descended from a common ancestral gene.

translation synthesis of protein using mRNA code

prevent translation of the gene, such as an insertion or deletion that stops **translation** at the beginning of the gene sequence. Pseudogenes also arise from mutation in a gene's promoter region. The promoter is the site at the beginning of the gene that attracts the enzyme called RNA polymerase. Without a functional promoter, the gene cannot be transcribed effectively, and so cannot lead to protein production.

Retroposition is a very common source of pseudogenes. Pseudogenes have been discovered because their sequences are similar to functional genes. In humans, pseudogenes are known to exist for topoisomerase (a gene that cuts DNA to prevent twisting), ferritin (an iron storage protein), two different forms of actin, and many other genes.

The Role of Transposable Genetic Elements

Transposable genetic elements are DNA segments that move around in the genome. Many biologists consider them to be a form of "selfish DNA," a kind of genetic parasite that serves no useful function for the host, but remains in the genome because it is efficient at getting itself copied. They can be present in large numbers of copies. In humans, more than a million copies of a single element, called Alu, account for about ten percent of the entire genome.

Insertion of a transposable element can disrupt a gene, creating a pseudogene. When a transposable element moves, it occasionally also takes a gene with it, placing it in a new position under the control of different regulatory elements. Alternatively, it may move an enhancer, thus affecting both the gene whose enhancer was removed and the gene (if any) it is now placed closer to. Some transposable elements themselves contain enhancers, further increasing the chances of altering gene expression when they are inserted in a new location.

Exon Shuffling

The coding portions of eukaryotic genes, termed "exons," are interrupted by noncoding regions, termed "introns." The evolutionary role of introns has been controversial since their discovery in 1977. Some scientists propose they are just another form of "junk DNA," and may be the relics of transposable elements or other forms of selfish DNA. Others suggest they may have played a central role in protein evolution.

The argument about the evolutionary importance of introns turns on exactly how they divide up the genes in which they are found. Proteins, which are encoded by genes, are not random strings of amino acids, but rather highly organized three-dimensional shapes, with different functions served by discrete parts, known as domains. A protein may contain half a dozen domains; one may bind a signaling molecule from outside the cell, another embeds the protein in a membrane, another binds an internal protein, and so on. It is often the case that each domain in a protein is folded up from a discrete segment of the amino acid chain.

Just as the domain's amino acids occur in sequence in the protein, the nucleotides that code for them occur in sequence in the gene. Those who propose that introns play a vital role in protein evolution suggest that exons correspond to the protein's domains, and that introns serve to divide the gene into these useful little bits of code. In this view, exons serve as "modules," or useful gene segments, that can be shuffled (via gene duplication and transposable elements, for instance) to create genes for new proteins with novel functions. For instance, a module for a membrane-embedding domain could be linked to a module for an oxygen-binding domain, allowing oxygen to be stored on a membrane, or a hormone-binding domain might be joined to a promoter-binding domain, allowing a hormone to control gene transcription.

One form of the disorder hemophilia is due to the insertion of a transposable element into a blood clotting gene.

The validity of this model of protein evolution depends on whether a gene's exons do indeed correspond to its protein's domains, and whether introns do actually separate domain-coding regions. So far the evidence is mixed, with some genes clearly divided this way, but many others showing complex or conflicting structures.

Because of this, scientists do not yet agree on the importance of exon shuffling in protein evolution. While it likely has occurred, it is unknown how widespread it may be. Also at issue is whether introns themselves arose early or late in life's evolution. If early, it may have been central to the development of all forms of life. The absence of introns in bacteria would then presumably be due to a streamlining of their genome by natural selection. If introns arose late, they were probably confined to eukaryotes and were therefore only important in their evolution.

While there is much that remains controversial, there is little disagreement about the importance of a related use of exons that occurs continually in many tissues. This is called alternative splicing. In this case, particular exons may be omitted, or they may be reassembled differently from tissue to tissue, creating tissue-specific variants, called isoforms, of the same protein. SEE ALSO ALTERNATIVE SPLICING; BIOINFORMATICS; CHROMOSOMAL ABERRATIONS; DEVELOPMENT, GENETIC CONTROL OF; GENE; GENE FAMILIES; GENETIC CODE; HEMOGLOBINOPATHIES; IMMUNE SYSTEM GENETICS; MUTATION; POLYPLOIDY; PSEUDOGENES; RNA PROCESSING; TRANSPOSABLE GENETIC ELEMENTS.

Richard Robinson

Bibliography

Alberts, Bruce, et al. *Molecular Biology of the Cell*, 4th ed. New York: Garland Science, 2002.

Cooper, David N. *Human Gene Evolution.* Oxford: BIOS Scientific Publishers, 1999.

Eickbush, T. "Exon Shuffling in Retrospect." *Science* 283 (1999): 1465–1467.

Extranuclear Inheritance *See Inheritance, Extranuclear*

Eye Color

When someone asks, "What color are your eyes?" he should really be asking "What color are your irises?" because it is the iris that contains the pigment that determines the color of your eyes. Despite the fascination eye color holds for us, the genes responsible for it in humans are not well-known and are more complex than most people think.

Iris Structure

The iris is the most visible portion of the uveal tract, which is the middle compartment of the eye. The iris is made up of blood vessels and connective tissue, in addition to melanocytes and other pigmented cells that are responsible for its distinctive color.

The iris contains muscles that control its movement and allow for changes in the size of the pupil, controlling the amount of light that enters the eye and that ultimately reaches the retina. Though its function is easily observed and obvious to any who look at it under varying lighting conditions, it is the appearance of the iris—specifically, its color—that is most striking and apparent. The structure itself takes its name from Iris, the Greek goddess of the rainbow and messenger of the gods.

anterior front

melanocytes pigmented cells

posterior rear

The iris consists of two layers of different embryological origin. The **anterior** border of the iris consists of the stroma, a loose and interrupted layer of connective tissue. It is composed of **melanocytes** and nonpigmented cells, as well as other types of cells and tissues. Melanocytes contain melanin, a brown or black pigment. The overall structure of the stroma is similar in irises of all colors. The iris pigment epithelium forms the densely pigmented **posterior** layer of the iris. It consists of two layers of tightly fused, pigmented cells.

Differences in iris color depends on the amount of pigmentation in the deep stroma, especially the anterior border layer, and on the density of the stroma, both of which influence how much light, and what wavelengths, are absorbed and reflected. As with other objects, the color we see is the result of reflected light. The stroma of brown irises is densely pigmented with melanin and absorbs much of the light that enters it. In many human populations, brown is the only eye color. Blue irises have lightly pigmented stroma, and light of longer wavelengths (red to yellow) readily penetrates the iris and is absorbed, while some light of shorter wavelength (blue) is reflected back and scattered by the iris stroma; hence the blue color.

alleles particular forms of genes

The inability to make melanin, as in albinism, leaves the iris without any pigment. The iris appears pink from the color of the blood flowing through it. Albinism is a recessive condition, requiring two defective **alleles** for melanin production, one inherited from each parent. Albinism also prevents pigment production in the hair and skin.

Genetics of Eye Color

Differences in iris color have been attributed to such causes as the temperature of the brain and eyes. Some people have stressed differences between dark-eyed and light-eyed populations and have suggested that eye color is related to general traits such as temperament or intellect. But, toward the middle of the nineteenth century, it had become clear that iris color was due to iris pigment, that this pigment developed soon after birth, and that the final quantity and distribution of the pigment was a hereditary trait.

Originally, iris color was thought to be a simple trait—one governed by a single gene with multiple forms, or alleles, corresponding to each color. In this scheme, blue was thought to be recessive, requiring two copies of

the blue allele in order to be displayed. Therefore, two blue-eyed parents could have only blue-eyed children, since each parent had only blue alleles. However, repeated observation of brown-eyed offspring from two blue-eyed parents showed this view to be wrong. Iris color is likely to be a polygenic trait—one governed by at least two genes and possibly more.

Brown versus blue eye color is believed to be controlled by two genes on chromosome 15, called *BEY1* and *BEY2*. Green versus blue eye color is believed to be controlled by a gene on chromosome 15, called *GEY*. In this system, blue is believed to be recessive to both brown and green. The protein products of these genes are unknown, however, as is the number of alleles possible for each. Furthermore, these three do not fully explain inheritance of all eye colors. More genes, which likely modify the action of these three, are probably involved.

Traditionally, iris color was felt to be stable throughout adulthood. However, iris color may change in response to disease. For example, there is a gradual unilateral (one-sided) loss of pigmentation in Horner's syndrome and in Fuchs' heterochromic iridocyclitis. There is also evidence for pigment loss in the iris as a result of aging, and changes in iris color may also occur spontaneously in normal people after adolescence. In addition, some commonly used drugs such as latanoprost (which lowers intraocular pressure) have caused hyperpigmentation in some irises. SEE ALSO COMPLEX TRAITS; INHERITANCE PATTERNS.

Eric A. Postel

Bibliography

American Academy of Ophthalmology. *Fundamentals and Principles of Ophthalmology: Basic and Clinical Science Course.* San Francisco: American Academy of Ophthalmology, 1995.

Imesch, Pascal D., et al. "The Color of the Human Eye: A Review of Morphologic Correlates and of Some Conditions That Affect Iridial Pigmentation." *Survey of Ophthalmology* 41, supplement 2 (1997): s117–s123.

McKusick, Victor A. *Mendelian Inheritance in Man: A Catalog of Human Genes and Genetic Disorders*, 12th ed. Baltimore: Johns Hopkins University Press, 1998.

Internet Resource

Online Mendelian Inheritance in Man. Johns Hopkins University and National Center for Biotechnology Information. <http://www.ncbi.nlm.nih.gov/Omim>.

Fertilization

Fertilization is the fusion of a female's egg cell (oocyte) and a male's sperm cell (spermatozoa) to form the first cell of a new and unique being. While on the surface this sounds like a simple process, there are many factors that make this possible.

Gametes

Gametes are unique from all other cells. Typically, each cell in the human body contains twenty-three pairs of chromosomes (for a total of forty-six). Mature egg and sperm cells contain only one copy of each chromosome (for a total of twenty-three). At fertilization the fusion of the two gametes will create a cell with the appropriate twenty-three pairs of chromosomes

gametes reproductive cells, such as sperm or egg

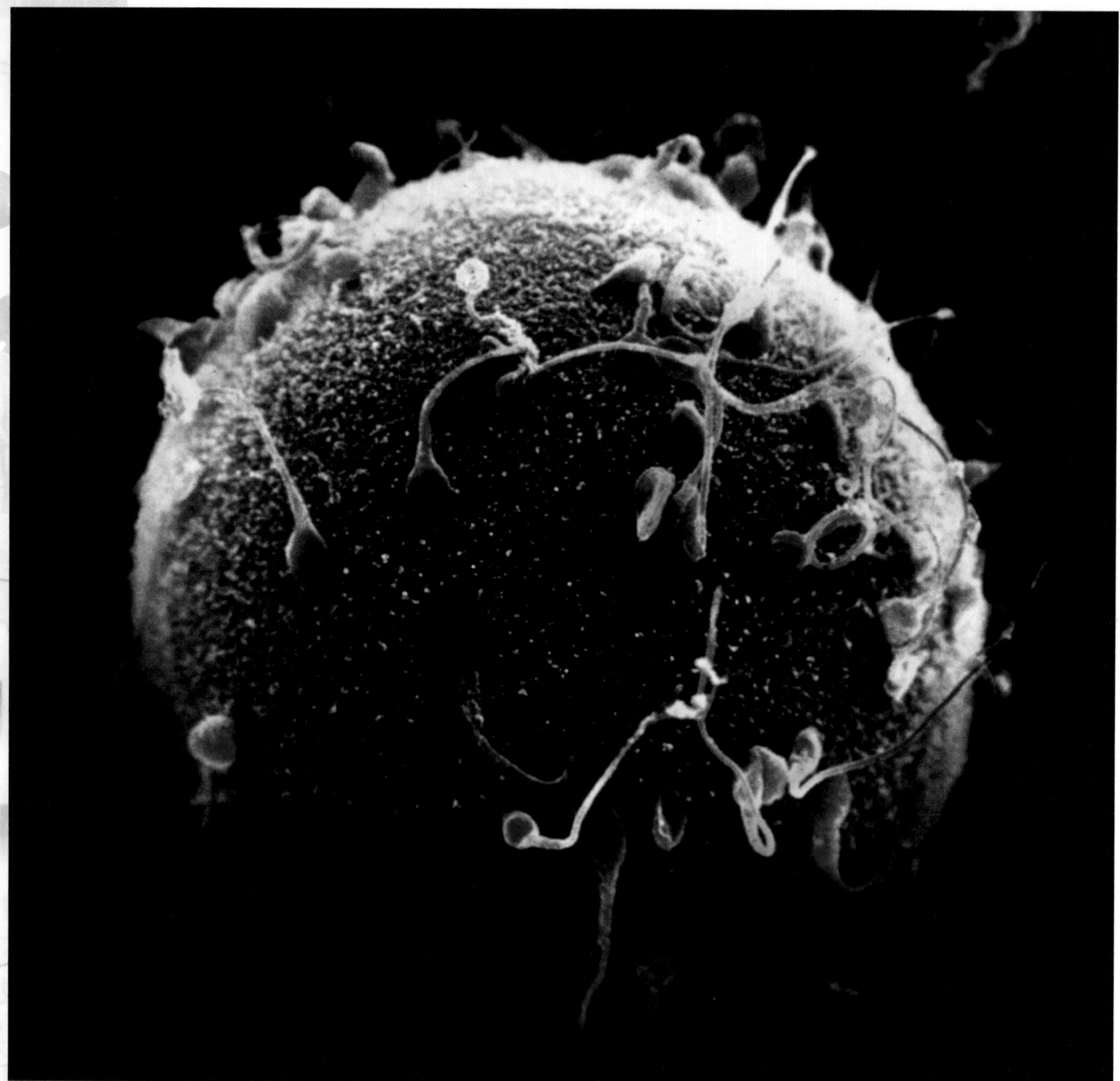

A scanning electron microscope (SEM) image captures many sperm attempting to penetrate an egg during the process of fertilization. If one of these sperm is to enter the egg it must create a way in; its "head" releases enzymes to break down the egg's outer layer, the "zona pellucida."

(forty-six individual chromosomes) necessary for human development. In this way, one chromosome of each pair will originate from each parent, making the new individual unique from any other person that came before. The specialized process by which the genetic material is shuffled and the chromosome number is cut in half, from forty-six to twenty-three chromosomes, is called **meiosis**.

Ovulation and Ejaculation

Sperm and egg cells are not only different from other cells, but are different from each other. A female is born with all the eggs she will ever have. At birth, the chromosomes of these eggs have only completed the beginning of meiosis (meiosis I) and will remain dormant (inactive) until the onset of menstrual cycles in puberty. Specific hormones produced during the menstrual cycle around day thirteen or fourteen trigger the continuation of meiosis in one egg each month. Meiosis is suspended for the second time in the middle of meiosis II around three hours prior to ovulation, and does not resume unless fertilization occurs.

meiosis cell division that forms eggs or sperm

During ovulation this egg, enclosed in two layers of protective material, is released from the ovary. The outer layer, the cumulus oophorus, is comprised of cells called cumulus cells, and the inner layer, the zona pellucida, is comprised of a jellylike coating made of protein and sugar. Once released from the ovary, the egg is swept into the fallopian tube. It is receptive to fertilization for only about one day. The sperm must reach the egg during this time, usually in the fallopian tubes, or fertilization will be impossible.

During sexual intercourse, millions of sperm are deposited into the vagina. They travel through the cervix and uterus to the fallopian tubes. Sperm can live within a woman's body for up to three days. Each sperm contain three distinct parts, the head, mid-piece, and tail. Each of these parts has a distinct purpose. The head of the sperm is composed of the nucleus (containing the chromosomes), an acrosome cap (containing **enzymes** crucial in fertilization), and an outer membrane. The mid-piece contains energy-producing mitochondria, and the tail is the mechanism for movement. Despite the ingenious design, fewer than 1 percent of the sperm released in an ejaculation ever make it to the egg. Factors inhibiting the success of sperm include abnormal formation and premature death from exposure to acidic vaginal secretions. Sperm can also be blocked by excess mucus covering the cervix, or they may travel to the fallopian tube that does not contain the egg. Fortunately, only one sperm is required to fertilize the egg.

enzymes proteins that control a reaction in a cell

In order to fertilize an egg, sperm must undergo the poorly understood process of capacitation. Capacitation involves changes to the acrosome, triggered by the cervical mucus, to prepare it to release the enzymes necessary to break through the zona pellucida. Upon reaching the surface of the zona pellucida, the sperm releases enzymes to break through. Once through the zona pellucida, the head of the sperm fuses to the egg's membrane, the tail of the sperm stops moving, and the egg engulfs the contents of the sperm.

It is crucial that only one sperm enters the egg. If an extra sperm passes through, a lethal condition known as polyspermy (many sperm) will occur. On the rare occasion this occurs, the fetus will be miscarried as a result of the extra set of chromosomes. To prevent this in most instances, a substance is released from the egg that changes the zona pellucida once it has been penetrated, blocking entry of any other sperm. Sperm penetration triggers the completion of the second meiotic division in the egg. With this division, the chromosomes of the sperm and egg come together in their own nucleus. The cell now officially becomes a **zygote**, the first cell of a new individual.

zygote fertilized egg

Variations

In some instances, an ovary releases more than one egg at one time, or both ovaries release an egg simultaneously. Each egg has the potential to be fertilized, resulting in multiple pregnancies. Since each conception originates

from a separate egg and sperm, individuals created in this way are as different as those conceived as separate births. These are referred to as fraternal twins. In rare instances, cells from the same embryo separate to create distinct embryos that are genetically identical and are referred to as monozygotic, or identical, twins.

The exact details of fertilization vary from animal to animal. Fertilization does not always take place inside an animal. For example, sea urchins, spiny animals attached to rocks on the ocean floor, release their eggs and sperm directly into the water. Large numbers of each (millions of eggs and billions of sperm) are necessary to ensure that enough eggs will be fertilized to maintain the population. Many other ocean creatures also release egg and sperm cells into the water. However, the eggs are fertilized only by sperm of the same species because of unique proteins on the surface of the egg. As in humans, fertilization immediately triggers a change in the surface of the egg, protecting it from penetration by other sperm, even sperm of the same species.

in vitro "in glass"; in lab apparatus, rather than within a living organism

Reproductive technology has introduced further variations in how eggs may become fertilized, permitting the process to occur outside the fallopian tubes. One of the most common is **in vitro** fertilization, in which eggs and artificially capacitated sperm are combined in a glass dish and the dividing embryos are later transplanted into the uterus. SEE ALSO MEIOSIS; REPRODUCTIVE TECHNOLOGY; TWINS.

Susan E. Estabrooks

Bibliography

Primakoff, Paul, and Diana G. Myles. "Penetration, Adhesion, and Fusion in Mammalian Sperm-Egg Interaction." *Science* (Jun. 21, 2002): 2183-2185.

Tobin, Allan, and Jennie Dusheck. *Asking about Life*, 2nd ed. Orlando, FL: Harcourt, 2001.

Wasserman, Paul. "The Biology and Chemistry of Fertilization." *Science* 235 no. 4788 (1987): 553–560.

Forensics *See DNA Profiling*

Founder Effect

alleles particular forms of genes

The term "founder effect" refers to the observation that when a small group of individuals breaks off from a larger population and establishes a new population, chance plays a large role in determining which **alleles** are represented in the new population. The particular alleles may not be representative of the larger population. As the new population grows, the allele frequencies will usually continue to reflect the original small group.

Genetic Characteristics of Founder Populations

genetic drift evolutionary mechanism, involving random change in gene frequencies

Because the founder population is small, **genetic drift** can play an important role in determining the genetic makeup of subsequent generations, and allele frequencies may fluctuate. For example, consider an extreme situation where a new population is founded by just two individuals, a male and a female, perhaps because they are stranded on an island. Assume that the

The Hutterite colony, to which these children belong, is a founder population with a high prevalence of asthma.

mother is **heterozygous** for a particular allele (*Aa*), while the father is **homozygous** (*AA*). If the couple has two children, there is a 25 percent chance that the mother will pass the *A* allele to both children.

heterozygous characterized by possession of two different forms (alleles) of a particular gene

homozygous containing two identical copies of a particular gene

If neither child inherits the allele, the *a* allele is effectively lost from the population. Even as they grow, many founder populations remain relatively genetically isolated, with little immigration into the population. Examples include founder populations that have remained isolated due to geographical location, such as Finland and Iceland, or due to religious customs, such as Amish and Hasidic Jewish groups.

Founder populations may have increased prevalence of certain genetic traits, including genetic disease. Disease alleles that happen to be present in the founders may be passed on to offspring, and, since the population is small, there may be a higher prevalence of the disease than in other, larger populations. Isolated founder populations, with little marriage outside of the populations, are especially likely to have a higher prevalence of recessive disorders, since parents are likely to share many genes, and there is an increased chance of inheriting two copies of a particular disease gene. Examples of rare genetic diseases that are prevalent in founder populations are Tay-Sachs disease in Ashkenazic Jewish populations and asthma in the Hutterian population.

In the nineteenth century, the rate of deafness on Martha's Vineyard (an island off the coast of Massachusetts) was thirty times that of the mainland population. Most deaf islanders were descendants of a small handful of English families who settled there around 1700.

Founder Populations Can Be Valuable for Genetic Studies

The same forces that lead to increased risk of disease also make founder populations particularly useful for identifying which genes are involved in genetic disease. Since the founder population is derived from a small number of individuals, it is likely that those individuals with a particular disease have a common genetic profile, rather than having multiple different disease mutations or susceptibility alleles. This genetic homogeneity is important, since genetic heterogeneity can make identification of any particular disease allele very difficult.

Linkage disequilibrium mapping is a powerful method for fine-mapping disease genes in founder populations. Linkage disequilibrium refers to the physical association between harmless but traceable marker alleles and a disease allele on a chromosome. The close proximity of the markers can help pinpoint the disease **locus**. Founder populations are particularly useful for linkage disequilibrium mapping since regions in linkage disequilibrium often span greater chromosomal distances than in general populations; that is, the disease gene will often be found with a larger set of common markers in a founder population than in a larger, more diverse population. This is expected because in founder populations, all chromosomes carrying a specific disease allele may be descended from a single ancestral chromosome, thus the disease allele will be in linkage disequilibrium with alleles at nearby markers. In a larger, more diverse population, the disease allele may have arisen on several different chromosomes, therefore the linkage disequilibrium, even for very close markers, may not be as great.

locus site on a chromosome (plural, loci)

One example of linkage disequilibrium mapping in founder populations is the identification of a region containing the diastrophic dysplasia gene in eighteen families from Finland. This condition causes bent or abnormal bone growth. The region to which the disease gene was localized was narrowed substantially because scientists were able to take advantage of the extensive linkage disequilibrium around this gene in the affected individuals, all of whom shared a series of alleles surrounding the disease gene.

A related method for mapping disease genes that is well-suited for founder populations is **haplotype** analysis. A haplotype is defined as the set of alleles that are inherited as a group from one parent. A haplotype forms an identifiable pattern that can be used to track inheritance of all the genes within it. There are only a small number of haplotypes among the founders. Recombination tends to break up haplotypes over time, with the alleles that are closest together remaining together the longest.

haplotype set of alleles or markers on a short chromosome segment

A haplotype that is constantly inherited with a disease can be analyzed to narrow the region in which the gene should be sought. This means that researchers can look for shared regions or segments of chromosomes among affected individuals to help identify the location of a disease gene. For example, genetic researchers were able to demonstrate that the majority of cases of idiopathic torsion dystonia (a neurological disease) in Ashkenazic Jews were due to a single mutation from a common ancestor, because the affected individuals shared common alleles (a consistent haplotype) on either side of the mutation.

Another advantage of genetic studies in founder populations is that good clinical and genealogical recordkeeping is often available. Many genetic studies have been successful in Finland because of the population history of this region. For instance, the current Finnish population is believed to have come from a small group of individuals who settled in the southwest part of the country about 2,000 years ago. Since the initial immigration, the population has continued to be relatively isolated, with little migration into it.

Genealogical records are available through church parishes and often go back six to twelve generations, allowing scientists to develop accurate and detailed family histories linking individuals together. Despite these advantages, for common diseases such as asthma, scientists must consider that

genes that cause asthma in Hutterites may or may not be relevant to other groups with asthma. Thus the scientist must weigh the advantages of performing genetic studies in small, historically isolated populations with the potential disadvantage of being unable to eventually generalize the studies' results. SEE ALSO GENETIC DRIFT; HARDY-WEINBERG EQUILIBRIUM; INBREEDING; LINKAGE AND RECOMBINATION; MAPPING; POPULATION BOTTLENECK; POPULATION GENETICS; TAY-SACHS DISEASE.

Eden R. Martin and Marcy C. Speer

Bibliography

Risch, Neil, et al. "Genetic Analysis of Idiopathic Torsion Dystonia in Ashkenazic Jews and Their Recent Descent from a Small Founder Population." *Nature Genetics* 9 (1995): 152–159.

Hastbacka, Johanna, et al. "Linkage Disequilibrium Mapping in Isolated Founder Populations: Diastrophic Dysplasia in Finland." *Nature Genetics* 2, no. 3 (1992): 204-211.

Strachan, Tom, and Andrew P. Read. *Human Molecular Genetics.* New York: Wiley-Liss, 1996.

Fragile X Syndrome

Fragile X syndrome is one of the most common causes of inherited mental retardation. Individuals with fragile X syndrome can exhibit moderate to severe mental retardation. Additional characteristics may include autistic-like behavior, hyperactivity, mitral valve prolapse (a heart valve defect), a large head circumference, a long face with a prominent forehead and jaw, protruding ears, flat feet, hyper-extensive joints ("double-jointedness"), and, in males, enlarged testicles. Fragile X syndrome is not restricted to any ethnic group. It was the first of the so-called triplet repeat diseases to be discovered, and study of it has led to a growing understanding of DNA instability and its role in disease.

Discovery of the Syndrome

The first family with fragile X syndrome was described by J. Purdon Martin and Julia Bell in 1943. This family had eleven severely retarded males, and the inheritance pattern of the mental retardation appeared to be X-linked. X-linked traits are inherited on the X chromosome and are more common in males, who have only one X chromosome, than in females, who have two.

In 1969, in a different family, Herbert Lubs observed a constriction near the end of the long, or q, arm of the X chromosome in four mentally retarded males and two of their mentally normal female relatives. This constriction made the X chromosome appear to be broken. Hence the name "fragile X."

For years, little attention was paid to Lubs's finding. Renewed interest in the observation emerged in the late 1970s, when additional families were identified with mental retardation and the same chromosome abnormality, or fragile site. Moreover, in 1977 Grant Sutherland discovered that the ability to detect this fragile site was dependent on the chemicals used to study patients' chromosomes. Sutherland's crucial observation helped develop the

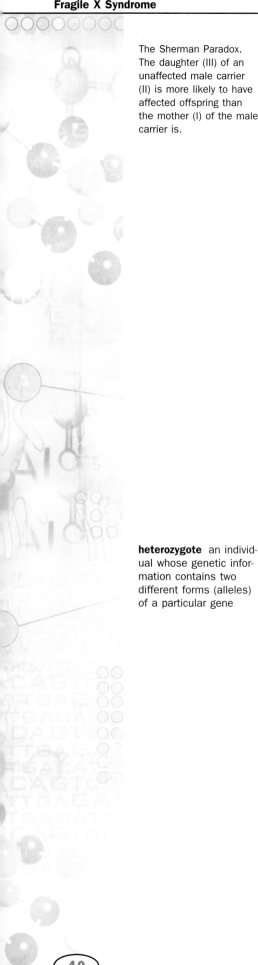

The Sherman Paradox. The daughter (III) of an unaffected male carrier (II) is more likely to have affected offspring than the mother (I) of the male carrier is.

first diagnostic test for fragile X syndrome. Using this knowledge, investigators reexamined chromosomes from the original fragile X family described in 1943 and demonstrated that, indeed, affected individuals in this family carried the characteristic fragile site.

Puzzling Inheritance Pattern

While the location of this fragile site established that fragile X syndrome was indeed X-linked, inheritance of this disorder was clearly not typical of other X-linked disorders. At first, it was believed that fragile X syndrome was an X-linked recessive genetic disorder. However, there were many observations inconsistent with this inheritance pattern.

heterozygote an individual whose genetic information contains two different forms (alleles) of a particular gene

If the disorder was truly inherited in an X-linked recessive manner, **heterozygote** carrier women would not display any characteristics of the syndrome, and all carrier males would. But there were reports of affected females, and of males who carried the fragile site but were unaffected. It was particularly difficult to reconcile that some male carriers could be so severely affected while others were completely unaffected.

Because of these puzzling observations, in 1985 Stephanie Sherman and her colleagues studied the inheritance pattern of fragile X syndrome more closely. They demonstrated that the risk of expressing mental retardation was dependent on the individual's position in the pedigree, with risk increasing in later generations. The daughter of an unaffected male carrier was more likely to have affected offspring than the mother of the unaffected male carrier was: something had changed on the X chromosome over the two generations. This observation became known as the "Sherman paradox" and was crucial to understanding the genetic mutation that causes fragile X syndrome.

To explain the unusual inheritance pattern, Sherman, her colleagues, and several other scientists hypothesized that the alleged gene for fragile X syndrome was mutated in a two-step process. They proposed that the first mutation caused a "premutation" state that produced no clinical symptoms, and that a second mutation was required to convert the premutation to a "full mutation" form that was associated with the characteristic symptoms

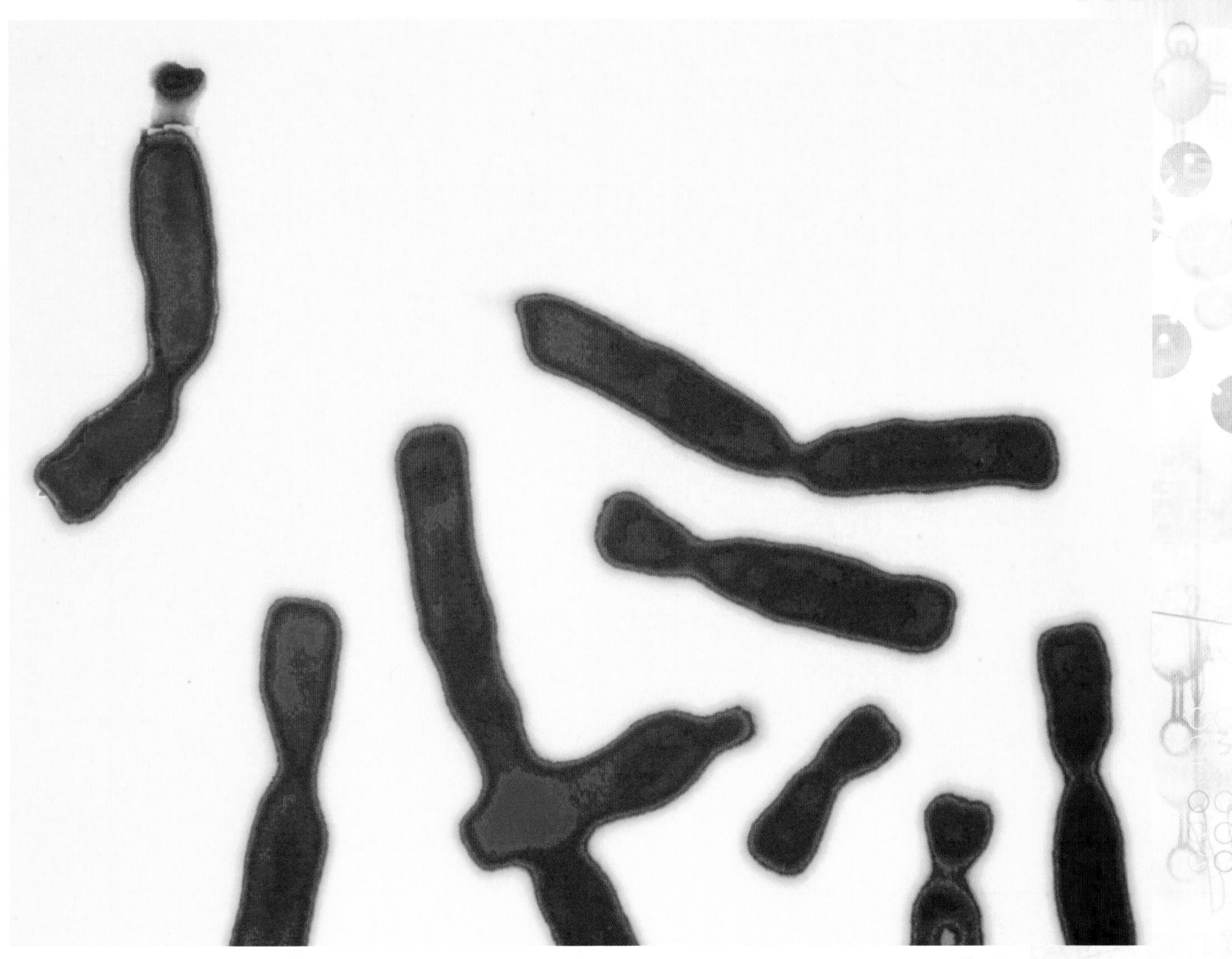

of fragile X syndrome. Moreover, conversion from a premutation to a full mutation was proposed to occur only when the premutation was transmitted from a carrier female.

The fragile X chromosome is stained with purple dye. The affected chromosome is described as "fragile" because of the extra constriction near the end of the long arm of the X chromosome. Mental disability presents in 80 percent of carrier males, and 50 percent of carrier females.

An Expanding Gene

In 1991 an international team of scientists identified the gene and mutation that causes fragile X syndrome. They found that in families with fragile X syndrome, there is a piece of the *FMR1* gene, called a CGG repeat, which is abnormally expanded.

In the general population, the repeat length can range from about six to fifty-four copies of the CGG, and the repeat is stable, or is passed from parent to child without change. In fragile X families, the premutation form of the repeat contains between fifty and two hundred copies of the CGG repeat, and the repeat is unstable.

Premutation alleles can expand to full mutation **alleles** (with more than two hundred copies of the CGG repeat) by transmission of the premutation from a mother to her child. A woman's risk of having a child with the

alleles particular forms of genes

full mutation correlates to her own repeat size. The larger her premutation, the more she risks having a child who carries the full mutation.

The CGG repeat is usually interrupted by a single AGG trinucleotide every ten CGG repeats, but this can vary from individual to individual. Because premutation alleles have fewer AGG interruptions compared with normal-size *FMR1* alleles, it is believed that the AGG interruptions are important for stability of the CGG repeat.

Individuals with a premutation do not express the clinical symptoms associated with fragile X syndrome, although it has been reported that premutation carrier females can experience premature ovarian failure. Individuals who carry the full mutation can express symptoms of fragile X syndrome because they are missing the protein produced by the *FMR1* gene. Males with a full mutation always exhibit some symptoms of the disorder. Due to X inactivation, females with a full mutation may or may not express symptoms.

Although there is currently no cure for fragile X syndrome, scientists are making great progress in understanding the biology of the disorder. In the mid- to late 1990s, Stephen Warren and colleagues determined that the *FMR1* gene product, named FMRP, is an RNA-binding protein that shuttles in and out of the nucleus and is involved in binding various messenger RNAs. Moreover, scientists successfully developed mice that lack the *FMR1* gene, which will greatly aid research. Symptoms of fragile X mice include learning disabilities, hyperactivity, and, in males, enlarged testicles. Prevailing hypotheses about FMRP suggest that this protein is involved in forming neural connections in the developing brain.

The identification of *FMR1* and the expanded CGG repeats was a landmark discovery in human genetics because it established a novel class of human genetic mutations, trinucleotide (or triplet) repeat expansions. Since the discovery of *FMR1* and the expanding CGG repeats, scientists have identified more than ten other human genetic disorders that are caused by expansions of trinucleotide repeats, including disorders such as Huntington's disease and myotonic muscular dystrophy. SEE ALSO INHERITANCE PATTERNS; INTELLIGENCE; MOSAICISM; TRIPLET REPEAT DISEASE; X CHROMOSOME.

Allison Ashley-Koch

Bibliography

Hagerman, Randi Jenssen, and Amy Cronister, eds. *Fragile X Syndrome: Diagnosis, Treatment, and Research*, 2nd ed. Baltimore: Johns Hopkins University Press, 1996.

Internet Resource

Online Mendelian Inheritance in Man: Fragile Site Mental Retardation 1; FMR1. Johns Hopkins University and National Center for Biotechnology Information. <http://www.ncbi.nlm.nih.gov/htbin-post/Omim/dispmim?309550>.

Fruit Fly: *Drosophila*

Drosophila melanogaster, a common fruit fly, was one of the first model organisms used in genetic research, and continues to be one of the most important. Thomas Hunt Morgan (1866–1945) developed *Drosophila* as a model system in 1909. Morgan, along with his students, Calvin Bridges, Alfred

Sturtevant, and Hermann Muller, made some of the most important discoveries in genetics through their work with *Drosophila.* Among these were the genetic explanation of sex linkage (the location of a gene on a sex chromosome); proof that genes are contained on chromosomes; and the demonstration that genes are arranged on a chromosome in a linear order with fixed, measurable distances between them, the principle that underlies genetic mapping.

Like other good model organisms, *Drosophila* is easy to rear in the laboratory. It has a short life cycle, lasting about two weeks, and produces many offspring. Each female can lay hundreds of eggs. These traits make it ideal for isolating mutants and carrying out many genetic crosses rapidly.

Mutants are the cornerstone of genetic analysis. To find a mutation one must be able to recognize an observable physical trait, or **phenotype**, such as a change in anatomical structure or behavior. At first glance, watching a tiny fruit fly landing on a rotting banana, one may be hard pressed to imagine that anyone could spot an anatomical variant, much less begin to study such a complex subject as behavior. Observed through a low-powered microscope, however, *Drosophila* is a sculptural masterpiece of bristles, segments, colors, and mosaic patterns. By studying *Drosophila* mutants, scientists have devised ways to genetically dissect the cellular bases of these phenotypes, as well as such startlingly complex behaviors as learning, memory, and even sleep.

phenotype observable characteristics of an organism

A feature of a model organism that aids geneticists is a small genome size and a small number of chromosomes, since the less DNA there is to sort through, the easier it is to find genes. *Drosophila*'s genome, containing about 180 million base pairs, is approximately one-twentieth the size of the human genome. There are four pairs of chromosomes: the X and Y sex chromosomes, and **autosomes** 2, 3, and 4. The complete nucleotide sequence of the gene-rich portion of the genome was determined in 2000. The genome is estimated to encode approximately 13,000 genes.

autosomes chromosomes that are not sex determining (not X or Y)

Drosophila molecular geneticists make wide use of **transposons**. These are short segments of DNA that, when injected into a cell, can insert themselves into the chromosomal DNA at random positions. Using recombinant DNA methods, a researcher can splice any gene into a transposon, which can then serve as a **vector** for introducing the gene into a fly.

transposons genetic elements that move within the genome

vector carrier

Alternatively, transposon insertion can be used to cause mutations in genes. While much of the chromosomal DNA consists of sequences that code for non-protein elements, such as introns and "spacer" sequences between genes, a transposon may become inserted directly into a protein-coding sequence. This usually alters the amino acid sequence of the protein encoded by a gene, rendering the gene product dysfunctional. Even without knowing which gene was mutated, or where in the genome it is located, a researcher can make use of the transposon insertion as a "molecular tag" to rapidly identify the gene. Since the sequence of the transposon is known, a DNA probe can be designed to detect it (and therefore find the gene which it has mutated) by molecular hybridization methods.

An unusual phenomenon of the chromosomes in certain of *Drosophila*'s tissues provides a powerful tool for determining the positions of individual genes. The chromosomes in the fruit fly's salivary gland cells replicate

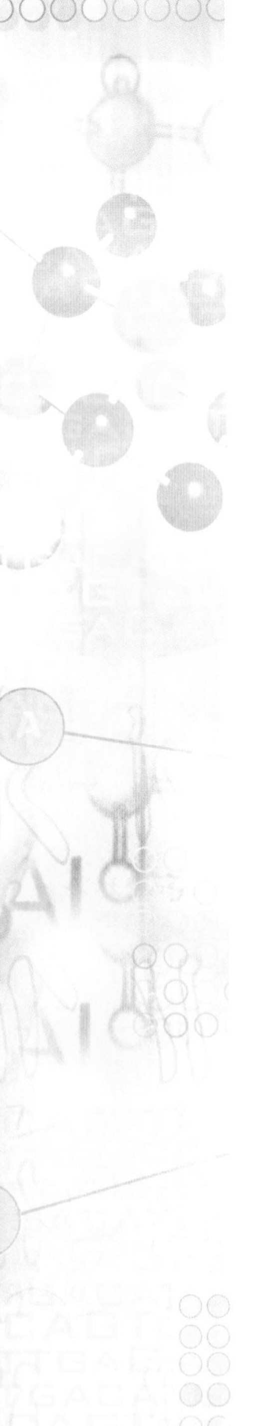

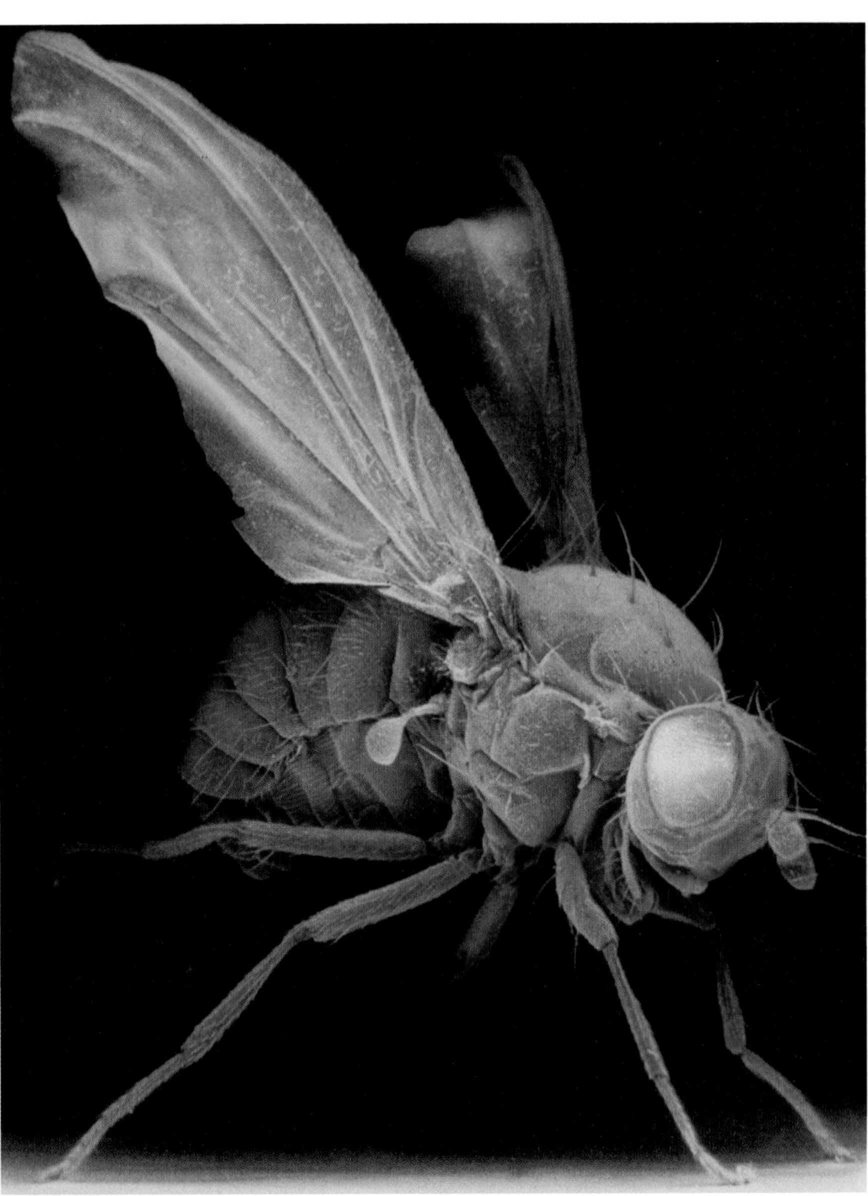

Because of their ablity to produce large amounts of offspring in a short amount of time, fruit flies were ideal specimens for early genetic experiments.

metaphase stage in mitosis at which chromosomes are aligned along the cell equator

several hundred times without separating from each other by mitosis and cell division. Instead, the newly replicated DNA strands line up parallel to one another to form a tight bundle, called polytene chromosomes. Polytene chromosomes are less condensed than normal **metaphase** chromosomes, and can be around 2 millimeters long, large enough to easily be examined in detail under a low-power microscope. When stained with certain dyes, polytene chromosomes display characteristic banding patterns along their length. *Drosophila* geneticists have made maps of the banding patterns and have learned to use them as landmarks to help them locate genes of interest. Polytene chromosomes make *Drosophila* an excellent organism for the sub-branch of genetics known as cytogenetics, which is genetic analysis through directly visualizing the chromosomes themselves.

One of the most important areas of research to come from studies on *Drosophila* is that of embryonic development. By analysis of mutants, developmental biologists have elucidated a complex and precise picture of how genes orchestrate the development of a fertilized egg into an adult fly. Many of the genes exercising the master control over these processes encode transcription factors, which are proteins that regulate when and where particular genes are transcribed to produce messenger RNA. The initial set of genes in the hierarchy acts in the **oocyte**, even before it is fertilized by a sperm. Their main function is to define the spatial polarity of the oocyte, determining which is the front and rear (posterior and anterior polarity), and which is the belly and back (dorsal and ventral polarity). These genes, in turn, activate genes that divide the embryo into segments and subsegments, which will eventually become the body segments of the adult animal. Later-acting genes, termed homeotic genes, act within each segment to define its identity, for example a wing or a leg. Remarkably, the genes and regulatory pathways involved in *Drosophila* development are highly conserved (that is, very similar genes and pathways are present), not only in other invertebrates, but also in mammals, including humans. SEE ALSO CHROMOSOME, EUKARYOTIC; DEVELOPMENT, GENETIC CONTROL OF; *IN SITU* HYBRIDIZATION; MODEL ORGANISMS; MORGAN, THOMAS HUNT; MULLER, HERMANN; MUTAGENESIS; MUTATION; TRANSCRIPTION FACTORS; TRANSPOSABLE GENETIC ELEMENTS.

oocyte egg cell

Paul J. Muhlrad

Bibliography

Alberts, Bruce, et al. *Molecular Biology of the Cell*, 3rd ed. New York: Garland Publishing, Inc., 1994.

Lawrence, Peter. *The Making of a Fly: The Genetics of Animal Design.* Oxford, U.K.: Blackwell Science, 1992.

Watson, James D., et al. *Recombinant DNA*, 2nd ed. New York: Scientific American Books, 1992.

Weiner, Jonathan. *Time, Love, Memory: A Great Biologist and His Quest for the Origins of Behavior.* New York: Alfred A. Knopf, 1999.

Internet Resource

"Drosophila." The WWW Virtual Library. <http://ceolas.org/VL/fly/index.html>.

Gel Electrophoresis

Gel electrophoresis is a widely used technique for separating electrically charged molecules. It is a central technique in molecular biology and genetics laboratories, because it lets researchers separate and purify the nucleic acids DNA and RNA and proteins, so they can be studied individually. Gel electrophoresis is often followed by staining or blotting procedures used to identify the separated molecules.

Basic Procedure

In electrophoresis, an electric field is generated to separate charged molecules that are suspended in a matrix or gel support. Negatively charged molecules move toward the **anode**, on one side of the gel, and positively charged molecules move toward the **cathode**, on the other side. The gel itself is a porous matrix, or meshwork, often made of carbohydrate chains. Molecules are pulled through the open spaces in the gel, but they are slowed down by the meshwork based on their differing properties.

andode positive pole

cathode negative pole

A solution containing DNA is added to the gel using a micropipette.

The parameters that determine the migration rate of these molecules through the meshwork are the strength of the electric field, the composition of the gel support or matrix, the composition of the liquid buffer solution the gel sits in, and the size, shape, charge, and chemical composition of the molecules being separated. Smaller molecules move faster than larger molecules, because they encounter less frictional drag in the gel. The size of the pores in the gel can be changed so this frictional drag is increased or decreased, allowing faster separation, or finer resolution.

The electrophoretic technique can analyze and purify a variety of biomolecules, but is mainly used to separate nucleic acids and proteins. A basic consideration for choosing this technique is the composition of the sample to be separated—for example, does it contain nucleic acids (DNA or RNA), or is it composed of proteins, or carbohydrates? What are the sizes of the molecules to be separated? Another important consideration is the purpose of the separation—is it qualitative, where the technique is being used to evaluate the composition of the sample, or is it quantitative, in that the separated materials are to be collected for further analysis? Cellulose or starch is used as a support medium for low molecular-weight biomolecules such as amino acids and carbohydrates, whereas separation of proteins and nucleic acids are almost always done in gels made of a porous insoluble material such as agarose or acrylamide.

Separation of Proteins

PAGE. Proteins are usually separated using vertical polyacrylamide gel electrophoresis (PAGE), a process that separates them on the basis of their electric charge and their size. Proteins with a greater negative charge will be attracted more strongly and move faster toward the anode. The **charge density** on the proteins would cause smaller molecules to move more quickly through the gel's pores.

charge density ratio of net charge on the protein to its molecular mass

The size of a gel's pores can be changed depending on the size range of the proteins being separated. This is done by raising or lowering the con-

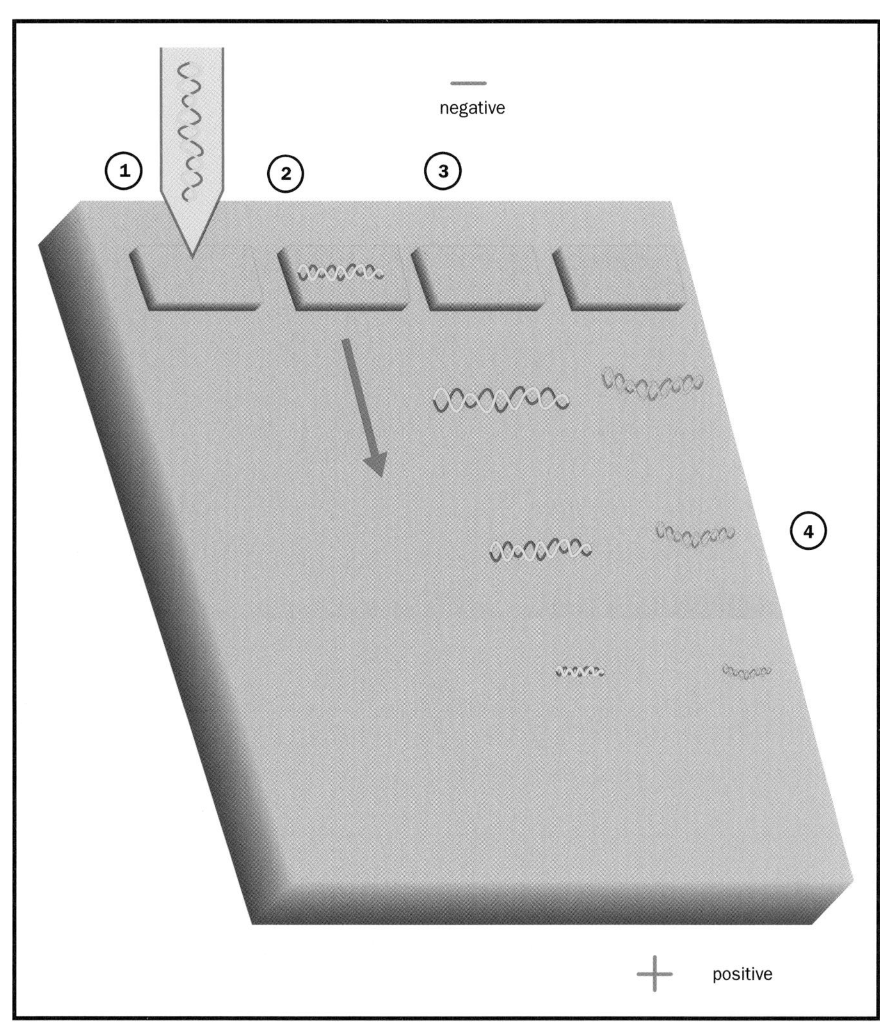

(1) DNA fragments are loaded into the gel. (2) Electric current is turned on, pulling the negatively charged DNA toward the positive electrode. (3) Smaller DNA fragments move faster through the gel. (4) The DNA is stained, revealing its position.

centration of acrylamide and bisacrylamide in the gel. Increasing the concentration results in more crosslinking between the two components, decreasing the pore size. Decreasing the concentration increases the pore size of the gel. Small proteins are separated better in a gel with large pores.

SDS-PAGE. Protein activity, such as for **enzymes**, can be determined once they are separated in the gel under conditions that do not **denature** the enzyme. Researchers can determine a purified protein's molecular weight by measuring how quickly the protein moves through a gel. The protein is first purified and denatured with heat and a reducing agent that disrupts disulfide bonding. It is then treated with an anionic detergent, sodium dodecyl sulfate (SDS), which disrupts the secondary, tertiary, and quaternary structure of the protein and coats it uniformly with negative charges. When run through a gel, the protein's migration rate is indirectly proportional to the logarithm of its molecular weight, so the smaller protein runs the fastest. The uniform negative charges ensure that the protein's migration rate is a function only of its molecular weight, not of whatever charges happen to be on it. Better resolution and separation can be obtained in an SDS-PAGE gel by first tightening the protein band before separating the proteins by size. This is accomplished by having the separating, or resolving, gel on the bottom and the larger pore gel, known as the stacking gel, on top. The

enzymes proteins that control a reaction in a cell

denature destroy the structure of

(1) A protein mixture is first separated by pH. (2) All proteins are made negative by the addition of a detergent, and are separated by size. (3) The final gel is stained to reveal the position of hundreds of unique proteins.

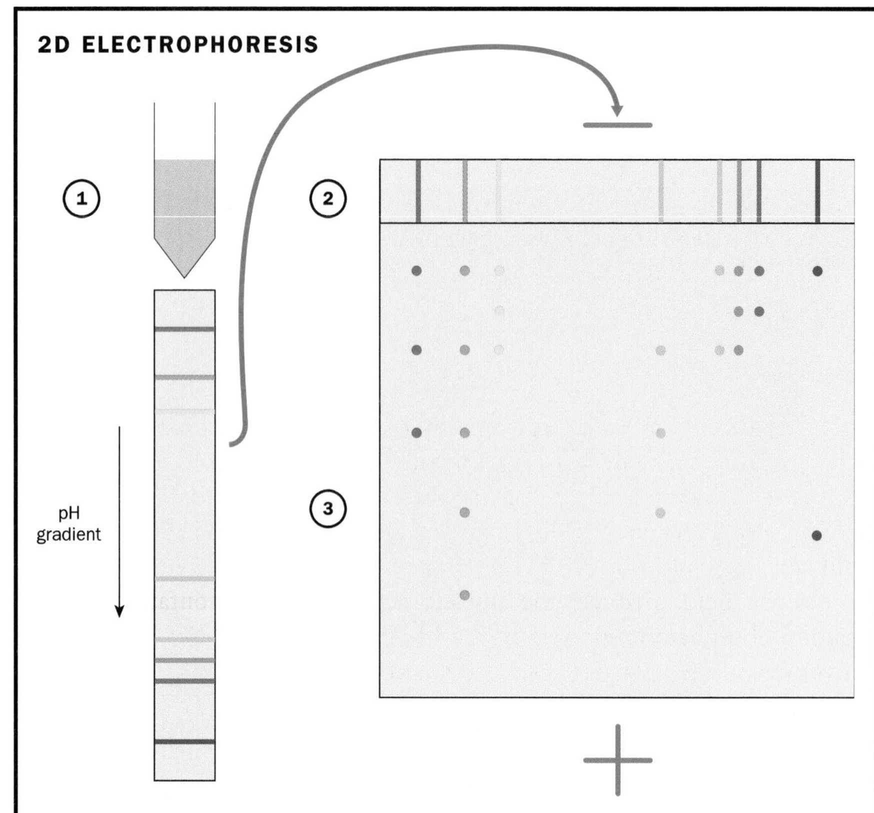

proteins enter the top gel, where they are maintained by in a tight zone between ions generated by the electric field. This is accomplished by having ions that run both slower and faster than the negatively charged proteins, so that ions sandwich the proteins between them, tightening the protein band. The proteins leave the stacking gel and enter the separating gel. In this gel the ions no longer sandwich the proteins because of a change in pH, and the pore size is smaller so that the proteins separate by size.

If the protein is run on a gel along with a ladder of proteins of known weight, then the molecular weight of the protein can be determined by comparing its migration rate to that of proteins whose molecular weights are known. This technique is known as "sodium dodecyl sulfate–polyacrylamide gel electrophoresis," or SDS-PAGE.

Isoelectric Focusing. Researchers can use gels to determine a protein's "isoelectric point," or the pH at which the protein's net charge is zero. Because pH changes the ionization state of several amino acid groups, the net charge on a protein is pH-dependent. By running proteins through a gel that has a pH gradient from one end to the other, this charge is gradually changed. At a certain pH, each protein's net charge will become zero, and the protein will stop moving. This procedure is known as "isoelectric focusing."

Two-Dimensional Electrophoresis. In "two-dimensional electrophoresis," a mixture of proteins is first separated in an isoelectric focusing tube gel. This tube is then placed sideways on an SDS-PAGE gel. In this way, proteins are separated based on two parameters: size and isoelectric point. Compared to techniques based on only one parameter, two-dimensional electrophoresis separates more proteins at once.

Two-dimensional electrophoresis is an important tool in **proteomics**. It can be used to separate large numbers of proteins that are isolated all at once after being expressed in response to a hormone, drug, or other stimulus. It can be combined with the use of DNA microarrays to allow a researcher to determine both what genes are expressed in response to a stimulus and what proteins are produced by these genes, which thereby determine an organism's physiological response to stimuli.

proteomics the study of the full range of proteins expressed by a living cell

Separation of DNA and RNA

Nucleic acids come in a very wide range of sizes, from several dozen base pairs to many millions. No single technique can be used to separate them all. Instead, researchers analyze the nucleic acid molecules using the overlapping electrophoretic techniques of polyacrylamide, agarose, and pulse-field gel electrophoresis. Each technique places DNA or RNA molecules in an electric field. Because the nucleic acid fragments contain negatively charged phosphate groups along the backbone of the DNA molecule, they move toward the positively charged anode. As with proteins, the migration rate of nucleic acids through a gel depends on their conformation, the buffer composition, the concentration of the gel support, and the applied voltage.

nucleic acids DNA or RNA

Agarose Gels. The techniques discussed so far are good for separating proteins and small nucleic acid fragments from 5 to 500 **base pairs**. The small pores of the polyacrylamide gels, however, are not appropriate for larger DNA fragments or intact DNA molecules such as plasmids. Gels made of agarose, a natural seaweed product, are used to characterize nucleic acids that are 200 to 500,000 base pairs long.

base pairs two nucleotides (either DNA or RNA) linked by weak bonds

Agarose gels, which can be purchased commercially, are prepared by dissolving purified agarose in warm electrophoresis buffer, cooling the solution to 50 °C (122 °F), and then pouring it into a mold, where it turns into a gel. Just as with polyacrylamide, the concentration of agarose in a gel determines the size of its pores. A comb placed in the gel before it sets produces the wells necessary for loading nucleic acid samples.

Nucleic acid fragments that are to be separated by size must be in "linearized" form. Plasmids, for example, must have their circular structure cut open using **restriction enzymes** before they are run on the gel. Otherwise, their rate of migration will depend on how supercoiled they are and whether they are nicked, instead of on their size. Nucleic acids that are separated in a gel can be seen with ethidium bromide or other stains.

restriction enzymes enzymes that cut DNA at a particular sequence

Pulse-Field Gel Electrophoresis. The conventional agarose gel electrophoresis described above separates nucleic acid fragments smaller than 50,000 base pairs (50 kilobase pairs). Pulse-field gel electrophoresis separates huge pieces of DNA that are between 200 and 3,000 kilobase pairs long. In this technique the electric field is not held constant during the separation. Instead, its direction and strength are repeatedly changed, with the molecules reorienting themselves every time the current changes. The molecules then slither like a snake through the gel matrix, in a process known as "reptation," with smaller fragments moving faster than larger ones. As the gel runs, it heats up and becomes more fluid, with the pulsing allowing

the larger pieces to move more easily the longer the gel runs. Typically such gels are run overnight.

Once separated, large DNA pieces, such as complete genes, can be isolated for further experiments. They can be cloned into a bacterium, sequenced, or amplified by polymerase chain reaction. SEE ALSO BLOTTING; CLONING GENES; DNA MICROARRAYS; POLYMERASE CHAIN REACTION; PROTEINS; PROTEOMICS; PURIFICATION OF DNA; SEQUENCING DNA.

Linnea Fletcher

Bibliography

Bloom, Mark V., Greg A. Freyer, and David A. Micklos. *Laboratory DNA Science: An Introduction to Recombinant DNA Techniques and Methods of Genome Analysis.* Menlo Park, CA: Addison-Wesley, 1996.

Gene

Genes are functional units of DNA that contain the instructions for making proteins or RNA. Genes also act as units of heredity, transferring the same instructions from parent to offspring. The nature, structure, and regulation of genes has been a central topic of scientific research for more than 100 years.

History of the Gene and Structure of DNA

Genes were first defined as units of hereditary transmission. The name "gene" was coined by Wilhelm Johannsen in 1909, although the concept of a discrete unit governing inherited characteristics goes back at least to Gregor Mendel in 1861. The work of Thomas Hunt Morgan and his colleagues established that genes were located on chromosomes, and in the mid-1940s Oswald Avery demonstrated that genes were composed of DNA (deoxyribonucleic acid). Since that time, some types of viruses have been discovered that use ribonucleic acid (RNA) instead of DNA, but here we shall concentrate on DNA genes. The discovery of the structure of DNA in 1953 by James Watson and Francis Crick set the stage for the next fifty years of research into gene structure, function, and regulation.

nucleotides the building blocks of RNA or DNA

DNA is a linear molecule composed of subunits called **nucleotides**. Each nucleotide is made of a sugar and phosphate group, plus a chemical base, of which there are four types: adenine, thymine, guanine, and cytosine (A, T, G, C). Nucleotides are typically referred to by the name of their base. DNA exists as a pair of strands, wound around one another into a double helix, with the bases directed into the center. The structure and charges of the bases dictate that A on one strand can match only up with T on the other, and C only with G. This complementarity provides the basis for faithful replication of the entire DNA molecule.

Genes Code for Protein and RNA

eukaryotic describing an organism that has cells containing nuclei

While all genes are made of DNA, not all stretches of DNA act as genes. Indeed, in **eukaryotic** organisms, most of the DNA does not function as genes, meaning it is not the code for making proteins or RNA. Some DNA outside of genes has a structural role, some are remnants of old genes that

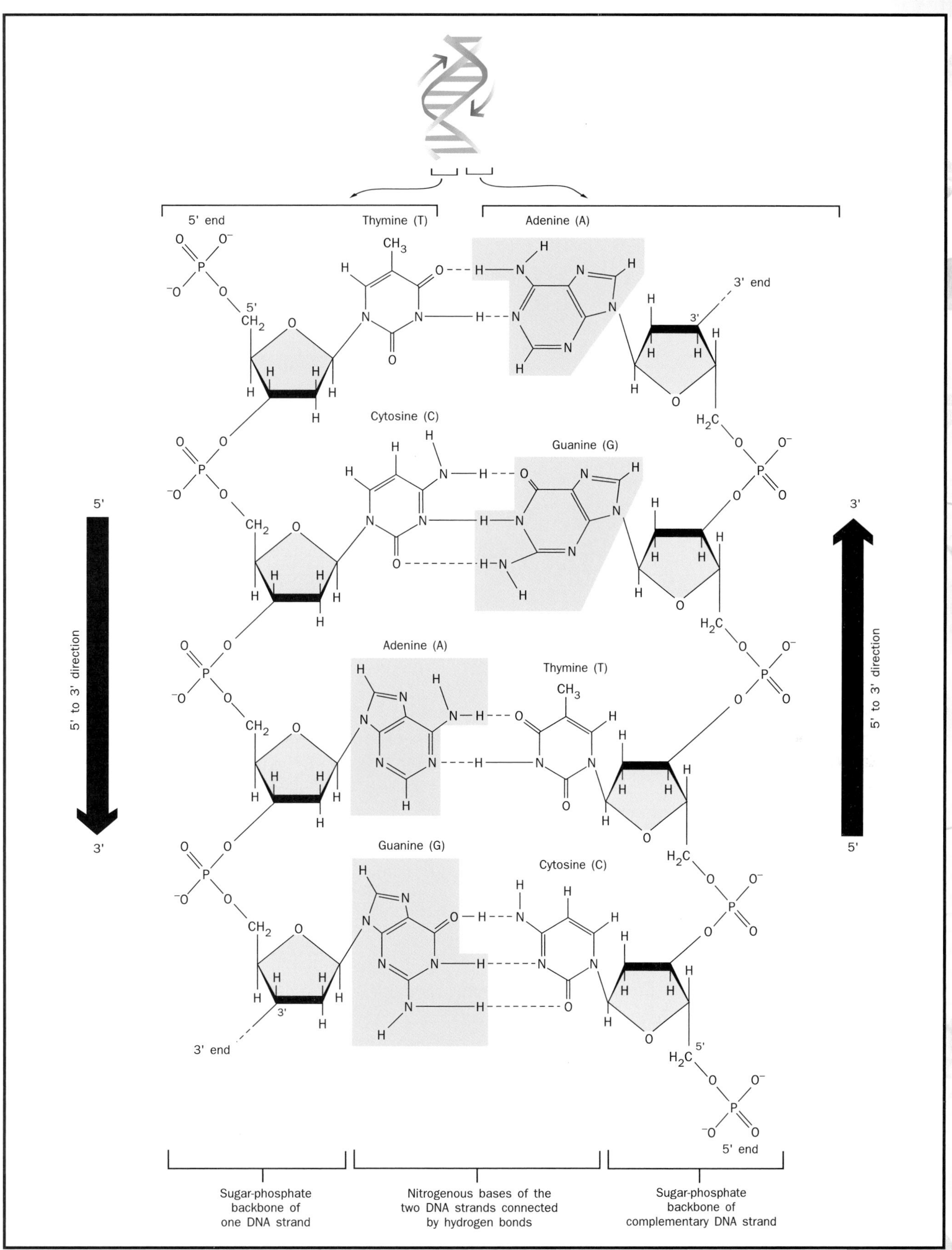

DNA nucleotides pair up across the double helix.

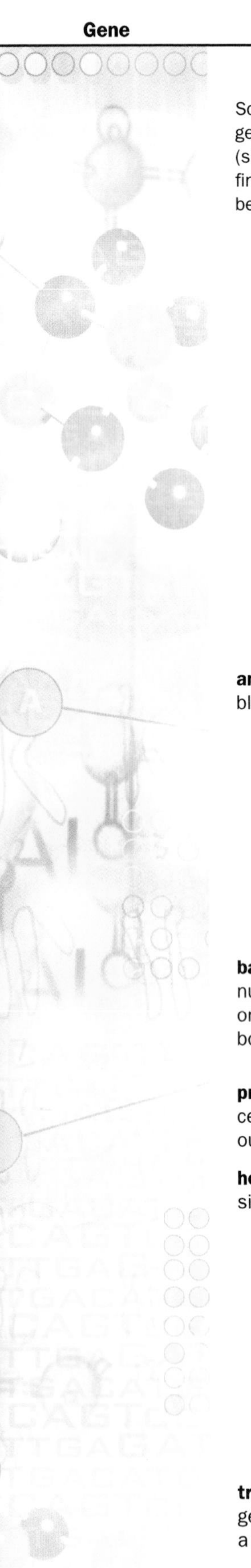

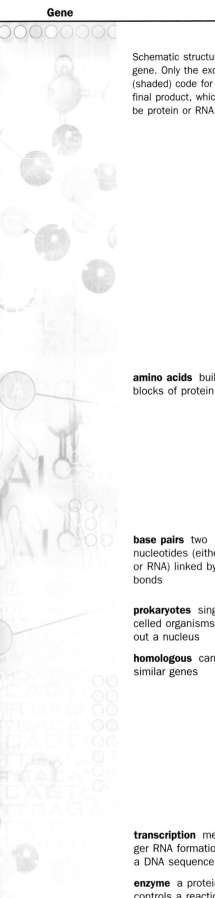

Schematic structure of a gene. Only the exons (shaded) code for the final product, which may be protein or RNA.

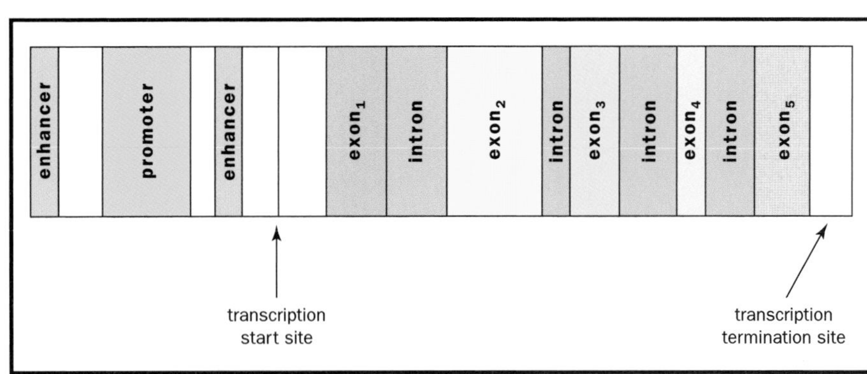

now are functionless, and much of it appears to be "junk," inserted and copied by viruslike sequences. Within a gene, usually only one side of the double helix actually codes for product; the other side is silent. Which side of the helix acts as code varies from gene to gene.

amino acids building blocks of protein

Almost all genes code for proteins. Proteins are strings of **amino acids**, and the sequence of nucleotides in the gene dictates the sequence of amino acids in the protein. Proteins perform almost all the functions in cells, and can be grouped into four major classes: they act as enzymes that control the rate of chemical reactions in the cell; they form structural components of organelles, membranes, and other cell components; they receive and transmit signals between and within cells; or they act as regulators of genes by latching onto DNA, thereby increasing or decreasing the rate at which the gene is used, or "expressed."

base pairs two nucleotides (either DNA or RNA) linked by weak bonds

prokaryotes single-celled organisms without a nucleus

homologous carrying similar genes

Genes vary in length. The largest human gene is 2.5 million **base pairs** in length, and codes for the muscle protein named dystrophin, which is more than 3,500 amino acids long. Eukaryotic genes generally produce proteins of about 150 to 3,000 amino acids in length. Some genes are relatively small, as in **prokaryotes**, which produce proteins of 50 to 300 amino acids. Most eukaryotic protein-coding genes are present in only two copies per genome, occurring in the same position on **homologous** chromosomes, one of which is received from each parent. If the two copies differ slightly they are called alleles. Changes in nucleotide sequences are termed mutations or polymorphisms, depending on their effect.

Some genes code not for protein but for RNA molecules that have their own functions within the cell. These include the transfer RNAs, ribosomal RNAs, and a variety of other smaller RNAs with roles in the nucleus. RNA-coding genes are usually present in multiple copies per eukaryotic genome.

Gene Expression

transcription messenger RNA formation from a DNA sequence

enzyme a protein that controls a reaction in a cell

Expression of protein-coding genes begins with the process of **transcription**. During transcription, the helix is unwound, and an **enzyme** (RNA polymerase) binds to the DNA. It then moves along the DNA, and beginning slightly "downstream" at the so-called initiation site, it copies one of the strands to form a molecule of RNA. Transcription ceases when the polymerase reaches a special DNA sequence called the termination site, usually a region high in G-Cs followed by A-Ts.

In prokaryotes, this RNA product is ready to use for protein synthesis, and is called messenger RNA (mRNA). After the mRNA of a gene is formed, it is used by the cell in protein synthesis (**translation**) at the ribosomes.

translation synthesis of protein using mRNA code

Thus, the prokaryotic gene consists of an RNA binding site (called the "promoter"), a transcription initiation site, the coding region, and a termination signal. The initiation site should not be confused with the start signal for protein synthesis, nor the termination site with the stop signal in protein synthesis. Each of the translation signals is within the coding region, or "open reading frame," of the gene.

Eukaryotic Genes

In eukaryotic cells, genes are more complex. It was discovered in 1977 that eukaryotic genes are functionally separated into coding segments called exons, which are interrupted by noncoding sequences of DNA called introns. The entire region between the initiation and termination sites is transcribed, including the introns, to form the primary transcript. This must then be processed by special enzymes that cut out the introns and splice together the exons to form an mRNA. The mRNA is then exported from the nucleus for translation.

The existence of introns allows for the creation of multiple proteins from one gene, by the use or exclusion of different exons. Such alternative splicing gives rise to protein "isoforms," highly similar but slightly different proteins, with functions that vary as well. Isoforms are typically tissue-specific. For example, the muscle enzyme creatine kinase exists in one form in the heart, and another form in the skeletal muscles (such as the biceps), which have different ends formed through use of different exons. Even though it codes for two or more proteins, most scientists call such a DNA sequence a single gene.

Eukaryotic genes also contain a sequence close to the termination site called the polyadenylation signal. After transcription, this sequence prompts a special enzyme, called poly-A polymerase, to cut the RNA chain and begin adding multiple adenine nucleotides, as many as 250, to the primary transcript. This poly-A tail helps transport the RNA out of the nucleus, stabilizes it in the **cytoplasm**, and promotes efficient transcription at the ribosome.

cytoplasm the material in a cell, excluding the nucleus

Thus, the eukaryotic gene consists of an RNA binding site (promoter), a transcription initiation site, the coding region including exons and introns, the polyadenylation signal, and a termination site.

Genes for RNAs are transcribed in the same way, but the RNA formed is not translated into protein. Details vary among different types, but most RNA-coding genes do not contain introns. Transcripts of the ribosomal RNA genes must be cut apart to form a number of smaller functional RNA molecules.

Controlling Gene Expression

The complexity of any living cell is due to the well-orchestrated interactions of its proteins. Just as an orchestra cannot have every instrument play at once, a cell cannot have all its proteins function at once. One method of regulating protein function is to control when the protein is made, which

is to say when the gene is expressed. Prokaryotic genes are usually controlled by operon systems, relatively simple systems that tie expression directly to metabolic activity in the cell. Eukaryotic genes are controlled by more complex regulatory systems that respond to hormones, growth factors, internal conditions, and many other influences.

To ensure that each gene is expressed when, and only when, it is needed, each eukaryotic gene has several control regions, termed the promoter and enhancer regions. These do not code for amino acids but are critical for proper gene expression. Mutations in these regions often change the rate at which a gene is expressed, or the factors in the cell or the environment to which it responds.

transcription factors proteins that increase the rate of gene transcription

The promoter region is a sequence of 20 to 200 nucleotides "upstream" of the coding region to which the RNA polymerase enzyme binds, permitting it to begin transcribing the DNA. Promoters differ in size and sequence in prokaryotic and eukaryotic genes. Promoters attract RNA polymerase by first binding a variety of other proteins, called **transcription factors**. In some eukaryotic genes, promoter sites also occur within the coding region, allowing alternative transcripts with fewer exons.

Enhancers, also called activation sites, are located either nearby or far away from the promoter. Because DNA is looped and coiled, however, these sites are actually physically close to the gene's promoter even when distant on the DNA strand. Enhancers are gene-specific, and attract a variety of transcription factors. All of these work together to increase the rate of transcription by increasing the likelihood of RNA polymerase binding. Controlling the availability of these proteins is an important factor in regulating expression of the gene. SEE ALSO CHROMOSOME, EUKARYOTIC; CHROMOSOME, PROKARYOTIC; CRICK, FRANCIS; DNA; EVOLUTION OF GENES; GENE EXPRESSION: OVERVIEW OF CONTROL; GENE FAMILIES; GENETIC CODE; MENDEL, GREGOR; MORGAN, THOMAS HUNT; MUSCULAR DYSTROPHY; MUTATION; NATURE OF THE GENE, HISTORY; NUCLEOTIDE; OPERON; PROTEINS; RNA POLYMERASES; RNA PROCESSING; TRANSCRIPTION; TRANSCRIPTION FACTORS; WATSON, JAMES.

Elof Carlson

Bibliography

Alberts, Bruce, et al. *Molecular Biology of the Cell*, 4th ed. New York: Garland Science, 2002.

Carlson, Elof. *The Gene: A Critical History*. Philadelphia, PA: Saunders Publishing, 1966.

Muller, H. J. "The Development of the Gene Theory." In *Genetics in the Twentieth Century*, L. C. Dunn, ed. New York: Macmillan, 1951.

Olby, Robert. *The Path to the Double Helix*. Seattle, WA: University of Washington Press, 1974.

Gene and Environment

phenotype observable characteristics of an organism

Questions of "nature versus nurture" have been asked of most human traits: Is it our genes, inherited from our parents, that make us the way we are, or is it the environment in which we live? A **phenotype** is a trait that can be observed and described in a population. Although some phenotypes may be totally controlled by genetic or environmental factors, most are influenced by a complex combination of the two. Genes and environmental factors may work independently, or they may interact with one another to cause the phenotype.

Classes of Human Genetic Phenotypes

Human phenotypes are often classified as either simple (or Mendelian) or complex. A "simple" or Mendelian phenotype is one that demonstrates a recognizable inheritance pattern (such as autosomal dominant or recessive, or X-linked). A Mendelian phenotype is caused by a particular genetic variant, or **allele**, of a gene. The expression of Mendelian phenotypes may vary by age, but, in general, the effect of a single gene is sufficient to cause the phenotype.

allele a particular form of a gene

In contrast, "complex" phenotypes do not adhere to simple Mendelian laws, and they are influenced by several factors (either genetic or environmental) acting independently or together. In complex phenotypes, alleles of particular genes increase the probability that the phenotype will develop, but do not determine with certainty whether a person will have the phenotype. They are neither necessary nor sufficient to cause the phenotype. Genes that act in this fashion are called susceptibility genes. The complex interaction of susceptibility genes with other genetic and environmental risk factors determines whether or not a person will develop a complex phenotype.

Gene-Environment Interaction in Phenylketonuria

Phenylketonuria (PKU) is a classic example of gene-environment interaction. PKU was originally described as an autosomal recessive metabolic disease, in which people with two defective copies of the phenylalanine hydroxylase gene are unable to convert phenylalanine into tyrosine. This inability leads to an accumulation of phenylalanine in the blood, causing problems with nerve and brain development that result in mental retardation.

The treatment of PKU by removing foods containing phenylalanine from the diet (and thus reducing the accumulation of phenylalanine) demonstrated that mutations in the phenylalanine hydroxylase gene cause mental retardation only in the presence of dietary phenylalanine. Since phenylalanine is very common in the diet, this gene-environment interaction was not detected at first. PKU serves as an illustration that phenotypes that are apparently Mendelian in nature may have complex interactions with other genes and with the environment. Removing the exposure to dietary phenylalanine prevents mental retardation, and phenylalanine does not cause mental retardation in the absence of **mutations** in the phenylalanine hydroxylase gene. Therefore, both factors are needed to cause mental retardation due to PKU.

mutations changes in DNA sequences

The identification of the gene-environment interaction in PKU has led to the effective treatment of this genetic disorder. Individuals who carry mutations in the phenylalanine hydroxylase gene, if placed on a low-phenylalanine diet, generally do not develop the symptoms of PKU. To identify individuals at risk of PKU, newborns are screened for elevated phenylalanine levels in the blood. Those infants with positive screening tests are then evaluated further. Those with PKU (about 1 in 10,000 live

births) are then placed on low-phenylalanine diets to prevent the development of mental retardation.

Methods for Identifying Gene-Environment Interactions

As the example of PKU demonstrates, studies that attempt to identify factors important in determining human phenotypes must simultaneously examine multiple genetic and environmental factors. It is generally not possible to experiment directly on humans to observe the effects of a gene or environmental factor on the expression of a phenotype. Human genetic studies are generally observational studies, where the researcher is limited to observing the combinations of exposures that naturally occur in the population. Genetic **epidemiologists** must apply statistical methods to these observational data to evaluate how genes and the environment affect the development of a **phenotype**.

epidemiologists person who studies the incidence and spread of diseases in a population

phenotype observable characteristics of an organism

The simultaneous analysis of genetic and environmental factors allows the identification of environmental factors that interact with each other. Researchers use statistical models to compare the joint effects of genetic and environmental factors in people with the phenotype and people without the phenotype. Such "case-control studies" are commonly used to examine the relationship between disease phenotypes and both genetic and nongenetic risk factors.

The strength of the association between a risk factor and the disease is described by the "relative risk," which is the probability of having the phenotype if exposed to the risk factor divided by the probability of having the phenotype if not exposed to the risk factor. A relative risk greater than 1 suggests that the risk factor increases the probability of developing the phenotype, whereas a relative risk less than 1 suggests the risk factor decreases the probability of the phenotype. An association between a risk factor and a phenotype exists if the relative risk is significantly different from 1.

An example of a risk factor and relative risk can be seen in Alzheimer's disease (AD). One gene that influences the risk of AD is the *APOE* gene. Three common alleles are known: ϵ-2, ϵ-3, and ϵ-4. Caucasian Americans with one or more copies of ϵ-4 are two and one-half times more likely to develop AD than are people with two ϵ-3 alleles. Interestingly, the ϵ-4 allele is not as strong of a risk factor for AD among African Americans or Hispanics, who nonetheless have higher risks of developing AD than Caucasians.

Patterns of Gene-Environment Interactions

There are many potential patterns of interaction that could exist between genetic and environmental factors for a phenotype. Several plausible statistical models of gene-environment interaction have been described. Phenotype expression can either be:

(1) Increased only in the presence of both the susceptibility genotype and the environmental factor;

(2) Increased by the environmental factor alone but not by the genotype alone;

(3) Increased by the genotype alone but not by the environmental factor alone;

(4) Increased by either, with joint effects being additive or multiplicative;

(5) Reduced by the genotype and not affected by the environmental factor alone, but increased in the presence of both; or

(6) Reduced by the genotype, increased by the environmental factor, and increased by the presence of both.

These models are simple and consider the effect of only one gene and one environmental factor. Interactions are likely to be much more complex, involving multiple genes, multiple environmental factors, genetic heterogeneity, and heterogeneity of exposure. However, finding statistical interaction between two factors is just the first step in unraveling a complex phenotype. Once statistical interactions are identified, other laboratory studies may be performed to establish what biological interactions, if any, exist.

The level of cholesterol in the bloodstream is an example of a trait that is caused by a complex set of genetic and environmental factors. In families with familial hypercholesterolemia (FHC), elevated cholesterol levels are inherited in a Mendelian, autosomal dominant pattern. However, only about 4 percent of all individuals in the top 5 percent of cholesterol levels in the population carry the gene responsible for FHC. Other genetic and environmental risk factors clearly influence cholesterol levels. For example, in people with and without FHC, consumption of cholesterol in the diet independently modifies cholesterol levels. Other factors such as exercise and medication use likely interact with dietary intake to determine blood levels of cholesterol. Other genetic factors may also be involved.

With the exception of a relatively small number of phenotypes that are completely genetically determined, almost all human phenotypes represent a combination of environmental and genetic factors. Understanding the ways in which genes and environment work together to impact human health is one of the great challenges in the study of complex phenotypes. SEE ALSO ALZHEIMER'S DISEASE; CARDIOVASCULAR DISEASE; COMPLEX TRAITS; DIABETES; INHERITANCE PATTERNS; METABOLIC DISEASE; STATISTICS.

William K. Scott

Bibliography

Khoury, Muin J., Terri H. Beaty, and Bernice H. Cohen. *Fundamentals of Genetic Epidemiology.* New York: Oxford University Press, 1993.

Thompson, Margaret W., Roderick R. McInnes, and Huntington F. Willard. *Thompson and Thompson: Genetics in Medicine*, 5th ed. Philadelphia: W. B. Saunders Company, 1991.

Vogel, Friedrich, and Arno G. Motulsky. *Human Genetics: Problems and Approaches*, 2nd ed. Berlin: Springer-Verlag, 1986.

Gene Discovery

Gene discovery is the process of identifying genes that contribute to the development of a trait or **phenotype**. Researchers often try to discover the genes that are involved in specific diseases. They also try to find the genes that contribute to many other traits.

phenotype observable characteristics of an organism

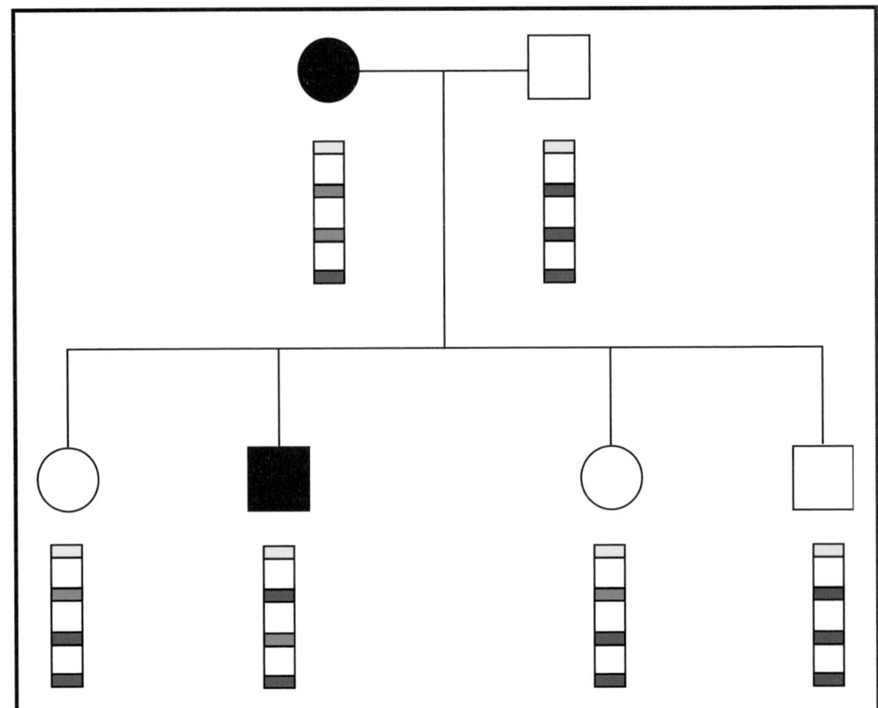

Polymorphic markers along the chromosomes (here shown as different colored bars) are examined to determine which is coinherited with the disease (shaded circle and square). The red marker in the third position is found only in the two family members with the disease.

Gene discovery begins with clearly defining a trait of interest and determining if that trait has a genetic and/or environmental basis. This is done using several approaches, such as sibling recurrence risk ratio, familial aggregation, and twin and adoption studies. The sibling recurrence risk ratio is the frequency of a disease among the relatives of an affected person, divided by the frequency of the disease in the general population. The greater the ratio, the stronger the genetic component of the disease.

A trait also is suspected of having a strong genetic component when familial aggregation, which is the clustering of patients in a single family, occurs. Familial aggregation can sometimes be misleading, however. Since families often share the same environment, it is difficult to know whether environmental or genetic factors are the cause of clustering.

In twin studies, concordance rates play a critical role. Concordance is the percentage of second twins that exhibit a trait when the trait occurs in the first twin. Twins generally also share environments, so concordance rates are often compared between monozygotic twins, who are genetically identical, and dizygotic or "fraternal" twins, who share on average 50 percent of their genetic material. The greater the difference in concordance rates between monozygotic and fraternal twins, the stronger the genetic contribution to the trait. The combination of evidence from all these approaches indicates whether a particular phenotype is likely to have a genetic basis.

Approaches for Identifying Genes

Once genetic influence has been established, two research approaches are commonly used to identify the specific genes involved. These are the candidate gene approach and the genomic screening approach.

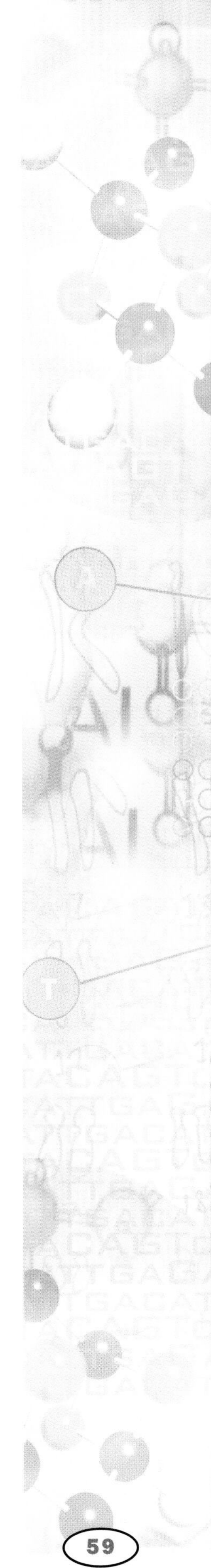

Candidate Gene Approach. In the candidate gene approach, genes are selected based on their known or predicted biological function and on their hypothesized relation to the disease or trait. These genes are subject to mutation analysis to determine whether they are really involved. The problem with the candidate gene approach is that it relies on assumptions about the molecular mechanisms underlying the development of a trait. However, diseases are usually studied because little is known about their causes, so initial ideas about these "molecular mechanisms" often prove to be wrong.

In addition, the candidate gene approach can be very time consuming, and it has been successful only infrequently. Chromosomal abnormalities, such as deletions, inversions, or **translocations**, in individuals exhibiting a trait, as well as animal models mimicking the trait, are especially important for a candidate gene approach, since they provide clues to the genetic basis of the trait.

translocations movements of chromosome segments from one chromosome to another

Genomic Screen Approach. A genomic screen is a systematic survey in which **polymorphic** DNA markers, evenly spaced along all the chromosomes, are used to determine if a marker is inherited along with the trait, indicating genetic linkage. This is performed taking the DNA from each individual in the study and identifying the type of marker each has on his chromosomes. These data are then analyzed using statistical programs to see if the marker and the trait that is being studied travel together through families significantly more often than would be expected just by chance. If a DNA marker is found to be linked with a trait, it suggests that the marker and the gene responsible for the trait are rarely separated by crossing over and are therefore near each other on a region of a chromosome. Further fine mapping of this region with more closely spaced markers can narrow the region where the gene of interest lies.

polymorphic occurring in several forms

Genomic screening does not require prior biological understanding of the pathophysiology of a disease. It requires large sets of data from families containing multiple members who are affected with the trait, and it tends to be the more expensive of the two approaches.

Genomic screening usually leads to the identification of one or several **loci**, or relatively large areas in the **genome**, that are linked with a trait but that contain many different genes. The genes in such regions need to be prioritized.

loci sites on a chromosome (singular, locus)

genome the total genetic material in a cell or organism

Genes are considered to be good candidates when their putative functions fit with the known or predicted pathway of the disease. If any known gene in the linkage region appears to be a good candidate gene, it is subjected to mutation analysis to determine if there is a potentially disease-causing mutation that segregates only with the affected individuals.

Both the gene candidate and the genetic screen approach require collecting data from a large number of families. Recruiting and medically evaluating affected and unaffected individuals for participation in a genetic study, and collecting their DNA samples, is a long and complicated phase of gene discovery.

Positional Cloning

Once one or more loci have been identified through a genomic screen as possibly containing a gene of interest, additional techniques are needed to

locate the exact gene responsible. Positional cloning is the process of identifying a disease gene based on its location on a chromosome. Genetic screening is the inital step of positional cloning. The usual steps are: (1) linkage (locates a chromosome area); (2) fine mapping (narrows down the initial genomic area to a smaller region); (3) candidate gene analysis (looks for mutations in genes lying in that small area). By using positional cloning, researchers can identify or "clone" a gene knowing only its location on a chromosome, as determined through linkage analysis. Once the location is identified, a physical representation of the genes and DNA in the linked region is constructed.

Before the Human Genome Project was completed, such a physical representation was constructed by using a "contig," a group of overlapping DNA fragments that together cover the linked region. The contig is a scaffold or platform on which to place genes and other sequences in the correct position. The geneticist continues to collect families looking for new recombinations that reduce the piece of DNA that all the affected family members share but that has not been inherited by any of the unaffected individuals.

polymorphisms occurring in several forms

Positional cloning is a laborious process if the region is large and if the genes and **polymorphisms** making up the contig are not known. This portion of the process has been greatly helped by the Human Genome Project, as it provides the contig all filled out and correctly mapped.

In most cases, the smallest inherited piece of DNA is still quite large molecularly, and can contain many genes. Only one gene causes the disease, though, so each gene must be tested for mutations that segregate with the trait. This can be very time consuming as well. If no mutations are found, the process is repeated with the next gene in the smallest shared region of DNA.

exons coding regions of genes

The third gene identified through positional cloning was the cystic fibrosis transmembrane conductance regulator gene. The gene, identified in 1989, regulates chloride ion transport across the plasma membrane and consists of twenty-seven **exons** spread over a 230-kilobase-pair region on chromosome seven. This gene was difficult to clone because of the lack of chromosome abnormalities that would have helped locate it. No human genome sequence was available at that time. Numerous technical problems also arose in constructing a physical map of the linkage region, creating the need to screen numerous libraries to obtain a full-length clone of the gene. More than 550 mutations were identified by late 2001.

Complex Diseases

bioinformatics use of information technology to analyze biological data

The sequencing of the human genome and the advancement of **bioinformatics** and molecular tools to identify genes has made the process of gene discovery much easier. Most of the human genome sequence is now available on the Internet, with known and predicted genes annotated. This has reduced the need to laboriously build contigs.

Many of the common single-gene diseases have already been associated with a gene, and research efforts are now shifting to the more difficult task of finding genes that give susceptibility to developing complex diseases. (Such diseases show familial aggregation but do not follow any clear Mendelian inheritance pattern.)

It is thought that a combination of several genetic predisposition factors interact with environmental factors to trigger complex diseases, and that it is not a single gene but multiple genes that contribute to such traits, and the identification of each of these genes is correspondingly more difficult.

Complex traits can constitute various other challenges for researchers. Genetic heterogeneity is where **alleles** at more than one locus trigger the same phenotype, or mutations in the same gene cause different phenotypes. Reduced penetrance is where a predisposing genotype does not necessarily cause the phenotype to manifest itself. Phenocopy is where a trait looks identical but has a different cause than the one being studied.

alleles particular forms of genes

To address these challenges, scientists use association studies, which are based on the principle that if a particular allele and trait occur simultaneously at a statistically significant frequency, the allele is likely to be involved in the development of the trait. (Linkage studies, by contrast, are based on finding DNA markers and traits that are linked within families.)

Alzheimer's disease is one example of a complex trait. Three genes have been found to contribute to the rare, early forms of the disease. Genetic screens have found that a fourth locus is linked to the common, late-onset form of the disease. Association studies have revealed that one allele of this fourth locus increases a patient's risk of developing Alzheimer's in a dose-dependent fashion, where the risk posed by having two alleles is greater than the risk posed by having just one. SEE ALSO ALZHEIMER'S DISEASE; BIOINFORMATICS; CLONING GENES; COMPLEX TRAITS; CYSTIC FIBROSIS; DNA LIBRARIES; HUMAN GENOME PROJECT; INTERNET; LINKAGE AND RECOMBINATION; MENDELIAN GENETICS; TWINS.

Sofia A. Oliveria and Jeffery M. Vance

Bibliography

Lewis, Ricki. *Human Genetics: Concepts and Applications*, 5th ed. Boston: McGraw-Hill, 2002.

Peltonen, Leena, and Victor A. McKusick. "Dissecting Human Disease in the Postgenomic Era." *Science* 291 (2001): 1224–1229.

Gene Expression: Overview of Control

The chromosomes of an organism contain genes that encode all of the RNA and protein molecules required to construct that organism. Gene expression is the process through which information in a gene is used to produce the final gene product: an RNA molecule or a protein.

Each cell in a multicellular organism such as a human contains the same genes as every other cell. Nonetheless, there are hundreds of distinct types of cells in the human body, each expressing a unique set of genes. Indeed, it is this unique constellation of expressed genes that makes each cell type distinct.

Cells may also change the genes they express over time, and they are constantly adjusting the amount of protein made in response to changing conditions. How does a cell express some, but not all, of the genes in its **genome**? How does it react to environmental changes to adjust the level of

genome the total genetic material in a cell or organism

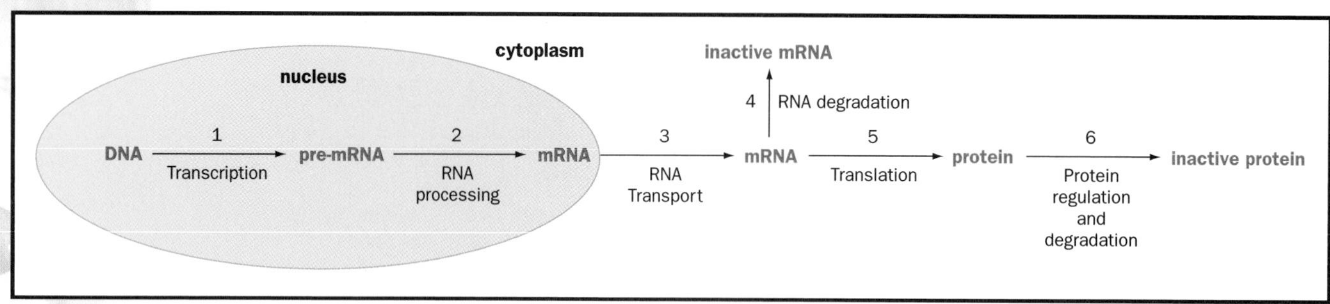

Figure 1. The flow of genetic information from DNA to proteins. Control can be exerted at each numbered step. Most control occurs at step one, transcription.

gene expression? These are the problems of control of gene expression. While the genes whose final products are RNA molecules are also regulated, this entry will focus on genes that encode proteins.

The Flow of Genetic Information from Genes to Proteins

transcription messenger RNA formation from a DNA sequence

cytoplasm the material in a cell, excluding the nucleus

phosphorylation addition of the phosphate group PO_4^{3-}

Cells can regulate gene expression at every step along the way, from DNA to the final protein, as shown in Figure 1. Genetic information in DNA is first copied to form an RNA molecule, in a process known as **transcription**. The RNA used to make proteins is called messenger RNA (mRNA) because it carries information from the DNA to the ribosome, where protein synthesis occurs.

The mRNA serves as a template to guide protein synthesis. Scientists refer to protein synthesis as "translation" because ribosomes translate an mRNA sequence into a protein sequence. Prokaryotic cells use the mRNA directly as a template for protein synthesis.

Eukaryotic cells, however, must modify the precursor mRNA in several ways before it can be used to guide protein synthesis. The two ends are chemically altered, and sections of the RNA that do not encode protein sequences, called introns, are spliced out. Together, these modifications are called "mRNA processing." After processing, the mature mRNA moves from the nucleus into the **cytoplasm**, where it binds to the ribosome and serves as a template for synthesis of a protein.

The most important stage for the regulation of most genes is when transcription begins. This is because it costs the cell less energy to regulate transcription than to regulate the steps after transcription. The second point where regulation occurs is during RNA processing. Cells can regulate the rate of processing. In addition, the final mRNA product can be altered through alternative splicing, as shown in Figure 2. Alternative splicing can regulate the types of proteins produced from a single gene.

Cells can also regulate mRNA transport out of the nucleus. Once the mRNA has moved into the cytoplasm, the abundance of mRNA can be regulated by RNA degradation. Cells can regulate translation, controlling the number of proteins each mRNA produces. Finally, even after a cell has generated a protein, it can regulate the abundance and activity of that protein. For instance, cells regulate the activity of many proteins by post-translational modifications such as **phosphorylation**. Cells can also regulate the abundance of most proteins by degrading them.

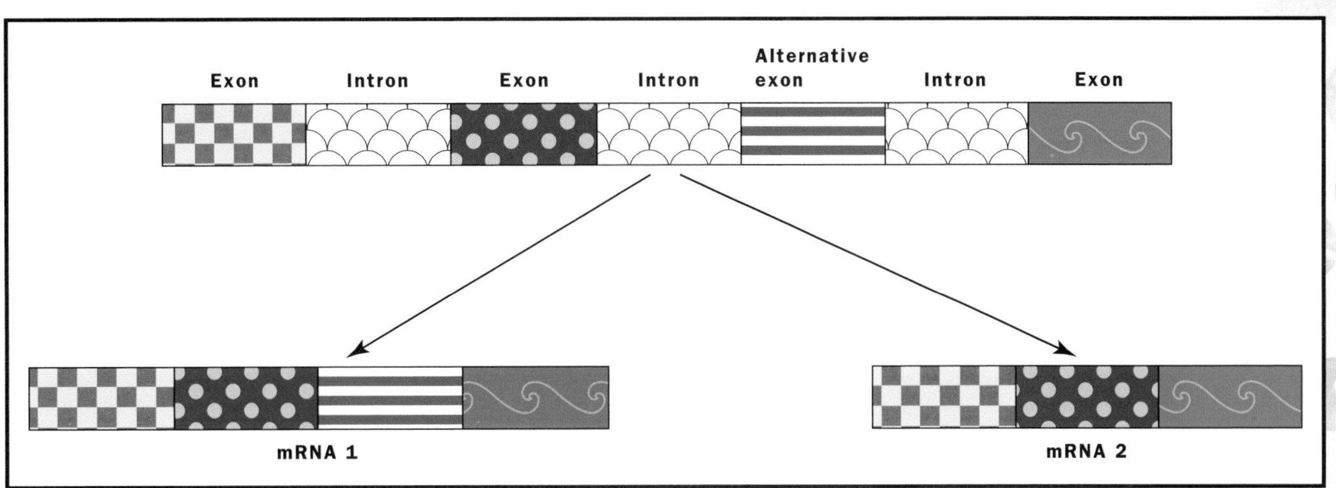

Figure 2. Alternative mRNA splicing leads to isoforms, or related proteins.

Gene Control Occurs at Several Levels

For a gene to be transcribed, **RNA polymerase** must first find the gene. This is made more difficult by the tight packing required to fit the entire genome within the nucleus. The cell uses this packing to its advantage, though, to prevent access to and expression of genes in some chromosomal regions.

RNA polymerase enzyme complex that creates RNA from DNA template

Regions that are tightly condensed are called heterochromatin and can be distinguished from more open regions (euchromatin) by their dense staining, when viewed under a microscope. In females, an entire X chromosome in each cell is kept condensed throughout life, to avoid a "double dose" of these genes (recall that females have two X chromosomes, while males have only one). This random X inactivation leads to mosaicism, in which some female cells express genes from one chromosome, while others express genes from the other.

Even within an active chromosome, some regions may be temporarily inactivated. Inactive heterochromatin can be converted to active euchromatin, and vice versa, by chemical modification of the histone proteins to which the DNA is attached in the chromosome. Negatively charged DNA is chemically attracted to the positively charged histones. By adding or removing chemical groups to the histones, this attraction can be modulated. A weaker attraction, as would occur by adding negatively charged groups to the histones, tends to open up the **chromatin**, favoring gene expression. A stronger attraction keeps it more condensed.

chromatin complex of DNA, histones, and other proteins, making up chromosomes

How Do Cells Regulate Transcription?

To understand transcriptional regulation, consider the structure of a typical eukaryotic gene, shown in Figure 3. The promoter of a gene is the binding site for a group of general **transcription factors** and for RNA polymerase. Transcription begins when a complex of proteins called TFIID binds to a promoter. The sequential binding of other general transcription factors and RNA polymerase follows. A protein tail tethers the RNA polymerase to the general transcription complex. When the general transcription factor TFIIH phosphorylates this tail, the RNA polymerase is released and moves along the DNA to begin transcription.

transcription factors proteins that increase the rate of gene transcription

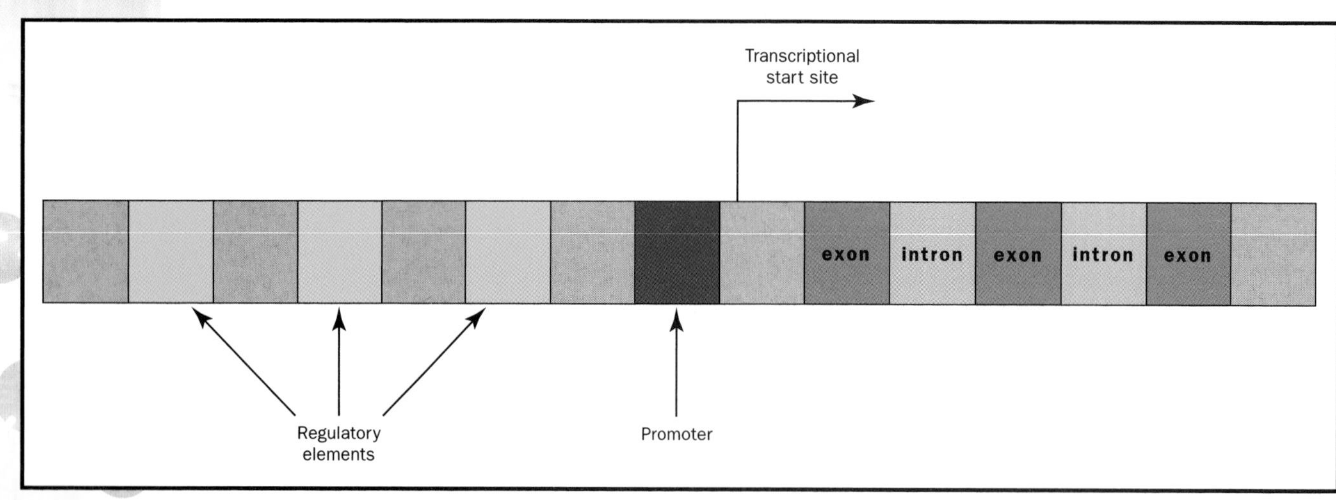

Figure 3. Structure of a typical eukaryotic gene.

These steps are identical for essentially all genes. However, the rate at which each of the steps occurs can be influenced by the presence or absence of other, gene-specific transcription factors. It is these other factors, often called gene regulatory proteins, that give the cell the ability to turn some genes on and others off. In contrast to the few general transcription factors, which assemble on the promoters of all genes, there are thousands of gene regulatory proteins. The regulatory proteins bound to a gene vary from gene to gene, and each is usually present at low levels in the cell.

Binding sites for these additional regulatory proteins are usually located upstream from the promoter. Surprisingly, these binding sites can be located some distance from the promoter and still regulate transcription. It is thought that this action at a distance can occur because the DNA between the regulatory sequence and the promoter can loop out, to allow the regulatory protein to contact the promoter, as shown in Figure 4.

Regulatory proteins are classified as either activators or repressors. Activators increase the rate of transcription, whereas repressors decrease it. The DNA sequences that bind activator proteins are called "enhancer elements," and those that bind repressor proteins are "repressor elements" or "silencer elements."

Many regulatory proteins have at least two distinct regions or domains, as shown in Figure 5. One domain binds to a specific DNA sequence. The other domain typically contacts the general transcription machinery assembled at the promoter. In one class of activating proteins, the activation domain contains a cluster of negatively charged (acidic) amino acids. Scientists believe acidic transcriptional activators accelerate the assembly of general transcription factors on the promoter. This is just one way a transcriptional activator can work. For instance, other regulatory proteins affect how tightly the gene is packaged within chromatin. Opening up the chromatin allows the transcription machinery to more quickly gain access to the promoter.

Gene regulatory proteins bind to the DNA when it is in a double-helical state. They recognize a specific DNA sequence by forming hydrogen bonds to chemical groups on the outside of the DNA. These proteins often contain common identifiable structural "motifs" that directly contact specific DNA sequences.

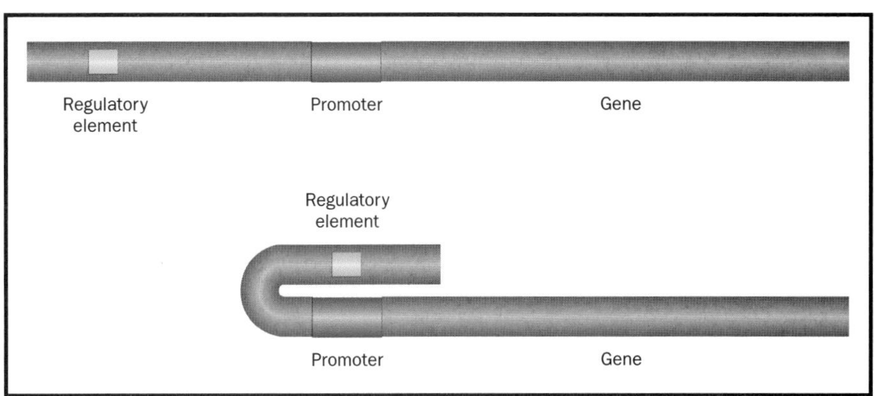

Figure 4. A regulatory element contacting the promoter by looping out the intervening DNA.

Combinatorial Regulation of Gene Expression

Some eukaryotic gene regulatory proteins work individually, but most act within a complex of proteins. Furthermore, a single gene regulatory protein may participate in multiple types of regulatory complexes. For example, a protein might function in a complex that activates the transcription of one gene and in a complex that represses transcription of another gene.

A gene that must be turned on at different times and in different tissues during development might have gene regulatory proteins clustered at multiple sites along its regulatory region. These complexes can then regulate the expression of the gene in a variety of developmental processes. The rate of RNA synthesis initiation will depend on the combination of regulatory proteins bound to the control regions of the gene. Thus, we can think of the regulatory region of the gene as an information processor, like a computer, that integrates input from all the regulatory proteins present and determines an appropriate level of RNA synthesis.

Regulation of Gene Expression during Development

During development, cells become different from one another because they synthesize and accumulate different proteins. Most of these differences come from changes in gene expression. Specialized cell types result from different genes being turned on or off in a coordinated manner. When a cell becomes a specific cell type, it continues in this role through many subsequent cell generations. This implies that cells remember the changes in gene expression involved in the choice of cell type. How is this achieved? One way is through the euchromatin-heterochromatin conversion, which can be faithfully inherited through cell division. Another way is for an important regulatory protein to activate its own expression as well as the expression of other genes. This gene, once expressed, will maintain its own expression.

As described above, it is usually a combination of gene regulatory proteins, rather than a single protein, that determines when and where gene expression occurs. Certain proteins can be more important than others, though. If all the other factors required for expression of a group of genes are present, a single gene regulatory protein can switch a cell from one developmental pathway to another.

fibroblasts undifferentiated cells that normally give rise to connective tissue cells

For example, forced expression of the MyoD protein in **fibroblasts** will cause these cells to form into muscle fibers. In an extreme example, homeotic

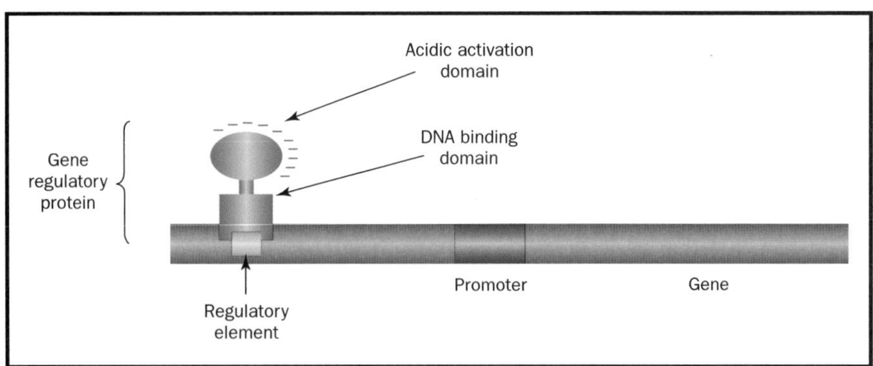

Figure 5. Domain structure of an acidic activation protein. This protein enhances the ability of general transcription factors to bind to the promoter.

genes specify large regions of an animal's body plan. Mutations in these genes can transform one body part into another. For instance, a mutation in the *antennapedia* gene of fruit flies will convert antennae to legs. Thus, expression of a single gene can trigger the expression of a whole battery of genes.

Combinatorial Control

An advantage of multiple gene regulatory proteins over single ones is that many different genes can be controlled with a handful of proteins. Consider the opposite situation, in which every gene would need a unique regulatory protein. The gene for each of those proteins would also need its own protein, and so on. Instead, a smaller set of proteins combines in different ways to make a large set of regulatory possibilities. Imagine that any regulatory element must be composed of two proteins. Four proteins (a, b, c, and d) can combine in pairs in ten different ways (aa, ab, ac, ad, bb, bc, etc.), and could thus, theoretically, control ten different genes. In reality, the situation is more complex, with literally thousands of regulatory proteins combining in ways researchers have not even begun to calculate.

With combinatorial control, therefore, a regulatory protein does not necessarily regulate a particular battery of genes or specify a particular cell type. Instead it might serve many purposes, and those purposes might overlap with those of other regulatory proteins. A regulatory protein might be switched on in many cell types, at different locations in the animal, and several times during development. Thus, combinatorial gene control makes it possible to generate a great deal of biological complexity with relatively few gene regulatory proteins.

Hormones and Growth Factors

During development, cells can change the expression of their genes when influenced by both external and internal signals. Signals from outside the cell that influence gene expression include contact with other cells, growth factors, and **hormones**.

hormones molecules released by one cell to influence another

Growth factors are extracellular molecules that stimulate a cell to grow or proliferate. Examples include epidermal growth factor and fibroblast growth factor. Growth factors regulate gene expression indirectly through a network of intracellular signaling cascades.

Hormones are signaling molecules that endocrine cells secrete into the bloodstream. Some hormones, such as insulin, bind to cell surface receptors

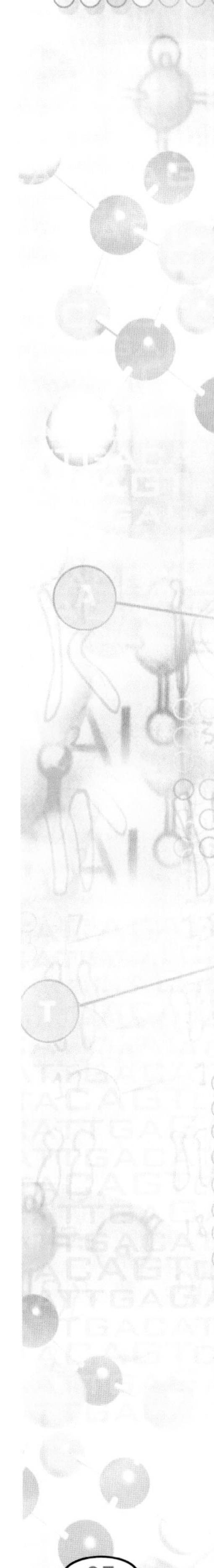

and affect gene expression through a network of intracellular signaling cascades. Other hormones, such as testosterone, pass through the cell membrane and bind to regulatory proteins in the cell that directly regulate transcription.

When the Regulation of Gene Expression Fails

When the control of gene expression fails, there can be serious consequences, such as death, birth defects, and cancer. Birth defects can result when the regulation of one or more genes important for development is lost. This often occurs because of a mutation, but it can also occur if the embryo or fetus is exposed to certain chemicals, such as alcohol. Mutations in the receptor for fibroblast growth factor, for instance, cause dwarfism. Cancer occurs when the regulation of genes that control growth and cell division, programmed cell death (apoptosis), and cell migration are lost. SEE ALSO ALTERNATIVE SPLICING; BIRTH DEFECTS; DEVELOPMENT, GENETIC CONTROL OF; GENE; HORMONAL REGULATION; POST-TRANSLATIONAL CONTROL; PROTEINS; RNA PROCESSING; SIGNAL TRANSDUCTION; TRANSCRIPTION; TRANSCRIPTION FACTORS.

Eric Aamodt

Bibliography

Alberts, Bruce, et al. *Molecular Biology of the Cell*, 4th ed. New York: Garland Science, 2002.

Lodish, Harvey, et al. *Molecular Cell Biology*, 4th ed. New York: W. H. Freeman, 2000.

Struhl, K. "Gene Regulation. A Paradigm for Precision." *Science* 293 (2001): 1054–1055.

Tjian, R. "Molecular Machines That Control Genes." *Scientific American* 272, no. 2 (1995): 54–61.

Gene Families

Gene families are groups of DNA segments that have evolved by common descent through duplication and divergence. They are multiple DNA segments that have evolved from one common ancestral DNA segment that has been copied and changed over millions of years.

The members of a gene family may include expressed genes as well as nonexpressed sequences. Such nonexpressed sequences include promoters, operators, transposable genetic elements, and pseudogenes, which are genes that are no longer functionally expressed.

Pseudogenes resemble other family members in their linear sequence of nucleotides. However, they usually either lack the signals that would allow them to be expressed or have significant deletions or rearrangements that prevent successful transcription or translation.

One well-studied gene family is that of the globins, shown in both Figures 1 and 2. The globin family contains many pseudogenes as well as many functional genes, including the genes coding for hemoglobins (α, β, γ, δ).

Gene families vary enormously in size and number, ranging, in the human **genome**, from just a few copies of very closely related sequences to more than a half-million copies of Alu sequences, which are transposable

genome the total genetic material in a cell or organism

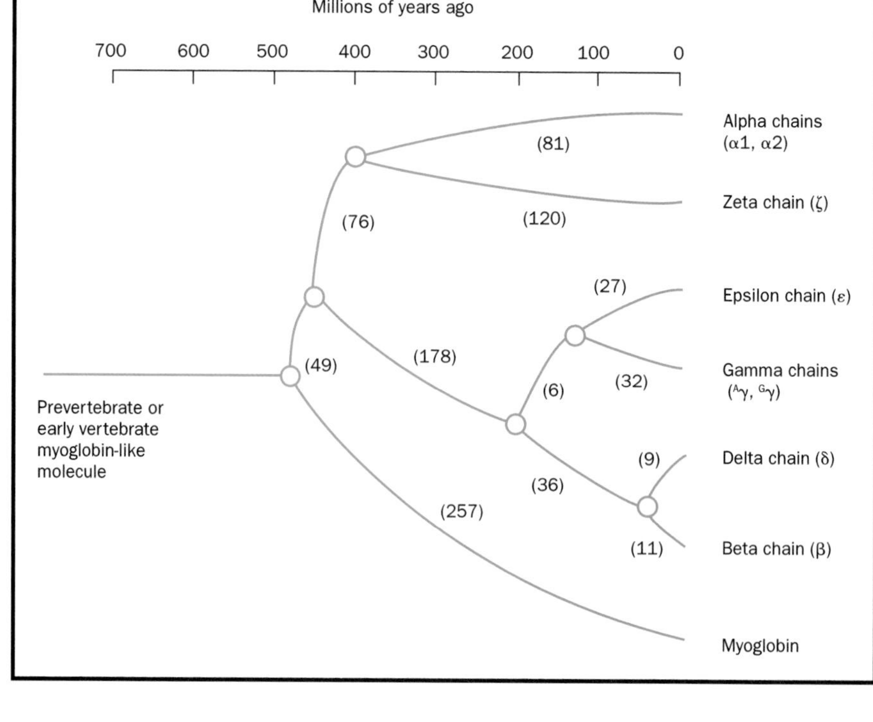

Figure 1. Evolution of the globin genes. The numbers in parentheses represent the estimated number of nucleotide changes needed to account for the observed amino acid differences. Adapted from <http://www.cord.edu/faculty/landa/courses/b315f99/sessions/phylogeny/globinPhylogeny.jpg>.

genetic elements with no known function. For genes that encode proteins, duplicate copies of genes have been found for over 2,000 proteins in a variety of genomes.

Members of gene families may be located in contiguous clusters on one chromosome, or they may be scattered throughout a genome. Homeotic genes, which lie in contiguous clusters on a few chromosomes, provide the best example of evolutionarily preserved gene order within a gene family. These genes play a role in the spatial development of the anterior-posterior axis of vertebrates and invertebrates. Four binary axes are laid down early in the **basal** body plan of most metazoans, including us: anterior-posterior, dorsal-ventral, left-right, and inside-outside. They affect locations on the body axis in roughly the same order as they are arranged along a chromosome, even though different clusters appear on different chromosomes. Amazingly, these genes work about the same in a fruit fly as they do in a human in establishing linear arrangements.

basal lowest level

In some gene families, related genes have stayed together over long periods of evolution, while in other gene families, members have become widely distributed within genomes. In ribosomal RNA genes, tandem arrays, in which multiple copies of the same gene occur one after another, have been observed. On the other hand, in the globin gene family, both the order and distribution of the genes, which are not identical, vary widely, even within one taxonomic group such as the mammals.

Members of a gene family usually have similar structures, but they may have diverged evolutionarily to such an extent that they are expressed in different ways. They may be expressed at different times in the development of multicellular organs or in different cells and tissues, and they may have acquired different functions. They may even have been transferred between organisms that are not closely related by evolution.

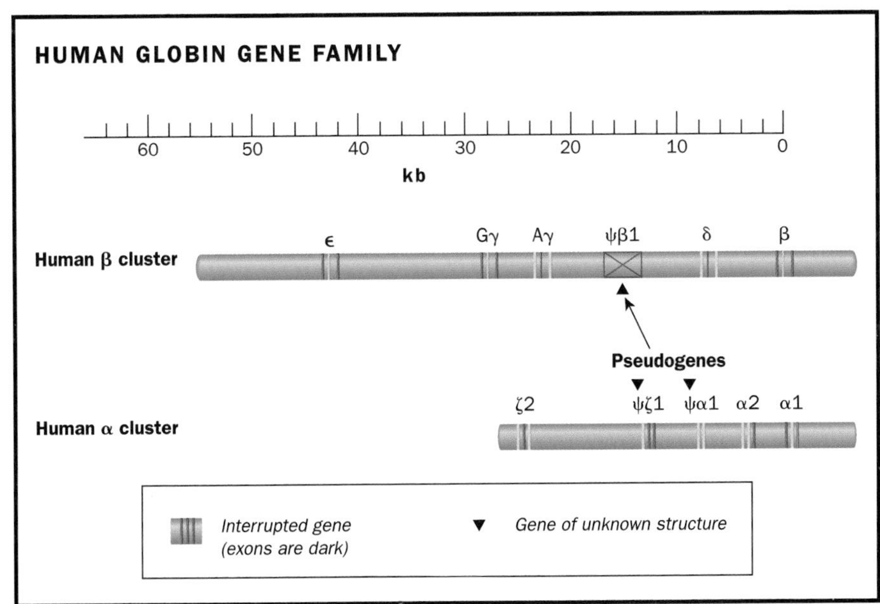

Figure 2. Representations of the two clusters of genes that code for human globin, the protein portion of hemoglobin. Adapted from <http://www.irn.pdx.edu/~newmanl/GlobinGeneEvolution.GIF>.

polyploidy presence of multiple copies of the normal chromosome set

One controversy about gene families involves whether they have arisen primarily by **polyploidy** or via tandem gene duplications. Polyploidy means that full genomes in an organism are duplicated either by mitosis or meiosis without cytokinesis or by matings between organisms with unequal numbers of chromosomes. This is followed by full copying of both parents' full genomes so each haploid set of chromosomes is now diploid. In tandem duplication, one or more copies of a gene lie on the same chromosome adjacent to one another. Polyploidy has been invoked to explain the evolution of complex new functions in taxa. Researchers give five reasons. Polyploids:

(1) have "higher levels of heterozygosity than do their diploid parents";

(2) "exhibit less inbreeding depression than do their diploid parents";

(3) "are polyphyletic . . . [which] . . . incorporates genetic diversity from multiple progenitor populations" and, thus, they have higher genetic diversity "than expected by models of polyploid formation involving a single origin";

(4) have genome rearrangements that are common; and

(5) are like duplicated genes, freed from intense selection pressure, which allows frequent evolution of new functions (Soltis and Soltis 2000, p. 310).

Austin Hughes has been the major critic of the often invoked polyploid hypothesis for origin of major animal groups because of the major substitutional load that would be involved and molecular phylogenetic evidence against it (1999, pp. 205–212). However, the results of most molecular evolutionary studies are more consistent with the gradualist view that new functions are generated primarily by tandem gene duplication and divergence of both sequence and function, spread over a long time. SEE ALSO DEVELOPMENT, GENETIC CONTROL OF; EVOLUTION OF GENES; POLYPLOIDY; PSEUDOGENES; TRANSPOSABLE GENETIC ELEMENTS.

John R. Jungck

Bibliography

Henikoff, Steven, et al. "Gene Families: The Taxonomy of Protein Paralogs and Chimeras." *Science* 278 (1997): 609–614.

Holmes, Roger S., and Hwa A. Lim, eds. *Gene Families: Structure, Function, Genetics and Evolution.* Singapore: World Scientific Publishers, 1996.

Hughes, Austin L. *Adaptive Evolution of Genes and Genomes.* New York: Oxford University Press, 1999.

Page, Roderic D. M., and Edward C. Holmes. *Molecular Evolution: A Phylogenetic Approach.* Malden, MA: Blackwell Science, 1998.

Patthy, Laszló. *Protein Evolution.* Malden, MA: Blackwell Science, 1999.

Soltis, Pamela S., and D. E. Soltis. "The Role of Genetic and Genomic Attributes in the Success of Polyploids." In *Variation and Evolution in Plants and Microorganisms,* Francisco J. Ayala, et al., eds. Washington, DC: National Academy Press, 2000.

Thorston, J. W., and R. DeSalle. "Gene Family Evolution and Homology: Genomics Meets Phylogenetics." *Annual Review of Genomics and Human Genetics* 1 (2000): 41–73.

Internet Resources

"Phylogenic Relationships between Globin-Type Proteins." Concordia College. <http://www.cord.edu/faculty/landa/courses/b315f99/sessions/phylogeny/globin Phylogeny.jpg>.

"Proposed Evolution of Globin Genes." Portland State University. <www.irn.pdx.edu/~newmanl/GlobinGeneEvolution.GIF>.

Gene Flow

Gene flow is the transfer of genetic material between separate populations. Many organisms are divided into separate populations that have restricted contact with each other, possibly leading to reproductive isolation. Many things can fragment a species into a collection of isolated populations. For example, a treacherous mountain pass may cut off one herd of mountain goats from another. In human beings, cultural differences as well as geographic separation maintain unique populations: It is more likely that a person will marry and have children with someone who lives nearby and speaks the same language.

WHAT IS AN ALLELE?

Alleles are forms of a gene that may differ between individuals or populations; brown and blue eye colors are due to different alleles for eye color.

Over time, reproductive isolation can lead to genetic differences between two populations. Gene flow between populations limits this genetic divergence, serving to inhibit the development of separate species out of the two separated populations.

gametes reproductive cells, such as sperm or eggs

The essential mechanism of gene flow is movement of individuals (or their **gametes**) between populations. For example, gene flow can occur in plant species when pollen is carried by bees or blown by the wind from one population of flowering plants to another.

Migration has been a significant feature of human history in both prehistoric and more recent times. No gene flow occurs if an individual migrates into a different population but does not reproduce. The migrant's genes must become part of the genetic makeup of the population into which it has migrated.

genetic drift evolutionary mechanism, involving random change in gene frequencies

In most populations, not all individuals contribute equally to the next generation. Because each individual can have different alleles, when only a subset of individuals reproduce, allele frequencies change from generation to generation, and some alleles may be lost. A change in allele frequency due to random chance is known as **genetic drift**, whereas a change due to differ-

ences in reproductive fitness is known as natural selection. Gene flow between isolated populations slows down their genetic drift from each other and reduces the power of natural selection to promote divergence between them. When there is a great deal of gene flow between populations, they tend to be similar; in this way, gene flow has a homogenizing effect. The opposite also tends to be true: If there is little or no gene flow between populations, the genetic characteristics of each population are more likely to be different.

Gene flow does not just occur between two populations. When a series of populations exists over a large area, gene flow may serve to keep even the most distant populations similar to one another. This can occur even if they do not exchange individuals or gametes as long as the alleles from one population eventually flow into the other population through a series of migrations or gamete movements. Similarly, other types of separation can also be overcome by this type of graded gene exchange. For instance, Great Danes and Chihuahuas cannot breed directly because of size incompatibility. But gene flow in both directions, through intermediate-sized dogs, keeps these two breeds from becoming separate species.

It is very difficult to assess gene flow directly, so population geneticists have devised a way to estimate gene flow by comparing allele frequencies. By determining allele frequencies in two different populations, the amount of gene flow between them, usually expressed as the number of migrants exchanged per generation, can be estimated. SEE ALSO GENETIC DRIFT; HARDY-WEINBERG EQUILIBRIUM; POPULATION GENETICS.

R. John Nelson

Bibliography

Avise, John C. *Molecular Markers, Natural History and Evolution.* New York: Chapman and Hall, 1994.

Futuyma, Douglas J. *Evolutionary Biology*, 3rd ed. Sunderland, MA: Sinauer Associates, 1998.

Mayr, Ernst. *Evolution and the Diversity of Life: Selected Essays.* Cambridge, MA: Belknap Press, 1976.

Weaver, Robert F., and Philip W. Hedrick. *Genetics*, 2nd ed. Dubuque, IA: William C. Brown, 1992.

Gene Targeting

Gene targeting is a method for modifying the structure of a specific gene without removing it from its natural environment in the chromosome in a living cell. This process involves the construction of a piece of DNA, known as a gene targeting **vector**, which is then introduced into the cell where it replaces or modifies the normal chromosomal gene through the process of **homologous** recombination.

vector carrier

homologous carrying similar genes

The Homologous Recombination Process

Homologous recombination is a process that occurs within the chromosome and which allows one piece of DNA to be exchanged for another piece. It is a cellular mechanism that is probably part of the normal process cells use to repair breaks in their chromosomes. Homologous recombina-

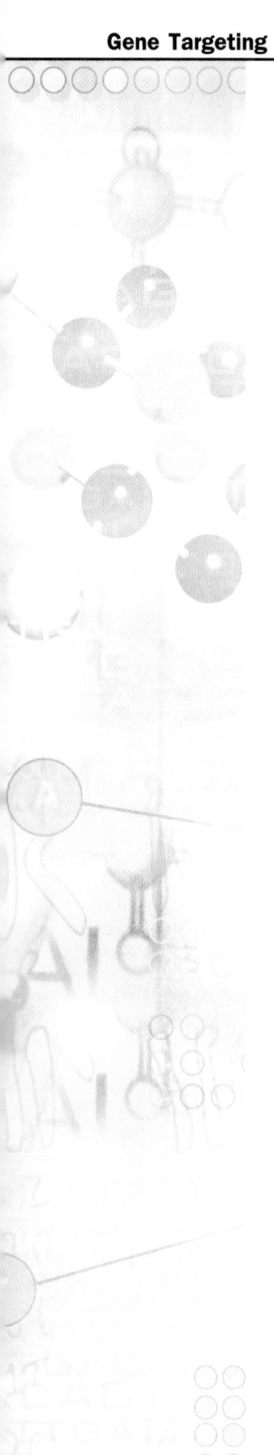

A modified version of the target gene replaces it in the chromosome. The target gene is removed and degraded. In this example, the gene is modified by insertion of an antibiotic resistance gene, which both inactivates the gene and allows efficient selection of transformed cells.

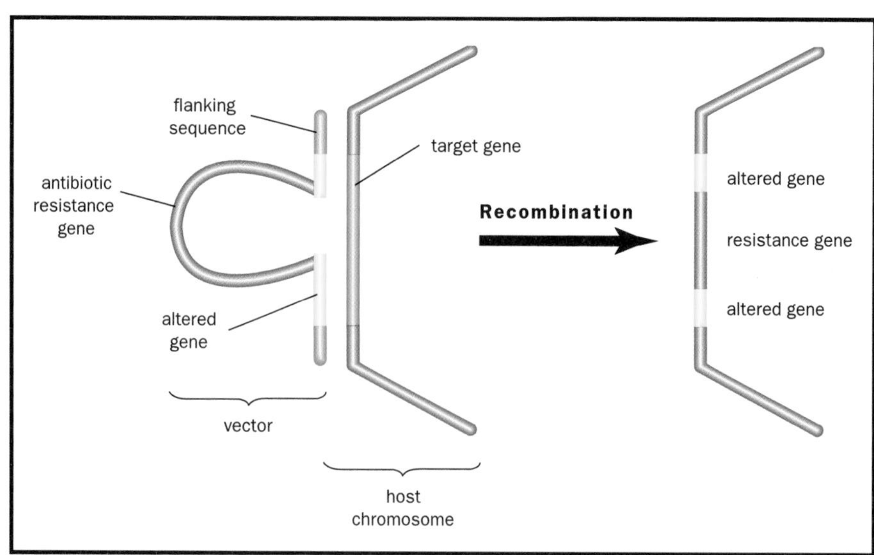

tion requires that the pieces of DNA undergoing recombination be almost identical (homologous) in sequence. In addition, sequences on either side of the target should be identical, to promote more efficient targeting and recombination.

By constructing a sequence that is homologous to a target sequence (such as a gene), laboratory researchers can replace one of the cell's own copies of a particular gene with a copy that has been altered in some way. It is also possible to replace only a part of a gene, such as one portion of its protein coding region. This permits the introduction of a **mutation** into specific cellular genes, which can either stop the gene functioning altogether (called a "**knock out**") or can mimic changes to genes that have been implicated in human diseases. The ability to target DNA constructs to particular locations in chromosomes is a very powerful tool because it allows the modification of more or less any gene of interest, in more or less any way desired.

mutation change in DNA sequence

knock out deletion of a gene or obstruction of gene expression

Homologous recombination of a DNA vector into a gene of interest can be done in almost any cell type but occurs at a very low frequency, and it is therefore important to detect the few cells that have taken up the gene. Gene targeting vectors are designed with this in mind. The simplest strategy is to include an antibiotic resistance gene on the vector, which interrupts the sequence homologous to the gene of interest and thus makes the inserted gene nonfunctional. This "selectable marker" gene makes the cells that possess it resistant to antibiotics, and can then be used to eliminate cells that are not genetically modified.

An example of a selectable marker that is commonly used for this purpose is the puromycin-N-acetyl-transferase (*pac*) gene, which confers resistance to the antibiotic puromycin, a drug that inhibits the function of **ribosomes**. After the introduction of the DNA construct, the cells are cultured with puromycin in the **medium**. This allows the selection of single cells that have incorporated the DNA construct into their own chromosomes. Cells lacking the *pac* gene will die in a culture medium containing puromycin. Once the puromycin resistant cells have been expanded into cell lines, the DNA of these cells can then be analyzed to select out a subset of the cells in which the introduced construct has integrated into the correct (target) gene.

ribosomes protein-RNA complexes at which protein synthesis occurs

medium nutrient source

For reasons that are not yet fully understood, the rate of homologous recombination in mouse embryonic stem (ES) cells is substantially higher than that of most other cells. Once a clone of ES cells with the correct targeting event has been identified, these cells can be used to introduced into the mouse via the process of **blastocyst** injection, which allows the study of gene function in the bodies of living, intact animals. Until very recently mice were the only organisms in which it has been possible to introduce targeted mutations into the germ line. The development of nuclear transfer (moving the nucleus from one cell to another), however, has allowed gene targeting to be done in other mammalian species, such as sheep and pigs.

blastocyst early stage of embryonic development

Adding or Deleting Genetic Material

As well as mutating or knocking out specific genes, gene targeting allows the introduction of novel pieces of DNA into a specific chromosomal location (this is often termed a "knock-in"). This allows researchers to examine the function of a gene in a variety of ways. For example, it is possible to examine where in the animal the gene is normally expressed by insertion (knock-in) of a fluorescent protein (such as green fluorescent protein, GFP) into the gene so that the cells expressing the gene begin to glow. In addition to changing single genes it is also possible to remove or alter large pieces of chromosomes.

Technologies also now exist that allow genes to be removed not just in a whole animal, as described above, but in a subset of cells or in a particular tissue. This can be achieved by modifying the vector to include target sites (termed loxP sites) for an enzyme called Cre recombinase. When the Cre enzyme is present in a mouse cell in which the target gene is surrounded by loxP sites, it will cut this gene out of the chromosome. This allows the function of this gene, which may be required for the mouse to normally develop, to be analyzed in a particular cell type or tissue where only the Cre recombinase is expressed.

Therapeutic Potential of Gene Targeting

It is hoped that gene targeting may eventually become useful in treating some human genetic disorders such as hemophilia and Duchenne muscular dystrophy. Treating human disease by the types of genetic approaches mentioned above is termed "gene therapy." This could, in principle, be achieved by replacing the defective gene with a normal copy of the gene in the affected cells of an individual undergoing treatment. In order to make this potential treatment effective it will be necessary to develop technologies that increase the frequency with which targeting occurs. This is currently the subject of much research.

The development of nuclear transfer technology also has opened up the possible alternative method of using homologous recombination for gene therapy based on cell transfer. Gene targeting would be used to replace the defective genes in selected **somatic** cells in culture, and their nuclei could then be transferred into stem cells. The stem cells can then be differentiated into the affected cell type (for example, into bone marrow cells for hemophilia) and these cells could then be transplanted to patients. SEE ALSO Embryonic Stem Cells; Gene; Gene Therapy; Marker Systems.

somatic nonreproductive; not an egg or sperm

Seth G. N. Grant and Douglas J. C. Strathdee

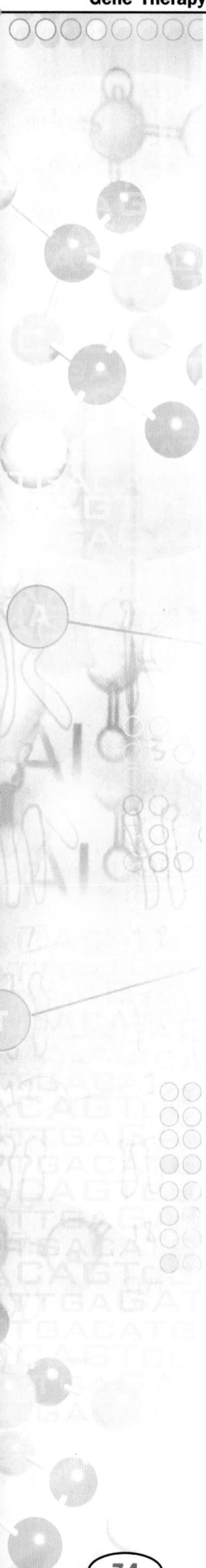

Bibliography

Hogan, B., et al. *Manipulating the Mouse Embryo: A Laboratory Manual,* 2nd ed. Cold Spring Harbor, NY: Cold Spring Harbor Laboratory Press, 1994.

Joyner, A., ed. *Gene Targeting: A Practical Approach,* 2nd ed. New York: Oxford University Press, 1999.

Torres, R. M., and R. Kuhn. *Laboratory Protocols for Conditional Gene Targeting.* New York: Oxford University Press, 1997.

Gene Therapy

Gene therapy is a new and largely experimental branch of medicine that uses genetic material (DNA) to treat patients. Researchers hope one day to use this therapy to treat several different kinds of diseases. While rapid progress has been made in this field in recent years, very few patients have been successfully treated by gene therapy, and a great deal of additional research remains to be done to bring these techniques into common use.

Disease Targets

Humans possess two copies of most of their genes. In a recessive genetic disease, both copies of a given gene are defective. Many such illnesses are called loss-of-function genetic diseases, and they represent the most straightforward application of gene therapy: If a functional copy of the defective gene can be delivered to the correct tissue and if it makes ("expresses") its normal protein there, the patient could be cured. Other patients suffer from dominant genetic diseases. In this case, the patient has one defective copy and one normal copy of a given gene. Some of these disorders are called gain-of-function diseases because the defective gene actively disrupts the normal functioning of their cells and tissues (some recessive diseases are also gain-of-function diseases). This defective copy would have to be removed or inactivated in order to cure these patients.

Gene therapy may also be effective in treating cancer or viral infections such as HIV-AIDS. It can even be used to modify the body's responses to injury. These approaches could be used to reduce scarring after surgery or to reduce restenosis, which is the reclosure of coronary arteries after balloon angioplasty. Each of these cases will be discussed in more detail below, but first we will deal with two technical issues of gene transfer: gene delivery and longevity of gene expression.

Gene Delivery

Whether given as pills or injections, most conventional drugs simply need to reach a minimal level in the bloodstream in order to be effective. In gene therapy, the drug (DNA) must be delivered to the nucleus of a cell in order to function, and a huge number of individual cells must each receive the DNA in order for the treatment to be effective. The situation is further complicated by the fact that a given gene may normally function in only a small portion of the cells in the body, and **ectopic expression** may be toxic. Thus, successful gene therapy often requires highly efficient delivery of DNA to a very restricted population of cells within the body.

ectopic expression expression of a gene in the wrong cells or tissues

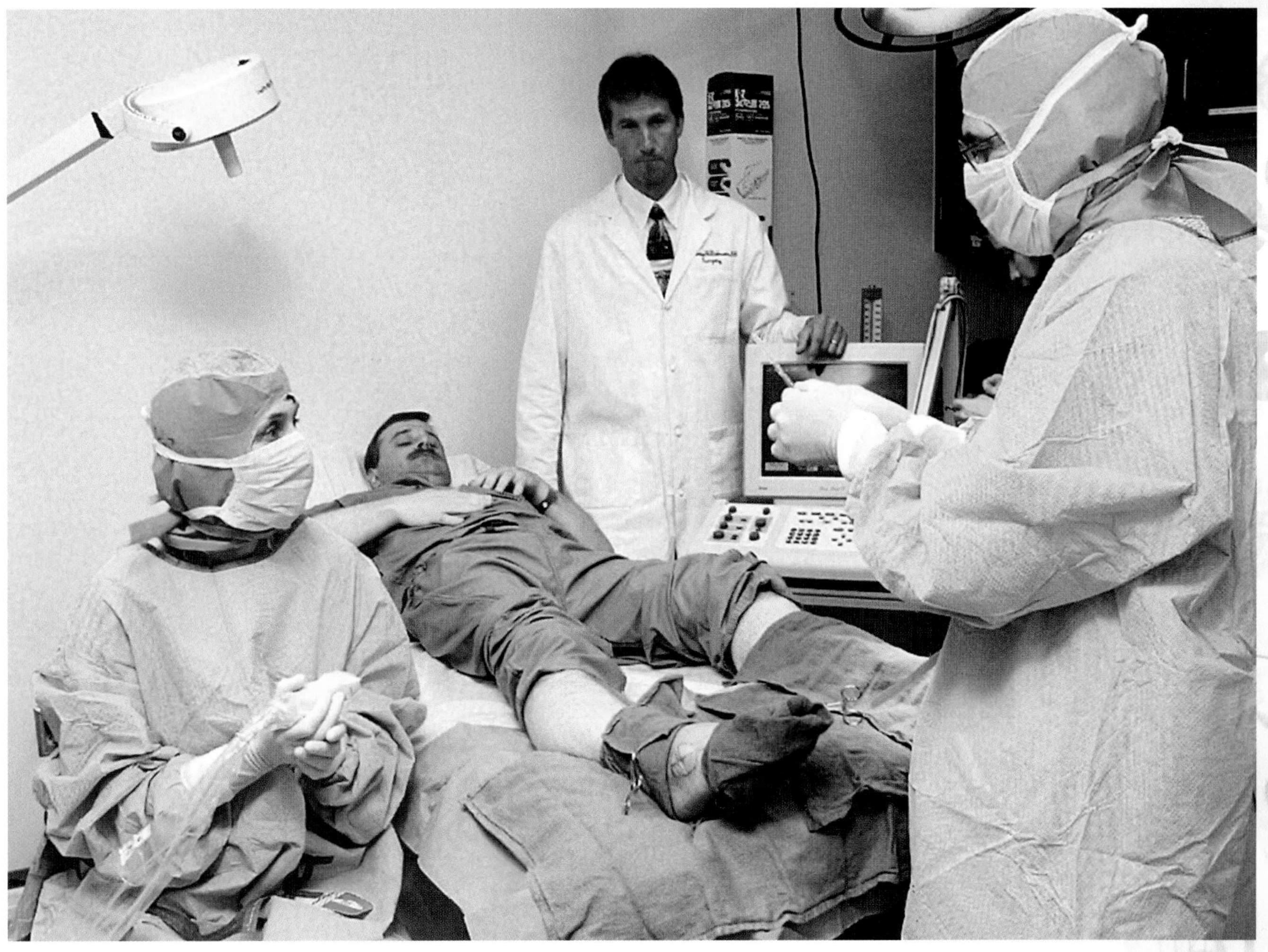

In 1999 doctors at the Ohio State University Medical Center prepare a gene therapy injection for 36-year-old Donovan Decker. He is the first patient to ever receive gene therapy for muscular dystrophy.

To achieve these goals, many researchers have turned to viruses. Viruses are parasites that normally reproduce by infecting individual cells in the human body, delivering their DNA to the nucleus of those cells. Once there, the viral DNA takes over the cell, converting it to a factory to make more viruses. The cell eventually dies, releasing more viruses to continue the cycle. Scientists can remove or disable some of the genetic material of the virus, making it unable to reproduce outside of the laboratory. This genetic material can then be replaced by the gene needed to treat a patient. The modified (or recombinant) virus can then be administered to the patient, where it will carry the therapeutic gene into the target cells. In this way, scientists can take advantage of the virus's ability, gained over millions of years of evolution, to deliver DNA to cells with tremendous efficiency. One of the most commonly used is a cold virus called adenovirus. Recombinant adenoviruses have been used in experimental gene therapy for muscle diseases, and can deliver genes to almost all of the cells in a small region surrounding the site of injection. Unfortunately, while adenoviruses excel at gene delivery, evolution is a double-edged sword, and the many mechanisms our own bodies have evolved to combat harmful viral infections are also used against therapeutic viruses, as will be discussed in more detail below.

Recombinant adenoviruses cannot be used to transfer DNA to all cell types, because they cannot reproduce themselves outside of the laboratory.

In April 2002 researchers announced that *ex vivo* gene therapy for severe combined immunodeficiency had been successful in five boys for up to 2.5 years.

When a cell with a recombinant adenovirus in it divides, only one of the two resulting cells contains the virus and the therapeutic gene it bears. The treatment of some diseases requires gene transfer to a stem cell, a cell that actively divides to create many new cells. For example, white blood cells live for only a short time, and must be constantly replenished by the division of **precursor** cells called hematopoietic stem cells. Gene therapy to treat an immune disease affecting white blood cells would thus require targeting these rapidly dividing cells. Researchers use a different kind of virus to accomplish this: retroviruses, so called because they contain RNA (a different kind of genetic material) rather than DNA.

precursor a substance from which another is made

When a retrovirus infects a cell, it converts its RNA to DNA and inserts it into the chromosome of the target cell. As the cell subsequently copies its own DNA during cell division, it copies the viral DNA as well, so that all of the progeny cells contain the retroviral DNA. At some later time, the viral DNA can liberate itself from the chromosome, direct the manufacture of many new viruses, and go on to repeat its life cycle. Recombinant retroviruses are engineered so that they can enter the target cell's chromosome, but become trapped there, unable to liberate themselves and continue their life cycle. Because all progeny cells still carry the recombinant retrovirus, they will also carry the therapeutic gene.

This is a great advantage over adenoviruses as a tool for gene delivery to dividing cells, but retroviruses have some drawbacks as well. They can only infect cells that are dividing quickly, and in most cases this infection must be carried out in the laboratory. Cells must be removed from the patient, infected with the recombinant retrovirus, grown for several weeks in the lab, and then reintroduced to the patient's body. This process, called **ex vivo** gene transfer, is extremely expensive and labor intensive. Nonetheless, this form of gene therapy has been used in one of the most successful clinical applications to date, the treatment of two patients with severe combined immune deficiency (SCID) caused by a defect in the adenosine deaminase gene.

ex vivo outside a living organism

Before treatment, these patients had essentially no immune system at all, and would have been required to live as "bubble children," completely isolated in a sterile environment. While their treatment did not completely cure their genetic disorder, it restored their immune systems enough to allow them to leave their sterile isolation chambers and live essentially normal lives. Many other viruses are being engineered for application to gene delivery, including adeno-associated virus, herpes simplex virus, and even extensively modified forms of the human immunodeficiency virus (HIV), to name just a few.

Many researchers are also exploring nonviral methods for gene delivery. One of the most successful of these methods consists of coating the therapeutic DNA with specialized fat molecules called lipids. The resulting small fatty drops called vesicles can then be injected or inhaled to deliver the DNA to the target tissue. Many different lipid formulations have been tested and different formulations work better in different tissues. These approaches have the great advantage that they do not stimulate the serious immune response that some viral **vectors** do. However, in general, these nonviral methods are not as efficient as viruses at transferring DNA to the target cells. No clearly superior method for gene delivery has yet emerged, and scientists are still

vectors carriers

actively developing both viral and nonviral methods. It is likely that many different methods will eventually be used, with each method specifically tailored to work best in a specific tissue or organ of the body.

Longevity of Gene Expression

One of the most challenging problems in gene therapy is to achieve long-lasting expression of the therapeutic gene, also called the transgene. Often the virus used to deliver the transgene causes the patient's body to produce an immune response that destroys the very cells that have been treated. This is especially true when an adenovirus is used to deliver genes. The human body raises a potent immune response to prevent or limit infections by adenovirus, completely clearing it from the body within several weeks. This immune response is frequently directed at proteins made by the adenovirus itself.

To combat this problem, researchers have deleted more and more of the virus's own genetic material. These modifications make the viruses safer and less likely to raise an immune response, but also make them more and more difficult to grow in the quantities necessary for use in the clinic. Expression of therapeutic transgenes can also be lost when the regulatory sequences that control a gene and turn it on and off (called **promoters** and enhancers) are shut down. Although inflammation has been found to play a role in this process, it is not well understood, and much additional research remains to be done.

promoters DNA sequences to which RNA polymerase bind to begin transcription

Examples of Gene Therapy Applications

There are many conditions that must be met in order to allow gene therapy to be possible. First, the details of the disease process must be understood. Of course, scientists must know exactly what gene is defective, but also when and at what level that gene would normally be expressed, how it functions, and what the regenerative possibilities are for the affected tissue. Not all diseases can be treated by gene therapy. It must be clear that replacement of the defective gene would benefit the patient. For example, a mutation that leads to a birth defect might be impossible to treat, because irreversible damage will have already occurred by the time the patient is identified. Similarly, diseases that cause death of brain cells are not well suited to gene therapy: Although gene therapy might be able to halt further progression of disease, existing damage cannot be reversed because brain cells cannot regenerate. Additionally, the cells to which DNA needs to be delivered must be accessible. Finally, great caution is warranted as gene therapy is pursued, as the body's response to high doses of viral vectors can be unpredictable. On September 12, 1999, Jesse Gelsinger, an eighteen-year-old participant in a clinical trial in Philadelphia, became unexpectedly ill and died from side effects of liver administration of adenovirus. This tragedy illustrates the importance of careful attention to safety regulations and extensive experiments in animal model systems before moving to human clinical trials.

Muscular Dystrophies. Duchenne and other recessive muscular dystrophies are well suited in many ways for gene therapy. These are loss-of-function recessive genetic diseases caused by mutations in the dystrophin gene or in genes for other structural muscle proteins. The normal levels of these pro-

teins are known, as are many of the ways that they function in the muscle cell. There is ample evidence in animal model systems that these diseases can be cured by delivery of functional copies of the gene. This is true in large part because muscle tissue has a tremendous capacity for repair and regeneration, so one could imagine that the heavily damaged muscle could repair itself after successful gene transfer. Muscle tissue is also an excellent target for gene transfer.

plasmid a small ring of DNA found in many bacteria

immunogenicity likelihood of triggering an immune system defense

Several different approaches have been used to transfer DNA to muscle. The most straightforward approach is the direct intramuscular injection of DNA in a circular form called a **plasmid**. The advantage to this approach is that it induces little to no immune response, although the overall number of cells expressing the gene is fairly low. In contrast, recombinant adenoviruses are extremely efficient at transferring genes to muscle, but give rise to a potent immune response that results in only short-term expression of the transferred genes. Because the efficiency of adenoviral transfer is so great, huge efforts are underway to reduce the **immunogenicity** of these vectors. These efforts have produced some significantly improved vectors, and research is now focusing on developing methods to prepare the large quantities necessary for clinical use. Adeno-associated virus combines the extremely high efficiency of adenoviral transfer with the very low immunogenicity of direct DNA transfer. However, this virus has a rather small capacity to carry DNA, so small that it cannot carry the dystrophin gene (one of the largest genes known), which is needed to treat Duchenne muscular dystrophy.

From these examples, it should be clear that many different approaches to gene therapy for muscular dystrophy have been tried, but that each approach suffers from one or more key shortcomings. In addition, all of these approaches to treat muscular dystrophy face one common problem: Although it is easy to transfer genes to a small part of a single muscle, simultaneously delivering a gene to all parts of all the muscles of the body is impossible with today's technology.

Hemophilia and Sickle Cell Disease. Because of the difficulty in treating diseases such as muscular dystrophy, many researchers have chosen to focus on genetic diseases that may be easier to treat, particularly those resulting from the lack of proteins freely dissolved in the bloodstream. Hemophilia is one such disorder, caused by a lack of blood-clotting proteins. Such patients have long been treated by the infusion of the missing clotting proteins, but this treatment is extremely expensive and requires almost daily injections. Gene therapy holds great promise for these patients, because replacement of the gene that makes the missing protein could permanently eliminate the need for protein injections. It really does not matter what tissue produces these clotting factors as long as the protein is delivered to the bloodstream, so researchers have tried to deliver these genes to muscle and to the liver using several different vectors. Approaches using recombinant adenoviruses to deliver the clotting factor gene to the liver are especially promising, and tests have shown significant clinical improvement in a dog model of hemophilia.

Gain-of-function genetic diseases present a very different sort of challenge because the mutant gene or genes create a new biological activity that actively interferes with the normal functioning of the cell. An example of such a disorder is sickle cell disease. Patients suffering from this disease have a defective hemoglobin protein in their red blood cells. This defective protein can cause their red blood cells to be misshapen, clogging their blood vessels and causing extremely painful and dangerous blood clots. Most of our genes make an RNA transcript, which is then used as a blueprint to make protein. In sickle cell disease, the transcript of the mutant gene needs to be destroyed or repaired in order to prevent the synthesis of mutant hemoglobin.

The molecular repair of these transcripts is possible using special RNA molecules called **ribozymes**. There are several different kinds of ribozymes: some that destroy their targets, and others that modify and repair their target transcripts. The repair approach was tested in the laboratory on cells containing the sickle cell mutation, and was quite successful, repairing a significant fraction of the mutant transcripts. While patients cannot yet be treated using this technique, the approach illustrates how biologically damaging molecules can be inactivated. Similar approaches are being developed to treat HIV-AIDS infections, and these may one day be used along with other antiviral therapies to treat this dreaded disease.

ribozymes RNA-based catalysts

Cancer. Very different strategies of gene therapy are used to treat cancer. When treating diseases such as muscular dystrophy, researchers try to deliver genes without detection by the patient's immune system. When treating cancer, the object is often precisely the opposite: to stimulate a patient's immune reaction to the tumor tissue and improve its ability to fight the disease. For this reason, tumor tissue is often transformed by the new gene to produce specific activators of the immune system, such as interleukins or GM-CSF (granulocyte monocyte colony stimulating factor).

Usually, cancer cells are not recognized by the immune system because they are in many ways identical to the patient's normal cells. These stimulating factors activate the immune system and help it recognize and attack the tumor tissue. In another approach, called "suicide therapy," a gene such as the herpes simplex virus thymidine kinase gene (*HSV-TK*) is transferred to the tumor. This gene normally does not occur in the human body, and it is not metabolically active. After several rounds of gene therapy have built up high levels of TK activity in the tumor, a drug called ganciclovir is given to the patient. This drug is inactive in normal cells, but the TK gene converts it into a potent toxin, killing the tumor cells. Even nearby tumor cells that do not have the TK gene can be killed by a phenomenon called the "bystander effect." This approach not only kills tumor cells directly, but also activates the immune system to further attack the tumor.

Anticancer gene therapy is a powerful adjunct to other more traditional forms of cancer treatment. Its advantages are that it can be beneficial even if only a portion of the tumor cells receive the transferred gene, there is no need for long-term gene expression, and it works with the immune system, rather than trying to defeat it. Anticancer gene therapy is already in significant use in the clinic, and is likely to become even more commonplace in the near future.

In summary, gene therapy covers several related areas of research and clinical treatment, all using the genetic material DNA as a drug. Gene therapy is currently being used, along with other techniques, to treat cancer.

One day, gene therapy may also be used to treat a variety of hereditary and nonhereditary diseases, ranging from loss-of-function disorders such as muscular dystrophy and hemophilia, to gain-of-function disorders such as sickle cell disease, to viral diseases such as HIV-AIDS. Active areas of research include improvements in the methods of gene delivery to the individual tissues and cells of the body and the modulation of the immune response to gene delivery. Many challenges remain to the successful maturation of gene therapy from the laboratory to the clinical setting. SEE ALSO CANCER; CYSTIC FIBROSIS; DISEASE, GENETICS OF; EMBRYONIC STEM CELLS; GENE DISCOVERY; GENE THERAPY: ETHICAL ISSUES; HEMOPHILIA; MUSCULAR DYSTROPHY; RETROVIRUS; RIBOZYME; SEVERE COMBINED IMMUNE DEFICIENCY; VIRUS.

Michael A. Hauser

Bibliography

Beardsley, T. "Working under Pressure." *Scientific American* 282 (2000): 34.

Clark, William R. *The New Healers: The Promise and Problems of Molecular Medicine in the Twenty-first Century.* New York: Oxford University Press, 1999.

Vogel, G. "Gene Therapy: FDA Moves against Penn Scientist." *Science* 290 (2000): 2049–2051.

Internet Resource

Institute for Human Gene Therapy. <http://www.yshs.upenn.edu/ihgt/>.

Gene Therapy: Ethical Issues

Gene therapy introduces or alters genetic material to compensate for a genetic mistake that causes disease. It is hoped that gene therapy can treat or cure diseases for which no other effective treatments are available. However, many unique technical and ethical considerations have been raised by this new form of treatment, and several levels of regulatory committees have been established to review each gene therapy clinical trial prior to its initiation in human subjects. Ethical considerations include deciding which cells should be used, how gene therapy can be safely tested and evaluated in humans, what components are necessary for informed consent, and which diseases and/or traits are eligible for gene therapy research.

Germ Line versus Somatic Cell Gene Therapy

somatic nonreproductive; not an egg or sperm

in vitro "in glass"; in lab apparatus, rather than within a living organism

Virtually all cells in the human body contain genes, making them potential targets for gene therapy. However, these cells can be divided into two major categories: germ line cells (which include sperm and eggs) and **somatic** cells. There are fundamental differences between these cell types, and these differences have profound ethical implications.

Gene therapy using germ line cells results in permanent changes that are passed down to subsequent generations. If done early in embryologic development, such as during preimplantation diagnosis and **in vitro** fertilization, the gene transfer could also occur in all cells of the developing embryo. The appeal of germ line gene therapy is its potential for offering a permanent therapeutic effect for all who inherit the target gene. Successful germ line therapies introduce the possibility of eliminating some diseases from a particular family, and ultimately from the population, forever. However, this also raises controversy. Some people view this type of therapy as unnatural, and liken it to "playing God." Others have concerns about the technical aspects. They worry that the genetic change propagated by germ line gene therapy may actually be deleterious and harmful, with the potential for unforeseen negative effects on future generations.

Somatic cells are nonreproductive. Somatic cell therapy is viewed as a more conservative, safer approach because it affects only the targeted cells in the patient, and is not passed on to future generations. In other words, the therapeutic effect ends with the individual who receives the therapy. However, this type of therapy presents unique problems of its own. Often the effects of somatic cell therapy are short-lived. Because the cells of most tissues ultimately die and are replaced by new cells, repeated treatments over the course of the individual's life span are required to maintain the therapeutic effect. Transporting the gene to the target cells or tissue is also problematic. Regardless of these difficulties, however, somatic cell gene therapy is appropriate and acceptable for many disorders, including cystic fibrosis, muscular dystrophy, cancer, and certain infectious diseases. Clinicians can even perform this therapy **in utero**, potentially correcting or treating a life-threatening disorder that may significantly impair a baby's health or development if not treated before birth.

in utero inside the uterus

Research Issues

Scientific and ethical discussions about gene therapy began many years ago, but it was not until 1990 that the first approved human gene therapy clinical trial was initiated. This clinical trial was conducted on a rare autoimmune disorder called severe combined immune deficiency. This therapy was considered successful because it greatly improved the health and well-being of the few individuals who were treated during the trial. However, the success of the therapy was tentative, because along with the gene therapy the patients also continued receiving their traditional drug therapy. This made it difficult to determine the true effectiveness of the gene therapy on its own, as distinct from the effects of the more traditional therapy.

Measuring the success of treatment is just one challenge of gene therapy. Research is fraught with practical and ethical challenges. As with clinical trials for drugs, the purpose of human gene therapy clinical trials is to determine if the therapy is safe, what dose is effective, how the therapy should be administered, and if the therapy works. Diseases are chosen for research based on the severity of the disorder (the more severe the disorder, the more likely it is that it will be a good candidate for experimentation), the feasibility of treatment, and predicted success of treatment based on animal models. This sounds reasonable. However, imagine you or your child has a serious condition for which no other treatment is available. How objective would your decision be about participating in the research?

Informed Consent

A hallmark of ethical medical research is informed consent. The informed consent process educates potential research subjects about the purpose of the gene therapy clinical trial, its risks and benefits, and what is involved in participation. The process should provide enough information for the potential

research subjects to decide if they want to participate. It is important both to consider the safety of the experimental treatment and to understand the risks and benefits to the subjects. In utero gene therapy has the added complexity of posing risks not only to the fetus, but also to the pregnant woman. Further, voluntary consent is imperative. Gene therapy may be the only possible treatment, or the treatment of last resort, for some individuals. In such cases, it becomes questionable whether the patient can truly be said to make a voluntary decision to participate in the trial.

clinical trials tests performed on human subjects

Gene therapy **clinical trials** came under scrutiny in September 1999, after the highly publicized death of a gene therapy clinical trial participant several days after he received the experimental treatment. This case raised concerns about the overall protection of human subjects in clinical testing, and specifically about the reliability of the informed consent process. In this case, it was alleged that information about potential risks to the patient was not fully disclosed to the patient and his family. It was further alleged that full information regarding adverse events (serious side effects or deaths) that occurred in animals receiving experimental treatment had not been adequately disclosed. Adverse events should be disclosed in a timely manner not only to the participants in these trials, but also to the regulatory bodies overseeing gene therapy clinical trials. Furthermore, participants had not been told of a conflict of interest posed by a financial relationship between the university researchers and the company supporting the research. Obviously, any conflicts of interests could interfere with the objectivity of researchers in evaluating the effectiveness of the clinical trials and should be disclosed during the informed consent process.

Appropriate Uses of Gene Therapy

How do researchers determine which disorders or traits warrant gene therapy? Unfortunately, the distinction between gene therapy for disease genes and gene therapy to enhance desired traits, such as height or eye color, is not clear-cut. No one would argue that diseases that cause suffering, disability, and, potentially, death are good candidates for gene therapy. However, there is a fine line between what is considered a "disease" (such as the dwarfism disorder achondroplasia) and what is considered a "trait" in an otherwise healthy individual (such as short stature). Even though gene therapy for the correction of potentially socially unacceptable traits, or the enhancement of desirable ones, may improve the quality of life for an individual, some ethicists fear gene therapy for trait enhancement could negatively impact what society considers "normal" and thus promote increased discrimination toward those with the "undesirable" traits. As the function of many genes continue to be discovered, it may become increasingly difficult to define which gene traits are considered to be diseases versus those that should be classified as physical, mental, or psychological traits.

ethicists a person who writes and speaks about ethical issues

germ line cells giving rise to sperm or eggs

To date, acceptable gene therapy clinical trials involve somatic cell therapies using genes that cause diseases. However, many **ethicists** worry that, as the feasibility of **germ line** gene therapy improves and more genes causing different traits are discovered, there could be a "slippery slope" effect in regard to which genes are used in future gene therapy experiments. Specifically, it is feared that the acceptance of germ line gene therapy could lead to the acceptance of gene therapy for genetic enhancement. Public debate

about the issues revolving around germ line gene therapy and gene therapy for trait enhancement must continue as science advances to fully appreciate the appropriateness of these newer therapies and to lead to ethical guidelines for advances in gene therapy research.

Promise of Gene Therapy

There was extensive media coverage and public excitement about the promise of gene therapy as the first clinical trials commenced in 1990. In fact, many individuals and families with genetic disorders expected an imminent cure for their diseases. Unfortunately, there are as yet few successes to report, even though hundreds of somatic cell gene therapy clinical trials for many different diseases have been attempted. The early hype of gene therapy might have been avoided with more open and honest communication about gene therapy and its expectations between the researchers, physicians, patients, and the public. The death of the participant in 1999 and deficiencies in the protocol (study design) used for that trial underscore the need for continued public discourse on gene therapy.

The promise of gene therapy has not diminished, even though its full therapeutic potential is not yet known. Scientists, physicians, patients, and families continue to look forward to many future successes for gene therapy. With the completion of the Human Genome Project and the accelerated discovery of human disease genes, the potential number of diseases for which gene therapy could be beneficial continues to increase. With further research into the technical aspects of gene therapy and continued public debate about the ethical issues involved in such treatments, it is hoped gene therapies will become standard, effective treatments in the next few decades. SEE ALSO CYSTIC FIBROSIS; EUGENICS; GENE THERAPY; GROWTH DISORDERS; MUSCULAR DYSTROPHY; PRENATAL DIAGNOSIS; SEVERE COMBINED IMMUNE DEFICIENCY.

Elizabeth C. Melvin

Bibliography

Anderson, W. French. "Human Gene Therapy." *Nature* 392 (1998): 25–30.

Smaglik, Paul. "Congress Gets Tough with Gene Therapy." *Nature* 403 (2000): 583–584.

Walters, Leroy. "The Ethics of Human Gene Therapy." *Nature* 320 (1986): 225–227.

Internet Resources

American Society of Human Gene Therapy. <http://www.asgt.org>.

"Gene Therapy." Human Genome Program of the U.S. Department of Energy. <http://www.ornl.gov/hgmis/medicine/genetherapy.html>.

Genetic Code

The sequence of **nucleotides** in DNA determines the sequence of amino acids found in all proteins. Since there are only four nucleotide "letters" in the DNA alphabet (A, C, G, T, which stand for adenine, cytosine, guanine, and thymine), but there are 20 different amino acids in the protein alphabet, it is clear that more than one nucleotide must be used to specify an amino acid. Even two nucleotides read at a time would not give sufficient combinations ($4 \times 4 = 16$) to encode all 20 amino acids plus start and stop signals. Therefore it would require a minimum of three DNA nucleotides

nucleotides the building blocks of RNA or DNA

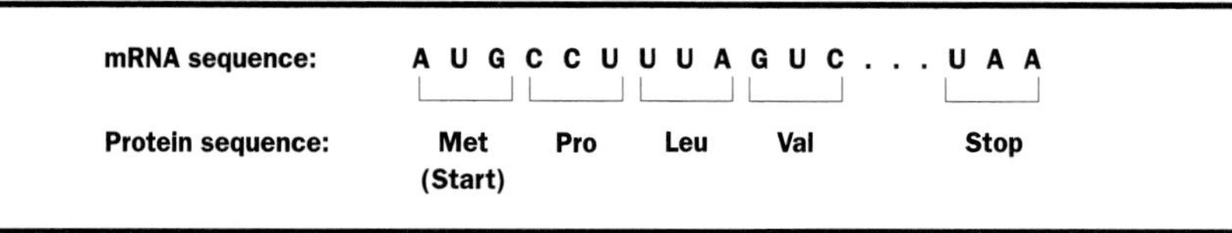

Figure 1. A messenger RNA is translated in triplets, beginning with the first AUG encountered by the ribosome. Translation stops at a stop codon, one of which is UAA.

to "spell out" one amino acid, and indeed this is the number that is actually used. RNA also uses a four letter alphabet when it reads and transcribes DNA instructions during protein synthesis, but its set of nucleotides is somewhat different, substituting U (uracil) for T (thymine).

codon a sequence of three mRNA nucleotides coding for one amino acid

polypeptide chain of amino acids

Any single set of three nucleotides is called a **codon**, and the set of all possible three-nucleotide combinations is called "the genetic code" or "triplet code." There are sixty-four different combinations or codons ($4 \times 4 \times 4 = 64$). We now know that three codons (UAA, UAG, and UGA) specify a "stop" signal, indicating the termination of the **polypeptide** chain being synthesized on the ribosome. Each of the remaining sixty-one codons encodes an amino acid. The "start" signal is the codon AUG, which also encodes the amino acid methionine. The codons are read from the messenger RNA molecule during protein synthesis, and, consequently, they are given in RNA bases rather than in the original DNA sequence. The reading of the codons is shown in Figure 1.

Translation

The gene is represented by the sequences of bases in the DNA molecule, which can, in a sense, be thought of as a "storage molecule" for genetic information. DNA is extremely stable, a property critical to the maintenance of the integrity of the gene. This stability is evidenced by the fact that DNA has been extracted from Egyptian mummies and extinct animals such as the woolly mammoth. It can be extracted from dried blood or from a single hair at a crime scene.

Each cell contains a complete set of genes, but only certain of these genes are active or "expressed" at any one time. When a gene is active, a "disposable" copy is transcribed from the gene into codons contained in a messenger RNA (mRNA) molecule. Unlike the DNA molecule, the mRNA molecule is relatively unstable and short-lived. This is so that when a gene is turned off, the mRNA does not remain in the cell forever, running off more proteins on the ribosomes that are no longer needed by the cell.

Another RNA molecule, called transfer RNA (tRNA), contains a specific region called the anticodon. The tRNA anticodon can base pair with the codon region of the mRNA during protein synthesis, using the base pairing rules of A-U, U-A, C-G, and G-C. Each tRNA carries a specific amino acid. Thus the tRNA carrying methionine has a UAC anticodon that pairs with the AUG codon of the mRNA bound to the ribosome. Similarly the tRNA for proline has a GGA anticodon.

In examining the table of codons (Table 1) you will see that there is more than one codon for each amino acid, except for methionine (AUG) and tryptophan (UGG). Different codons that code for the same amino acid are said to be "synonyms," and the code is said to be "degenerate" in the sense that there is not a single, unique codon for each of the twenty amino acids.

Table 1. Genetic code.

First base	Second base: U	C	A	G
U	UUU, UUC: Phe UUA, UUG: Leu	UCU, UCC, UCA, UCG: Ser	UAU, UAC: Tyr UAA, UAG: STOP	UGU, UGC: Cys UGA: STOP UGG: Trp
C	CUU, CUC, CUA, CUG: Leu	CCU, CCC, CCA, CCG: Pro	CAU, CAC: His CAA, CAG: Gln	CGU, CGC, CGA, CGG: Arg
A	AUU, AUC, AUA: Ile AUG: Met	ACU, ACC, ACA, ACG: Thr	AAU, AAC: Asn AAA, AAG: Lys	AGU, AGC: Ser AGA, AGG: Arg
G	GUU, GUC, GUA, GUG: Val	GCU, GCC, GCA, GCG: Ala	GAU, GAC: Asp GAA, GAG: Glu	GGU, GGC, GGA, GGG: Gly

The "Wobble" Hypothesis

Even before the genetic code had been elucidated, Francis Crick postulated that base pairing of the mRNA codons with the tRNA anticodons would require precision in the first two nucleotide positions but not so in the third position (the precise conformation of **base pairs**, which refers to the **hydrogen bonding** between A-T (A-U in RNA) and C-G pairs is known as Watson-Crick base pairing). The third position, in general, would need to be only a purine (A or G) or a pyrimidine (C or U). Crick called this phenomenon "wobble."

base pairs two nucleotides (either DNA or RNA) linked by weak bonds

hydrogen bonding weak bonding between the H of one molecule or group and a nitrogen or oxygen of another

This less-than-precise base pairing would require fewer tRNA species. For example, $tRNA^{Glu}$ could pair with either GAA or GAG codons. In looking at the codon table, one can see that, for the most part, the first two letters are important to specify the particular amino acid. The only exceptions are AUG (Met) and UGG (Trp) which, as indicated above, have only one codon each.

The Code Has No Gaps or Overlaps

The 1960s were an exciting time for molecular biologists, for it was then that the genetic code was broken. Two possibilities had to be considered for the genetic code. It was possible that the code had gaps, that is, some sort of punctuation mark or a "spacer" nucleotide or nucleotides between coding groups. Second, the code could be either overlapping or nonoverlapping. These possibilities are illustrated in Figures 2 and 3. An overlapping code would have the advantage that more information could be contained in a smaller space.

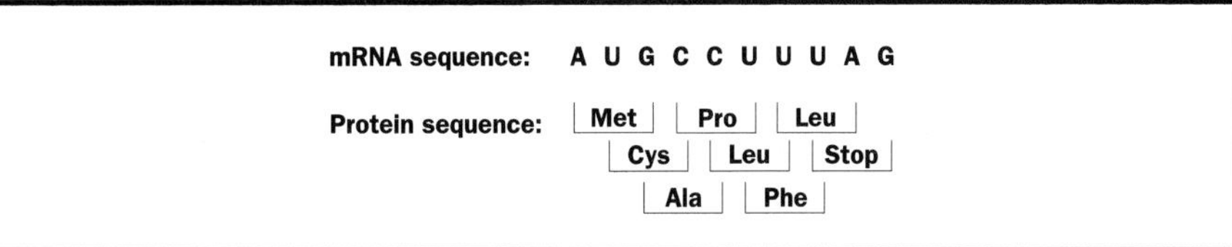

Figure 2. Any stretch of messenger RNA has three different reading frames, which can be translated to give different amino acid sequences. Only one of them is the correct sequence, dictated by the start point of the first triplet, AUG.

Figure 3. Models of mRNA showing the effect of hypothetical "punctuation marks" separating codons.

Punctuated mRNA:	A U G p C C U p U U A p G U C p . . .
Protein sequence:	Met Pro Leu Val
Punctuated mRNA:	A U G p C G C U p U U A p G U C p . . .
Insertion of a base:	▲
Protein sequence:	Met Arg Leu Val

However, in overlapping code a mutation that changed one base would lead to the changing of three consecutive amino acids in the protein sequence. Genetic evidence, available even before the code had been deciphered, indicated that a single point mutation, that is, a change in a single nucleotide, affected only one amino acid and thus suggested a nonoverlapping code.

Another possibility was that the code had punctuation marks, that is, a base (indicated by "p" in Figure 3) acting as a comma that would separate each codon. In this situation, if an additional base were inserted into a codon, then only that codon would be affected. In a code without punctuations or gaps, however, insertion of a single nucleotide would result in all codons from that point on being affected. This would in turn change the amino acid sequence in the protein from that point on. Again, genetic evidence ruled out a punctuated code, as base insertions do, in fact, affect the entire protein from the insertion point on, rather than just a single amino acid. This effect is called a frameshift mutation.

In the late 1970s DNA sequencing techniques were developed. A number of proteins had already been sequenced by protein sequencing methods. When the genes for these proteins were cloned and sequenced, the predicted protein sequence could be deduced. Agreement between the DNA and protein sequences confirmed the accuracy of the genetic code.

Exceptions to the Universal Genetic Code

E. coli the bacterium *Escherichia coli*

After the original genetic code of ***E. coli*** was completed in 1968, the genetic code was subsequently determined for many other organisms ranging from bacteria to mammals, including humans. The codons were found to be the same for all organisms, leading to the idea that the genetic code is "universal." Furthermore, it also suggested that life on Earth had a single evolutionary origin, otherwise there would have been numerous genetic codes. The code was established during evolution, probably by chance, as there are no compelling reasons one codon should prevail over another. After it was established, any subsequent changes in the code would prove to be lethal, for if one codon changed, then all similar codons in the entire organism's genome would have to change simultaneously—a highly unlikely possibility.

endosymbiotic a type of symbiosis in which one partner lives within the other

genome the total genetic material in a cell or organism

Thus, it was surprising to find that there are, in fact, a few rare exceptions to the universal code. These exceptions are listed in Table 2. Most of these exceptions are found in the mitochondrial genome. The mitochondrion is thought to have evolved from an **endosymbiotic** bacterium at the time when the eukaryotic cell first arose. The mitochondrial **genome** is small, and most of the genes of the original endosymbiont have migrated to the nucleus.

Table 2.

EXCEPTIONS TO THE UNIVERSAL GENETIC CODE

Organism	Normal codon	Usual meaning	New meaning
Mammalian mitochondria	AGA, AGG AUA UGA	Arginine Isoleucine Stop codon	Stop codon Methionine Tryptophan
Drosophila mitochondria	AGA, AGG AUA UGA	Arginine Isoleucine Stop codon	Serine Methionine Tryptophan
Yeast mitochondria	AUA UGA CUA, CUC, CUG, CUU	Isoleucine Stop codon Leucine	Methionine Tryptophan Threonine
Higher plant mitochondria	UGA CGG	Stop codon Arginine	Tryptophan Tryptophan
Protozoan nuclei	UAA, UAG	Stop codons	Glutamine
Mycoplasma capricolum bacteria	UGA	Stop codon	Tryptophan

In examining the exceptions to the universal genetic code in Table 2, you can see that there are only a few changes, most notably the use of a standard "stop" codon to encode an amino acid. For example, UGA normally is a stop codon. But in the mitochondria of the fruit fly *Drosophila melanogaster*, it encodes the amino acid tryptophan.

A few additional exceptions to the universal genetic code have also been identified. These include the nuclear genome of a few protozoan species and also in the bacterium *Mycoplasma capricolum*. These exceptions, however, do not imply multiple evolutionary origins of life. What is most striking is that the "exceptional" meanings of most of the codons are identical across all the organisms in which they are found, not different. Had there been multiple origins, we would expect to see drastically different genetic codes in these exceptional organisms. SEE ALSO CRICK, FRANCIS; *ESCHERICHIA COLI* (*E. COLI* BACTERIUM); NUCLEOTIDE; READING FRAME; RIBOSOME; TRANSCRIPTION; TRANSLATION.

Ralph R. Meyer

Bibliography

"The Genetic Code." *Cold Spring Harbor Symposia on Quantitative Biology*, vol. 31. Cold Spring Harbor, NY: Cold Spring Harbor Press, 1966.

Kay, Lily E. *Who Wrote the Book of Life? A History of the Genetic Code.* Stanford, CA: Stanford University Press, 2000.

Nirenberg, M. W., and J. H. Matthaei. "The Dependence of Cell-Free Protein Synthesis in *E. coli* upon Naturally Occurring or Synthetic Polyribonucleotides." *Proceedings of the National Academy of Sciences* 47 (1961): 1588–1602.

Genetic Counseling

Over the last half-century, our understanding of genetic disorders has increased spectacularly. When facts about inherited disorders first came to light, health professionals began to inform families about probable inheritance patterns and recurrence risks (the likelihood that offspring or other relatives might also inherit the disease).

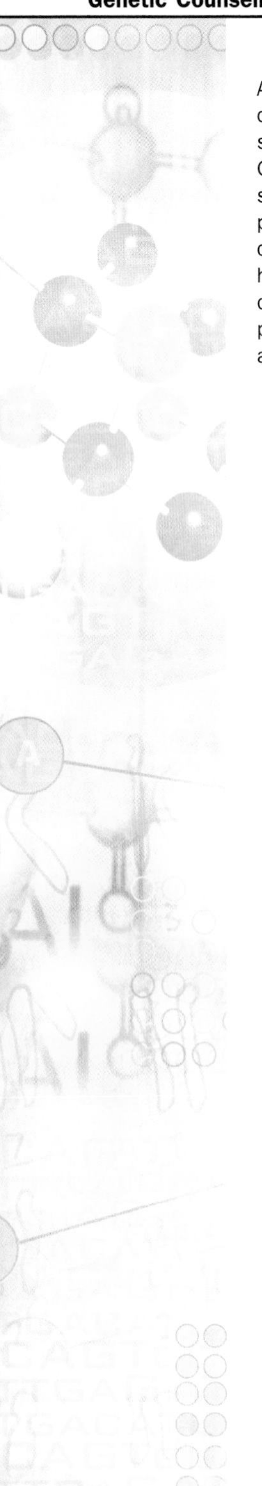

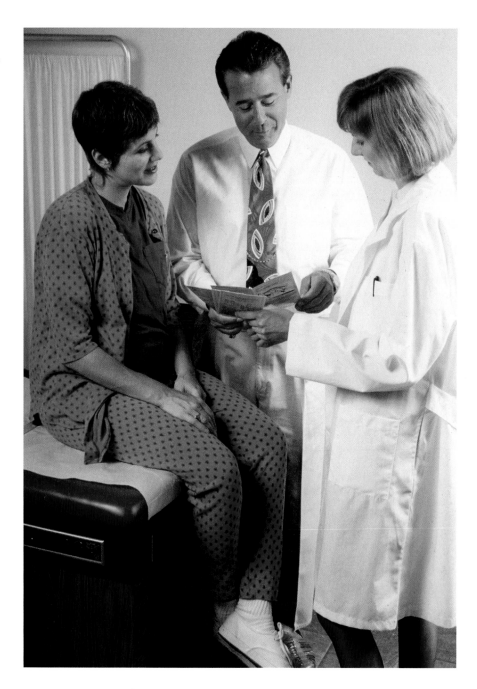

A couple receives genetic counseling during their sixth month of pregnancy. Genetic counselors are specially trained to provide support to couples whose families have histories of genetic diseases, or who are pregnant and carrying an affected child.

The Need for Genetic Counseling

Receiving a diagnosis of a genetic disorder can have profound impact for both patients and their family members, and it quickly became clear that aside from the need for medical and genetic information, families affected by genetic disorders had educational, social, and psychological needs that required attention. And though families were afflicted with different disorders passed on by different modes of inheritance (**autosomal** recessive, autosomal **dominant**, complex, or some other type), certain reactions were observed again and again.

autosomal describes a chromosome other than the X and Y sex-determining chromosomes

dominant controlling the phenotype when one allele is present

recessive requiring the presence of two alleles to control the phenotype

For instance, some parents of children with an autosomal **recessive** disorder felt profound guilt at having transmitted an inherited disorder to their child. On the other hand, a family member who was spared a genetic disorder that other family members developed frequently suffered "survivor guilt." Health-care providers also noted that family members at risk for developing a late-onset disorder live with intense anxiety about the future and often needed support and counseling. Finally, counseling was seen to be of potential help for family members who incorrectly inferred that they were at risk for having a child with a genetic disorder. Attempts to meet these varied needs and help give people a sense of control over their situation resulted in the emergence of a model of genetic education and support that came to be called genetic counseling.

An Evolving Field

In time, genetic counseling evolved into a profession. Since the early 1970s genetic counselors have been members of health-care teams providing comprehensive and consistent medical genetic services, while also tending to the social and emotional welfare of the patients and their families. In the United States, the first master's degree training program for genetic counseling was established in 1971. Since then the profession has grown tremendously. There are now more than 2,500 genetic counselors in the United States and 25 genetic counseling training programs. Many industrialized countries have adopted the United States's model of training for genetic counselors, and master's-level training programs now exist in Canada, Australia, Great Britain, and South Africa.

As the profession grows, the definition of genetic counseling also continues to evolve. Genetic counseling is currently defined as "a communication process, which helps an individual and/or family in a variety of ways." For instance, genetic counselors help patients and their families to comprehend the medical facts, including the diagnosis, probable course of the disorder, and available treatment options. Genetic counselors also help educate their clients about the way heredity contributes to the disorder and the risk of recurrence in relatives, and to understand the options available for dealing with this risk of recurrence.

Genetic counselors also teach their clients the medical facts relating to a disorder, enabling them to make informed, independent decisions. They understand that only if their clients possess the necessary facts about available medical care and genetic testing can their decisions be free of coercion. Finally, genetic counselors provide information helpful in accessing local and national support resources.

A key aspect of the genetic counselor's work is educational: helping clients to comprehend the genetic implications of their disorder. In addition, the diagnosis of a genetic disorder in an individual often leads to identification of other family members who may be at risk for having or passing on a genetic disorder, so genetic counselors often work with entire families. For instance, if the genetic tests of a female patient with two sisters disclose that she has a genetic change (mutation) in the BRCA1 gene, then her two sisters are at risk for carrying this same genetic change, which can cause breast cancer. Once the patient has been notified and has given her permission, these sisters would then be contacted and given the chance to learn about their own risk of carrying a disease-causing gene.

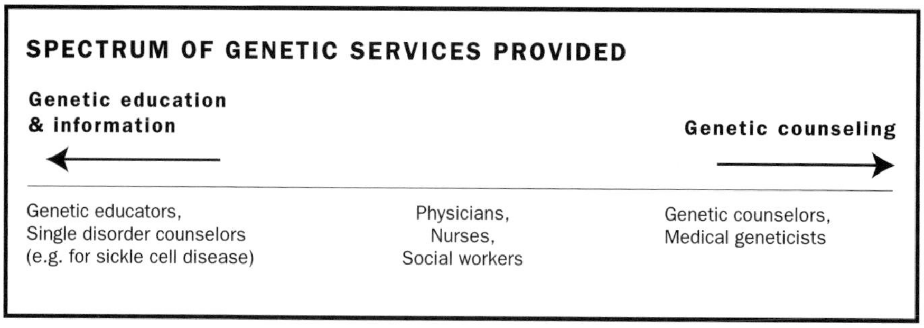

A visual representation of the various services offered by genetic counselors.

SPECTRUM OF GENETIC SERVICES PROVIDED

Genetic education & information ←		Genetic counseling →
Genetic educators, Single disorder counselors (e.g. for sickle cell disease)	Physicians, Nurses, Social workers	Genetic counselors, Medical geneticists

This second round of counseling is important. Armed with the information about their susceptibility for breast cancer, the sisters might choose to undergo genetic testing themselves, or they might begin early detection screening evaluations. Their new knowledge might also lead them to adopt lifestyle changes that could reduce their risk of developing breast cancer. Even when genetic testing is not available, early identification of at-risk patients and their family members can be valuable and quite possibly lifesaving.

osteoporosis thinning of the bone structure

The demand for genetic education and counseling will likely increase as knowledge accumulates about the genetic component of commonly occurring disorders such as breast cancer, Alzheimer's disease, heart disease, diabetes mellitus, and **osteoporosis**. As a result, a variety of professional specialists, such as genetic educators, physicians, nurses, social workers, medical geneticists, and genetic counselors, will increasingly be called upon to provide genetic education and counseling. Of this group, however, genetic counselors and medical geneticists are the most qualified to perform comprehensive genetic counseling.

Protection as Well as Education

eugenics movement to "improve" the gene pool by selective breeding

At the same time that they provide beneficial genetic counseling to patients and their families, professionals providing such a service must have a full understanding of the dangers of **eugenics**. The abuse of genetic information has led to many atrocities in the past. In Germany, the Nazis murdered nearly 7 million "genetically defective" people during World War II and forcibly sterilized nearly half a million others, all in the name of "eugenics"—a policy that calls for the systematic elimination of "unfit" members of the population. The United States also has a checkered past with respect to eugenics. In the early twentieth century, the United States passed laws allowing sterilization of the mentally handicapped and limiting the number of "genetically inferior" ethnic groups that were allowed to immigrate.

informed consent knowledge of the risks involved

In order to prevent such abuses from ever occurring again, the genetic counseling profession has followed in the footsteps of other health-care professions by establishing a code of ethics guiding professional behavior. Policies such as nondirectiveness, prevention of genetic discrimination, respect for patients' beliefs, complete disclosure, and **informed consent** are components of these ethical principles. Nondirectiveness, one of the major tenets of genetic counseling, is defined by the National Society of Genetic Counselors as enabling "clients to make informed independent decisions, free of coercion, by providing or illuminating the necessary facts and clarifying the alternatives and anticipated consequences."

Genetic counseling has become a vital part of medical genetics. With the knowledge gained from the past and the tools to help patients choose their paths, genetic counseling will continue to be invaluable in the rapidly growing field of human genetics. SEE ALSO EUGENICS; GENETIC COUNSELOR; GENETIC TESTING; INHERITANCE PATTERNS; PRENATAL DIAGNOSIS.

Chantelle Wolpert

Bibliography

Epstein C. J., et al. "Genetic Counseling." *American Journal of Human Genetics* 27 (1975): 240–242.

Fine, B., and M. Koblenz. "Conducting Pre-Test Patient Education." In *Humanizing Genetic Testing: Clinical Applications of New DNA Technologies.* Evanston, IL: Northwestern University, 1994.

Kessler, S. "Psychological Aspects of Genetic Counseling VI: A Critical Review of the Literature Dealing with Education and Reproduction." *American Journal of Medical Genetics* 34 (1989): 340–353.

———. "Process Issues in Genetic Counselling." *Birth Defects* 28, no. 1 (1992): 1–10.

National Society of Genetic Counselors. "Genetic Counseling as a Profession." In *National Society of Genetic Counselors.* Wallingford, PA: National Society of Genetic Counselors, Inc., 1983.

Reed, S. "A Short History of Genetic Counseling." *Social Biology* 21 (1974): 332–339.

Genetic Counselor

Genetic counselors are health professionals trained in genetics, genetic disorders, genetic testing, molecular biology, psychology and psychosocial issues, and the ethical and legal issues of genetic medicine. Most genetic counselors have a master's degree from a genetic counseling training program. The very first class of genetic counselors was graduated from Sarah Lawrence College in 1971. There are about 2,000 genetic counselors in the United States. Most are women under the age of forty, but the field is becoming more diverse.

Genetic counselors are board-certified by the American Board of Genetic Counseling. Board eligibility or certification is required for employment in many positions, and some states are beginning to license genetic counselors. While salaries vary significantly by geographic location, years of experience, and work setting, according to a Professional Status Survey conducted by the National Society of Genetic Counselors, Inc. (NSGC) in 2000 the mean salary for a full-time master's-level genetic counselor was $46,436. The NSGC, incorporated in 1979, is the only professional society dedicated solely to the field of genetic counseling. Its mission is "to promote the genetic counseling profession as a recognized and integral part of health care delivery, education, research, and public policy."

The role of the genetic counselor has evolved greatly since 1971. Initially, genetic counselors worked almost exclusively in the clinical setting under physician supervision, seeing clients who had been diagnosed as having a genetic disorder, were at risk for developing a genetic disorder, or were at risk for having a child with a genetic disorder. They would assess genetic risk, provide information, discuss available testing options, and provide

appropriate supportive counseling. The variety of patients and the information and testing options offered by genetic counselors was greatly restricted by the limited technology and genetic knowledge of the time.

Today, as a result of the Human Genome Project and other advances, genetic counselors are now able to offer more services and options. They are able to specialize in a particular area of interest, such as cancer, prenatal, pediatric, assisted reproduction, and metabolic or neurogenetic disorders. Most genetic counselors (more than 80 percent) still work in the clinical setting, either in a hospital or in private practice. However, advances in genetics have enabled genetic counselors to work in a variety of other settings including research, public health, education, and industry.

patient advocate a person who safeguards patient rights or advances patient interests

As a **patient advocate**, the genetic counselor also remains informed of ethical and legal issues regarding the use of information generated by the Human Genome Project and incorporates pertinent information into the counseling session. For example, the decision to undergo genetic testing may involve controversial issues. Depending on the type of test and the disorder present, testing may have implications for other family members, insurance eligibility or coverage, employment, and quality of life. It is the role of the genetic counselor to ensure that clients are aware of concerns relevant to their situation.

Opportunities for the genetic counselor also exist to consult on research projects, guest lecture, publish articles and books, and teach. Broad training makes genetic counselors highly adaptable to virtually any setting where genetic information is utilized. Overall, genetic counseling is a dynamic and evolving profession. SEE ALSO GENETIC COUNSELING; GENETIC TESTING; GENOMIC MEDICINE; POPULATION SCREENING.

Susan E. Estabrooks

Bibliography

Baker, Diane L., Jane L. Schuette, and Wendy R. Uhlmann, eds. *A Guide to Genetic Counseling.* New York: Wiley-Liss, Inc, 1998.

Internet Resource

National Society of Genetic Counselors, Inc. <http://www.nsgc.org>.

Genetic Discrimination

The potentially stigmatizing nature of genetic information and its history of abuse necessitate special provisions for its protection. Although the U.S. Supreme Court has established a basic right to privacy (in the case of *Roe* v. *Wade*), the nonspecific nature of this privacy protection fails to adequately guard against the unauthorized disclosure of personal genetic information.

There have been reports that genetic discrimination has been practiced by employers and insurers who are afraid of being required to cover the potentially high medical expenses of individuals who have a family history of a genetically inheritable disease, even if those individuals exhibit no symptoms of the actual disease. However, recent debate suggests that the actual occurrence of genetic discrimination has not yet been proven and that the perception of risk is exaggerated. Nevertheless, researchers working in human genetics should proceed under the assumption of potential harm and should therefore keep confidential all information that they collect or generate as part of any family study they conduct.

Antidiscrimination Legislation

The Rehabilitation Act of 1973 was the first law to prohibit employment discrimination by federally funded agencies and institutions based on physical disability. The 1992 reauthorization of this legislation defines an "individual with a disability" as any person who "is regarded as having impairment." Such a definition is broad enough to include an **asymptomatic** gene carrier who is perceived as being "sick" by an employer or insurer. For example, carriers of the trait for sickle cell disease may be "regarded" as having the disease, even if they show no symptoms of it. In some cases, such individuals have been denied health insurance because they were inappropriately viewed as having a preexisting condition, that is, as actually having the disease, rather than simply carrying the gene for it.

asymptomatic without symptoms

The 1990 Americans with Disabilities Act (ADA) significantly broadened the scope of the 1973 legislation by prohibiting discrimination against disabled individuals in most areas of employment and with regard to access to public transportation. While the ADA's definition of disability excludes people with a "characteristic predisposition to illness or disease," some professionals believe that the legislation allots sufficient protection against genetic discrimination in the workplace.

Nonetheless, there is disagreement about whether people with a genetic predisposition to disease are adequately protected under the ADA. However, a recent ruling by the Equal Employment Opportunity Commission has interpreted the ADA to include the protection of individuals from employer discrimination based on genetic test results. The debate has not yet been fully resolved. For instance, there is still the potential for employers to institute genetic screening programs prior to offering employment. Thus, the municipal, state, and federal judiciaries will be called on to define the extent of the ADA legislation in their respective precincts in future cases.

Although the ADA prohibits employer discrimination, this law, along with the majority of state laws, does not adequately protect those with genetic disorders, whether symptomatic or asymptomatic, against discrimination by insurers. Thus, even though employers are prohibited from discriminating against workers with genetic disorders, the employment opportunities for such persons may still be limited. This is because employers could refuse to hire such individuals on the grounds that their insurance premiums would be too high.

Several states have attempted to address this problem by passing laws protecting individuals from genetic discrimination by health insurers. However, these laws generally fail to offer comprehensive protection to all people at risk for genetic discrimination. Depending on the way an individual law is worded, healthy gene carriers; people who are predisposed to certain genetic disorders such as cancer; and people who are at a high risk of developing a genetic disorder due to family medical history, rather than because of a detected gene mutation may all be excluded from the protection the law is intended to provide. For instance, Section 514 of the federal Employee

Retirement Income Security Act (ERISA) exempts self-insured health benefit plans from state insurance laws, so the employees of companies that participate in such plans have no protection. Many employers, especially small businesses, are self-insured, and rely upon commercial insurance companies to administer their health plans. These employer-funded self-insurance plans qualify for the ERISA exemption, and therefore do not have to comply with state mandates for services or state laws regarding genetic discrimination, involuntary testing, or privacy.

Present Problems and Potential Solutions

There is one federal act related to genetic discrimination that overrides ERISA and therefore applies to all health plans, even those offered by small businesses. This federal act is called the Health Insurance Portability and Accountability Act (HIPAA). HIPAA specifies that if employees are covered by a group health insurance policy offered through their employer, they must be offered a similar policy when they change jobs. The new insurance company does not have to offer this coverage at the same rate as was offered by the old plan, but it cannot deny coverage by declaring the genetic disorder to be a "preexisting condition." In contrast, if a person is either unemployed or self-employed, there is no requirement that an insurer offer him or her a policy.

As information about genetic disorders rapidly expands, the potential harm from genetic discrimination also becomes magnified. As a result, there is increasing recognition of the need for federal legislation guaranteeing a right to genetic privacy. Working groups have been formed to evaluate the impact of genetic information on individual insurability and on the insurance industry. Among these groups are the American Society of Human Genetics' Ad Hoc Committee on Genetic Testing/Insurance Issues, and the Working Group on Ethical, Legal, and Social Implications of the Human Genome Project, which is jointly sponsored by the National Institutes of Health and the Department of Education. At the end of the twentieth century, congressional committees held hearings on these issues, and in February 2001 a bill was introduced that mandates genetic nondiscrimination in health insurance. SEE ALSO GENETIC COUNSELING; GENETIC TESTING; GENETIC TESTING: ETHICAL ISSUES.

Chantelle Wolpert

Bibliography

Internet Resource

"The Potential for Discrimination in Health Insurance Based on Prescriptive Genetic Tests." U.S. Congress. <http://energycommerce.house.gov/107/hearings/07112001Hearing322/hearing.htm>.

Genetic Drift

Genetic drift is the random change in the genetic composition of a population due to chance events causing unequal participation of individuals in producing succeeding generations. Along with natural selection, genetic drift is a principal force in evolution.

Allele Frequencies

Different forms of a gene are called **alleles**. Individual members of a population have different alleles. Together, all the alleles for all the genes in a population constitute the "gene pool" of the population. Through reproduction, individuals pass their genes on to the next generation. If considering only the effect of genetic drift, the larger the population is, the more stable the frequency of different alleles in the gene pool will be over time. In small populations allele frequencies are likely to change rapidly and dramatically over very few generations, or "drift," because of chance events. This rapid change can occur in small populations because each individual's alleles represent a large fraction of the gene pool, and if an individual did not reproduce it could have a much larger effect than in the case of an individual in a large population not reproducing. Also, alleles that are found infrequently are more likely to be lost due to random chance.

alleles particular forms of genes

After many generations, if only genetic drift is operating, populations (even large populations) will eventually contain only one allele of a particular gene, becoming "monomorphic," or fixed for this allele.

Random Events

Many types of random events that can affect the likelihood of alleles being passed to future generations can be imagined. An adult may fail to mate during mating season due to unusually adverse weather; a pregnant mother may discover a rich food source and produce unusually strong or numerous offspring; all the offspring of one parent may be consumed by predators. Many other scenarios are possible.

To see how such events affect allele frequencies, imagine a population that contains four individuals of an organism that reproduces once and dies. Let us examine how allele frequencies change for a gene that has two alleles, A and a. As with other genes, each individual has two alleles, one inherited from each parent. Imagine that three of the individuals are aa **genotype**, and one is Aa genotype. Thus, of the population's eight copies of the gene, one is A, and seven are a. Now imagine that because of random chance, the Aa individual does not reproduce. Therefore, only aa offspring are produced and the A allele is lost to the population. The A allele goes from a frequency of one-eighth to zero through the process of genetic drift.

genotype set of genes present

A large reduction in population size can lead to a situation known as a genetic bottleneck. After a genetic bottleneck the population is likely to have different allele frequencies. When only a very small number of individuals are left after a population decline, the population will have only the alleles present in these few individuals. This is known as the "founder effect." The founder effect can be viewed as an extreme case of a genetic bottleneck. If a population decline affects all individuals in the population without respect to the alleles they carry, genetic drift will have an effect on all genes.

Genetic drift has important implications for evolution and the process of **speciation**. When a small group of individuals becomes isolated from the majority of individuals of a species, the small group will genetically drift from the rest of the species. Because genetic drift is random and the smaller group will drift more rapidly than the larger group, it is possible that, given

speciation the creation of new species

enough time, the small group will become different enough from the large group to become a different species.

The fact that small populations are more subject to genetic drift has important implications for conservation. If the number of individuals of a species becomes small, it becomes increasingly influenced by genetic drift, which may result in the loss of valuable genetic diversity. Conservation biologists seek to maintain populations at sufficient numbers to counteract genetic drift. SEE ALSO CONSERVATION BIOLOGY: GENETIC APPROACHES; FOUNDER EFFECT; GENE FLOW; HARDY-WEINBERG EQUILIBRIUM; POPULATION BOTTLENECK; POPULATION GENETICS.

R. John Nelson

Bibliography

Avise, John C. *Molecular Markers, Natural History and Evolution.* New York: Chapman and Hall, 1994.

Futuyma, Douglas J. *Evolutionary Biology*, 3rd ed. Sunderland, MA: Sinauer Associates, 1998.

Mayr, Ernst. *Evolution and the Diversity of Life: Selected Essays.* Cambridge, MA: Belknap Press, 1976.

Weaver, Robert F., and Philip W. Hedrick. *Genetics*, 2nd ed. Dubuque, IA: William C. Brown, 1992.

Genetic Engineering *See Biotechnology*

Genetic Testing

mutations changes in DNA sequences

Genetic testing involves examining a person's DNA in order to find changes or **mutations** that might put an individual, or that individual's children, at risk for a genetic disorder. These changes might be at the chromosomal level, involving extra, missing, or rearranged chromosome material. Or the changes might be extremely small, affecting just one or more of the chemical bases that make up the DNA. In a broader sense, genetic testing includes other types of testing that provide information about a person's genetic makeup, such as enzyme testing to diagnose or identify carriers for a genetic condition such as Tay-Sachs disease.

With hundreds of genetic tests available, determining who should be offered testing and under what circumstances testing should occur is relatively complicated. In general, testing is offered to those at highest risk based on their ethnic background, family history, or symptoms. However, just because genetic testing is possible and a person is at risk, this does not mean it should be offered or will be useful to that person. Genetic testing is unlike other medical tests in that individual results may also provide information about relatives, may be able to predict the likelihood of a future illness for which treatment may or may not be available, may put the person at risk for harm such as discrimination, or may have limited accuracy. There are a number of settings in which genetic testing occurs and within each setting there are a variety of indications and considerations for testing.

Prenatal Genetic Testing

Prenatal (before birth) genetic testing refers to testing the fetus for a potential genetic condition. The pregnant woman is considered the patient and makes decisions regarding prenatal testing. There are a variety of circumstances under which a woman might be offered prenatal genetic testing. The parents of the fetus may have a genetic disorder, or they may be what is known as **carriers**. An abnormality or birth defect may be detected on ultrasound that could indicate a genetic condition. The fetus may be identified to be at increased risk for a chromosome abnormality, such as Down syndrome, or a birth defect, such as spina bifida, based on the result of a maternal serum screening test performed on the mother. This is a test that looks at several proteins made by the fetus that are found in a woman's bloodstream while she is pregnant. Or, the mother might be at increased risk for having a baby with a chromosome abnormality because of her age. While all women are at risk for having a baby with a chromosome abnormality, women who are age thirty-five or older are offered prenatal chromosome testing because the chance their fetus has a chromosome abnormality is equal to or higher than the chance she will have a miscarriage due to the sampling procedure.

carriers people with one copy of a gene for a recessive trait, who therefore does not express the trait

As with most genetic testing, prenatal genetic testing should occur in conjunction with genetic counseling. The genetic counselor provides supportive, nondirective counseling and information. Nondirective counseling means that while the counselor will try to facilitate decisions regarding testing and future pregnancy management, she will not make specific recommendations. Because the decision to undergo testing is personal and must take into account differences in beliefs, life circumstances, and the risk of the procedure, the decisions regarding testing and pregnancy management must be made by the patient. This encounter is also likely to include information about risk of the fetus being affected, the disorder in question, and available testing options.

Most genetic tests are performed on tissue or a blood sample. For obvious reasons, obtaining a sample from a fetus is not the same as obtaining one from a child or adult. Prenatal testing procedures are invasive, and there is a risk of miscarriage with every procedure. For this reason, specially trained physicians perform these tests. Prenatal testing can be accomplished using three different methods: amniocentesis, chorionic villus sampling, and **percutaneous** umbilical blood sampling. These tests differ in the type of fetal tissue studied, the timing of the testing during pregnancy, and in their risks and benefits.

percutaneous through the skin

Amniocentesis is the most common, and it carries the lowest risk of miscarriage (about one in two hundred pregnancies). It is typically performed between sixteen and eighteen weeks into the pregnancy and involves collecting a small amount of amniotic fluid that contains cells of the developing fetus, which can be used for testing. Chorionic villus sampling is performed earlier than amniocentesis, typically between ten and twelve weeks of pregnancy, but about one in one hundred pregnancies are miscarried as a result of this procedure. It involves obtaining a small sample of chorionic villi (fingerlike projections of the chorion, a membrane that will later develop into the placenta), which should contain cells of the fetus. Percutaneous umbilical blood sampling, typically performed after eighteen

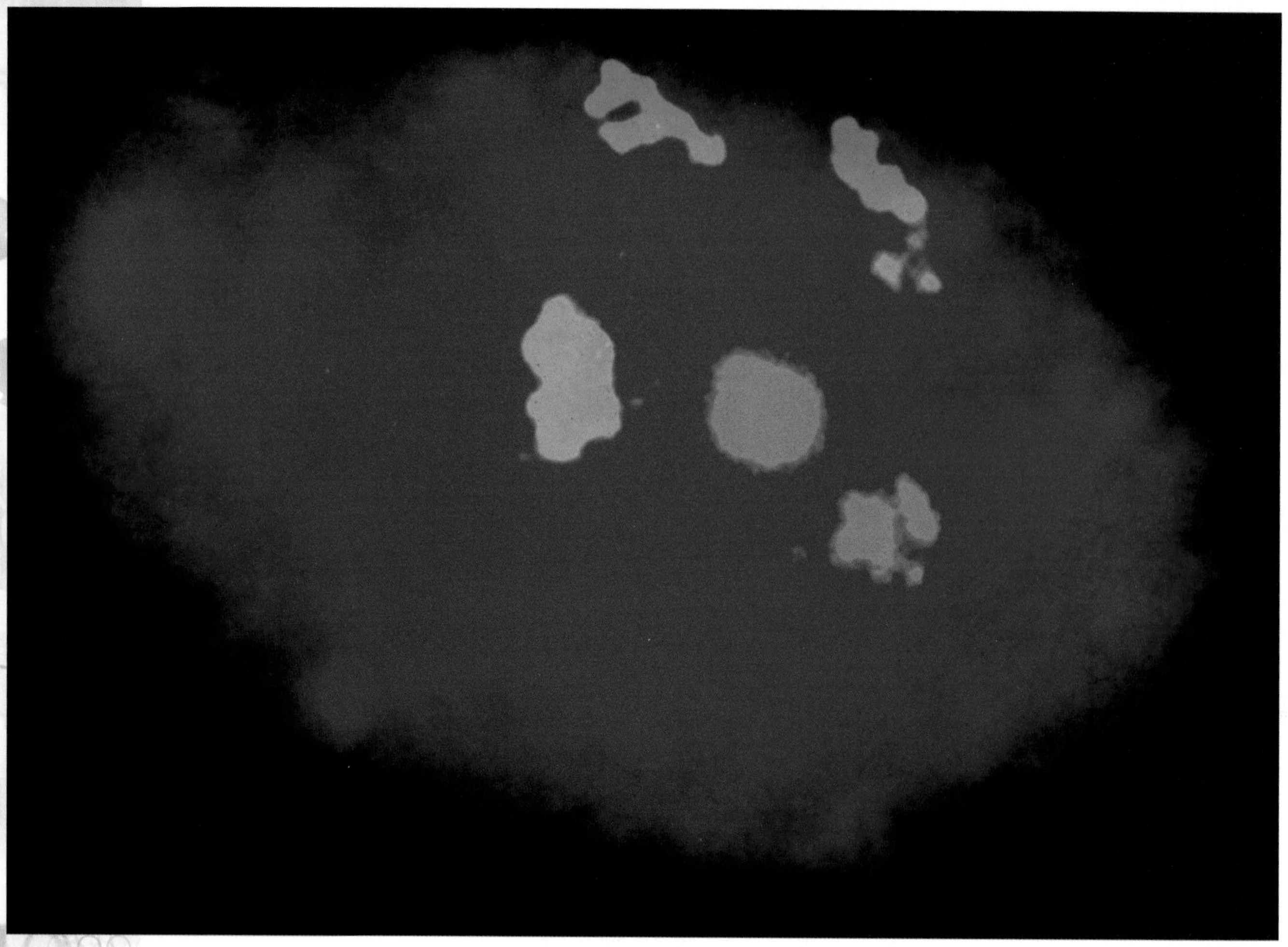

This genetic test is positive for Trisomy 18. Fluorescence *in situ* hybridization reveals three copies of chromosome 18 (green) and female sex chromosomes (fuschia) within a nucleus (blue).

in vitro "in glass"; in lab apparatus, rather than within a living organism

weeks, is the most difficult to perform and carries the highest risk of miscarriage (about one in fifty pregnancies). It involves withdrawing blood from the umbilical cord and is primarily used when results are needed extremely quickly, or when only a fetal blood sample can provide a given answer about the fetus. For example, it may be used to test the fetus if the mother has been exposed to an infectious organism known to cause birth defects.

Assisted Reproduction

While **in vitro** fertilization has been available for over two decades, more recently it has become possible to test the resulting embryos for genetic disorders when the embryos are between eight and sixteen cells in size. In this procedure, one to two cells are removed, and the section of DNA containing the gene in question is replicated and tested. Only those embryos identified to be free of risk (based on the DNA results) for developing the genetic disorder are implanted in the uterus. This technique is not widely available, however, and it is both expensive and time consuming. Thus, it is used only infrequently.

Newborn Screening

Newborn screening is unique in being the only genetic testing that it is mandated by the state. The premise of newborn screening is that, for some dis-

orders very early detection and initiation of treatment will prevent health problems, often mental retardation. All newborn infants are tested for a variety of genetic disorders. Each state determines for itself for what disorders to test their newborns. Disorders are chosen based on severity, incidence, ease and accuracy of testing, cost, and benefit of early diagnosis. All states test newborns for phenylketonuria (PKU), a metabolic disorder that is almost never evident at birth. Individuals with PKU are missing an enzyme called phenylalanine hydroxylase, which results in the buildup of phenylalanine. If left untreated, severe mental retardation develops. However, infants with PKU who are placed on a diet low in phenylalanine immediately after birth are expected to develop normally, making PKU an excellent candidate for newborn screening.

Symptomatic Genetic Testing

Genetic testing of individuals who are exhibiting symptoms of a genetic disorder is relatively straightforward. Testing is necessary to either make or confirm a diagnosis, which may improve treatment and establish risk estimates for other family members.

A concern related to symptomatic testing is the duty to recontact the patient in the future if more information becomes available. Often, symptomatic individuals, usually children, present with a variety of symptoms for which no diagnosis can be made clinically and for which there is no genetic test. However, with the completion of the Human Genome Project and the wealth of research being conducted, new genes are discovered regularly, which may result in new testing possibilities. Most physicians inform their patients that more information may be available in the future and ask their patients to contact the clinic periodically to inquire about such updates. It is unclear whether this is sufficient to fulfill the physicians' obligation; however, no clear standards exist on this issue.

Carrier Testing

Another common testing situation is carrier testing for autosomal recessive disorders. Autosomal recessive disorders are caused by the inheritance of two nonfunctioning genes, one from each carrier parent. The parents are referred to as carriers because they carry only one nonfunctioning gene and are, therefore, not affected by the disorder. Every individual is thought to be an unaffected carrier of some autosomal recessive disorder. This is only a problem, however, if two individuals who both carry the same recessive disorder conceive a child together. Under this circumstance, the child would have a one in four (25%) chance of inheriting a nonworking copy from each parent, thereby inheriting the disorder.

There are hundreds of genetic tests available, but it is not practical to perform every available test on each person. Carrier testing is typically offered only to those individuals who are at increased risk based on family history or ethnic background. While family history bears an obvious correlation, ethnic background is important because those who descend from the same group of ancestors are more likely to carry the same genetic changes. For example, individuals of Ashkenazi Jewish descent have about a one in thirty chance of being a carrier for a condition called Tay-Sachs disease,

whereas the carrier frequency is only about one in 300 in other populations. In instances such as this, population screening is often recommended.

Presymptomatic Testing

Presymptomatic testing (that is, testing a healthy person before symptoms appear) may be considered for a genetic disorder for which there is a family history. The decision to undergo this type of testing is not usually straightforward and should always be accompanied by genetic counseling. There are a number of considerations to take into account when deciding whether to proceed with testing. The first is the usefulness of the information. How will knowing the genetic information benefit the person? Testing is more favorable when preventive treatment is available, when results might have a significant impact upon life decisions, such as having children or getting married, or if it will ease extreme anxiety to learn one's genetic status. If no treatment is available, as in the case of Huntington's disease and other triplet repeat diseases, the information may be of less benefit. In some cases it may even be psychologically harmful.

mutations changes in DNA sequences

The second consideration is accuracy, not only of the actual test result, but also of its ability to predict the development of the disorder. Some disorders are caused by more than one gene, or by multiple changes or **mutations** in the same gene. A given test might not be able to look at all mutations or every gene that causes a disorder, leaving a person who tests negative with doubt as to whether they are truly mutation-free. Some genetic tests, particularly those for complex disorders, are for susceptibility genes. As the name implies, these are genes that make a person susceptible to developing a disorder, but do not guarantee it. An example of this is breast cancer. When deciding whether or not to test for such a disorder, it is important to ask how it would feel to test positive for a susceptibility gene for a serious genetic disorder that may never develop.

The third consideration is risk of personal harm. Testing positive for a disorder may put a person at risk of economic or social harm. Although rare, there have been instances where individuals have been denied insurance, employment, or both based on the results of genetic testing. Also, a person may be at risk of experiencing psychological or emotional problems after undergoing genetic testing. There have been instances, particularly with Huntington's disease testing, where individuals have committed suicide following a positive test result. All of these factors must be presented to, discussed with, and weighed by the individual considering testing, in the context of genetic counseling and prior to making a decision about testing.

Presymptomatic Testing of Children

Presymptomatic testing of children has somewhat different considerations. It is typically considered only when the onset of the disorder occurs in childhood, or when knowing the genetic status will significantly benefit the child, for example by enabling him to receive early preventive treatment. For example, children at risk for inheriting the gene that causes retinoblastoma (cancer of the retina) may be tested because the disease usually presents before age five. With early treatment, the long-term outcome is favorable. Know-

ing whether the child has inherited the gene will allow physicians to know whether to aggressively screen the child for signs of cancer development.

Many genetic professional organizations have developed position statements regarding genetic testing of children that discourage testing for disorders that do not pose a risk in childhood and for which early identification poses no benefit to the child. This includes adult-onset disorders, such as Huntington's disease, but also pertains to carrier testing of females for X-linked recessive disorders, such as muscular dystrophy. In most X-linked (sometimes referred to as sex-linked) recessive disorders, females who inherit a mutation on one of their two X chromosomes are usually unaffected carriers because the second X chromosome is able to compensate for the loss. However, because males have only one X chromosome (the other sex chromosome is a Y), they will be affected if they inherit a mutation. In genetic medicine, personal autonomy is a priority. Individuals have the right to make their own decision regarding genetic testing. If a child is tested for an adult-onset disorder or to determine carrier status, their right to make their own decision as an adult has essentially been taken away. SEE ALSO ALZHEIMER'S DISEASE; BREAST CANCER; CHROMOSOMAL ABERRATIONS; CYSTIC FIBROSIS; DOWN SYNDROME; GENETIC COUNSELING; GENETIC COUNSELOR; GENETIC DISCRIMINATION; GENETIC TESTING: ETHICAL ISSUES; INHERITANCE PATTERNS; METABOLIC DISEASES; MUSCULAR DYSTROPHY; POPULATION SCREENING; PRENATAL DIAGNOSIS; TAY-SACHS DISEASE; TRIPLET REPEAT DISEASES.

Susan E. Estabrooks

Bibliography

Holtzman, Neil A., et al. "Predictive Genetic Testing: From Basic Research to Clinical Practice." *Science* 24, no. 278 (1997): 602–605.

Martindale, Diane. "Pink Slip in Your Genes." *Scientific American* 284 (2001): 19–20.

Ostrer, Harry, Richard H. Scheuermann, and Louis J. Picker. "Benefits and Dangers of Genetic Tests." *Nature* 392 (1998): 14.

Ponder, Bruce. "Genetic Testing for Cancer Risk." *Science* 278 (1997): 1050–1054.

Rennie, John. "Grading the Gene Tests." *Scientific American* 270 (1994): 88–96.

Internet Resources

"Secretary's Advisory Committee for Genetic Testing." <http://www4.od.nih.gov/oba/sacgt.htm>.

"Understanding Gene Testing." U.S. Department of Health and Human Services. <http://rex.nci.nih.gov/PATIENTS/INFO_TEACHER/bookshelf/NIH_gene-testing/gene00.html>.

Genetic Testing: Ethical Issues

Genetic testing is the name given to a variety of laboratory techniques, all of which ultimately provide information about a person's underlying genetic makeup, also called their genotype. Genetic information is of interest because it can help in the diagnosis of a current health condition or provide insight into future health.

Advances in Genetic Science

Early understanding of genetic disease was first obtained about "single-gene disorders." These disorders include such diseases as sickle cell disease and

cystic fibrosis. They occur because of the strong influence of a mutation in a single gene. Using the tremendous amounts of information about the human genome made available by the Human Genome Project, scientists are now dissecting the genetic components of complex disorders, such as diabetes and heart disease, that come about as a result of both genetic and environmental factors. Information about the genetic factors in complex disorders can help to predict the probability of future development of disease.

Understanding the issue of probability is important for the ethical use of genetic testing technology. "Genetic determinism" describes the idea that genes hold the map of an individual's future health. Although the detection of some **mutations** can diagnose the presence of a disease, many other tests provide only information about risk for disease, not a certainty that disease will develop. For example, genetic testing to detect mutations in the gene for breast cancer (BRCA1) can give information about an individual's susceptibility to developing breast and ovarian cancer. However, the probability of developing cancer will depend not only on the status of the gene, but also on factors related to family history and environmental exposures.

mutations changes in DNA sequences

A Range of Ethical Issues

Ethical issues that arise within the context of genetic testing are similar to those that arise for any personal medical information. For example, there are concerns about protecting the privacy of the patient and the confidentiality of information, whether it is genetic data or any other item in the medical record. Possibilities of stigmatization or discrimination occur because of social perceptions of some diseases, whether the primary cause is genetic, infectious, or environmental. Furthermore, genetic testing requires informed consent; that is, patients need to be educated about the purpose of the testing (diagnosis, prognosis, or susceptibility assessment), the potential test outcomes and what they mean, and what the options are once the results are known.

Nevertheless, the perspective of "genetic exceptionalism" places genetic information in a category of different, special, or greater concern than other medical information. Knowledge of the genetic information of one person can have an influence on his family members because it implies information about their health status, disease risk, or the possibility of passing a condition on to their offspring. However, while concerns arise from genetic testing because of the obvious familial nature of genetic information, similar concerns arise from the simple knowledge of shared family history or environmental exposures, even without detailed genetic information.

Evaluation of the risks and benefits of genetic testing is an important factor in the process of considering the ethics of its use. The physical risks of genetic testing are generally minimal since, in most cases, DNA can be tested from easy-to-access cells: blood, buccal cells (inside of the cheek), and even hair follicles. However, there is a range of nonphysical risks, including loss of privacy and discrimination.

Consideration of the benefits of a genetic test is sometimes difficult. For example, when a therapy is available for a genetic disease, the availability of this benefit favors conducting the genetic test. If a therapeutic benefit is not available, however, the value of having this information becomes more questionable. Doctors and patients must ask if the infor-

mation will help in some way, such as in life or reproductive planning. Although it can be difficult to compare the value of the information to the possible risks associated with conducting the genetic test, it should be carefully and explicitly considered.

Prenatal and Childhood Testing

Testing in children raises a different set of risks and issues. Adults are able to choose for themselves about genetic testing after hearing the relative advantages and risks. Children, however, must rely on their parents to decide for them. On the one hand, deciding to test children to diagnose a disease seems appropriate in most situations, particularly when a treatment can be provided. On the other hand, there is disagreement about testing children for disorders with adult onset and for which there is no intervention. In this situation, respect for the child's own decision (autonomy) and preserving privacy are ethical arguments for waiting to offer the genetic test. Upon reaching adulthood, individuals may then choose whether to learn their own genetic information.

Genetic testing requires additional considerations when done in the prenatal setting. In this case, parents are deciding to obtain genetic information about a third party, the developing fetus. In cases where there is no available medical therapy for a diagnosed genetic condition, the availability of limited options results in tough ethical choices for parents. The options may range from continuing the pregnancy, preparing for a child with special medical needs, or terminating the pregnancy. To make these decisions, prospective parents need clear information regarding the meaning of test results. Issues to consider include the predictive value of the test, the likely severity and age of onset of the predicted disorder, and the probability that the genetic alteration detected will actually result in a disorder.

Presymptomatic Testing: Huntington's Disease

Presymptomatic genetic testing when no treatment is available stands in contrast to genetic testing done for the diagnosis of a current disorder. Huntington's disease (HD) is an example of a genetic disorder in which clinical manifestations begin in adulthood. Although some symptoms can be managed with medications, no treatment is yet available to alter the gradual loss of muscle control, psychiatric changes, and progressive **dementia**.

dementia neurological illness characterized by impaired thought or awareness

HD is an autosomal dominant disorder, which means that an individual with only one abnormal copy of the HD gene will develop the disease, and the children of affected individuals have a 50 percent chance of inheriting the genetic mutation. The gene abnormality occurs when a **polymorphic** CAG repeat sequence is expanded beyond the normal number of 10 to 29 copies. Diagnosis of HD is almost 100 percent sure when the number of CAG repeats is in the range of 36 to 121.

polymorphic occurring in several forms

Individuals at risk for HD may want to undergo genetic testing to end the uncertainty of not knowing whether they will be affected. This knowledge may allow for career and life planning. However, concerns that presymptomatic diagnosis of HD would lead to serious psychological distress or suicide led to the development of testing guidelines. The standard was established that all individuals who sought HD testing should receive pretest counseling to explain the test, assess their psychological status, and

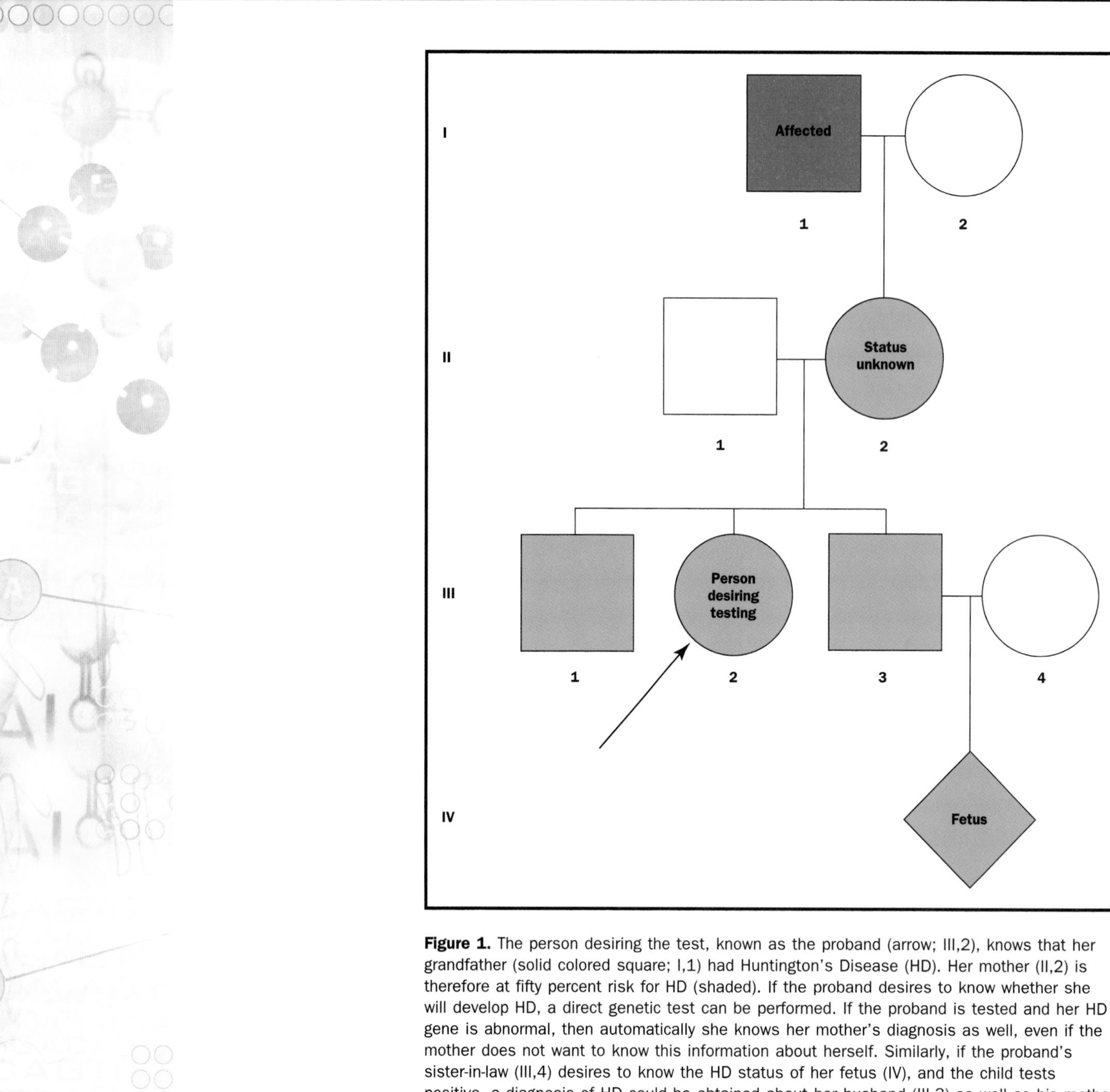

Figure 1. The person desiring the test, known as the proband (arrow; III,2), knows that her grandfather (solid colored square; I,1) had Huntington's Disease (HD). Her mother (II,2) is therefore at fifty percent risk for HD (shaded). If the proband desires to know whether she will develop HD, a direct genetic test can be performed. If the proband is tested and her HD gene is abnormal, then automatically she knows her mother's diagnosis as well, even if the mother does not want to know this information about herself. Similarly, if the proband's sister-in-law (III,4) desires to know the HD status of her fetus (IV), and the child tests positive, a diagnosis of HD could be obtained about her husband (III,3) as well as his mother (II,2) regardless of their preferences for knowing or not knowing their status.

prepare them to consider their possible reactions to the results, whether positive or negative. In addition, results should be given in person, and counseling and support should be available following the test.

Before the identification of a specific gene for HD, genetic testing was done by linkage studies within a family. Linkage analysis required the voluntary participation of affected and unaffected family members across several generations. Identification of the specific gene defect now allows direct genetic analysis, without examining the DNA of any other family member. However, because of the inheritance patterns of HD, information about one family member can give diagnostic information about another. Figure 1 diagrams the ethical complexities arising from situations in which one family member's genetic test results in necessary implications for other family members. Note that it is not clear whether the individual desiring testing has decisive rights, or if those rights belong to those family members who do not desire such testing.

Susceptibility Testing: Apolipoprotein E

Susceptibility testing is another area that raises possible ethical issues for patients and clinicians. Recall that susceptibility implies that the genetic test does not provide the final answer regarding disease, but rather that additional factors are involved. For example, cardiovascular disease is a major source of **morbidity** and mortality. Polymorphic variants in the apolipoprotein E gene (APOE) have been found to be associated with variations in blood levels of lipids, lipoproteins, and apolipoproteins, which may in turn influence risk for heart disease. Clinicians and patients may wish to learn which APOE variants they have, to help determine their susceptibility to cardiovascular disease and the likelihood that altering their lifestyle risk factors or using pharmaceutical therapy might help prevent cardiovascular disease.

morbidity incidence of disease

When deciding to test for APOE gene variants, however, there are additional consequences that must be considered, particularly the fact that the gene may play multiple roles in health and disease (**pleiotropy**). Studies have shown that one version of APOE, the e4 **allele**, is associated with an increased risk of Alzheimer's disease. When deciding to undertake genetic testing for the intended immediate medical benefit of understanding cardiovascular disease risk, the individual should therefore also consider the possibility of learning potentially unwanted additional information.

pleiotropy genetic phenomenon in which alteration of one gene leads to many phenotypic effects

allele a particular form of a gene

Support Mechanisms

Several support mechanisms exist for dealing with the complex decisions and ethical issues arising from the medical application of genetic testing. In the past, much genetic testing has been carried out by medical specialists with disease-specific expertise who have an understanding of the medical, psychosocial, and family implications of diagnoses of particular genetic diseases. Genetic support groups have also been established for education and mutual support among patients and families with genetic disorders. An organization called the Genetic Alliance provides a referral source to connect individuals with appropriate support groups and information about genetic disorders.

Typically, the best resource for an integrated approach to support for patients and families is a genetic counselor. Genetic counselors provide factual education about genetic disease inheritance, how to understand risk and medical probabilities, and the range of options for care. They also provide counseling to help patients and families to clarify and articulate their own values and motivations in order to make the most appropriate decisions. These may be decisions about whether to undergo a genetic test or how they may act on receiving the results of a test. SEE ALSO BREAST CANCER; GENETIC COUNSELING; GENETIC COUNSELOR; GENETIC DISCRIMINATION; GENETIC TESTING; HUMAN GENOME PROJECT; INHERITANCE PATTERNS; PLEIOTROPY; PRENATAL DIAGNOSIS.

Carol L. Freund and Jeremy Sugarman

Bibliography

Burgess, M. M. "Beyond Consent: Ethical and Social Issues in Genetic Testing." *National Review of Genetics* 2 (2000): 147–151.

Ross, L. F., and M. R. Moon. "Ethical Issues in Genetic Testing of Children." *Archives of Pediatrics and Adolescent Medicine* 154 (2000): 873–879.

Scriver, C. R., et al., eds. *The Metabolic and Molecular Bases of Inherited Disease*, 8th ed. New York: McGraw-Hill, 2001.

Welch, H. G., and W. Burke. "Uncertainties in Genetic Testing for Chronic Disease." *Journal of the American Medical Association* 280 (1998): 1525–1527.

Wilfond, B. S. "Genetic Testing." In *Ethics in Primary Care*, J. Sugarman, ed. New York: McGraw-Hill, 2000.

Internet Resources

Genetic Alliance. <http://www.geneticalliance.org/>.

Secretary's Advisory Committee on Genetic Testing. <http://www4.od.nih.gov/oba/sacgt.htm>.

Genetically Modified Foods

A genetically modified (GM) food is a plant that has a genetic change in each of its cells that a researcher has introduced. The modification may add a gene from a different species and thereby create a transgenic plant, or it may overexpress or silence a preexisting plant gene. Overexpression is accomplished by altering the **promoter** region of a gene, which controls how rapidly and in which cells the encoded protein is synthesized, thus directing a plant to manufacture more of a natural product. Conversely, a gene may be "silenced" (directed not to synthesize a protein) through the use of antisense technology, which applies a complementary nucleic acid to messenger RNA, halting expression of the encoded protein.

promoter DNA sequence to which RNA polymerase binds to begin transcription

Genetic Modification in Animals and Plants

Animals have not yet been genetically modified to provide foods. Transgenic animals can, however, produce certain pharmaceuticals, but this approach is still experimental. One possible future use of transgenic animals is to create herds of cattle or sheep that are genetically resistant to developing transmissible spongiform encephalopathies, such as scrapie in sheep and "mad cow disease" in cattle.

Genetic modification in plants produces the same types of changes that result from traditional agricultural techniques, such as controlled breeding. However, genetic modification alters one gene at a time in a controlled manner, and typically has faster results than breeding plants with particular combinations of traits. With standard breeding techniques, it may take a generation to introduce, or remove, a single gene. Breeding a polygenic trait (a trait that involves more than one gene) into apples, which have a generation time of four years, could take two decades or longer.

GM traits that have already been introduced into plants include resistance to insects, insecticides, and herbicides; larger fruits; salt tolerance; slowed ripening; additional nutrients; easier processing; insecticide production; and the ability to take its own nitrogen from the air, lowering reliance on fertilizer. Specific products of genetic manipulation include insect-resistant corn, frost-resistant strawberries, rice that makes beta-carotene (a vitamin **precursor**), frost- and salt-tolerant tomatoes, delayed-ripening pineapples and bananas, canola with a healthier oil profile, and cotton and trees altered to make it easier to process fabric and paper. Some transgenic combinations are strange. Macintosh apples that have been given a gene from a *Cecropia* moth that encodes an antimicrobial protein, for example, are resistant to a bacterial infection called fire blight.

precursor a substance from which another is made

Regulatory Concerns

Whether a new variety of crop plant presents a hazard to human health depends upon the nature of the trait, not how the plant received that trait. For example, the U.S. Department of Agriculture found that a variety of potato obtained through conventional breeding was very toxic, and so it was never developed as a food. However, a potato developed through genetic modification at about the same time did not contain the toxin and was apparently safe to eat. This is why U.S. government regulatory agencies do not evaluate crops on how they were developed, but on their effects on the digestive tracts of animals.

Even after government agencies approve the marketing of a GM crop, consumer acceptance is crucial to its success. The FlavrSavr tomato, for example, was introduced in the 1980s. It ripened later, while in the supermarket, which extended its shelf life while providing an attractive product. However, the developers had focused only on this characteristic, and the tomatoes just did not taste very good. Consumer objection to GM foods also contributed to the FlavrSavr's failure. However, a high-solids GM tomato sold in England before the anti-GM movement began was popular with consumers, largely because it was priced lower than other tomatoes.

The Technique of Genetic Modification

The first step in developing a transgenic plant is to identify a trait in one type of organism that would make a useful characteristic if transferred to the experimental plant. The components of an experiment to create a transgenic plant are the gene of interest, a piece of "vector" DNA that delivers the gene of interest, and a recipient plant cell. Donor genes are often derived from bacteria, and are chosen because they are expected to confer a useful characteristic, such as resistance to a pest or pesticide.

To begin, the donor DNA and vector DNA are cut with the same **restriction enzyme**. This creates hanging ends that are the same sequence on both of the DNA molecules. Some of the pieces of donor DNA are then joined with vector DNA, forming a recombinant DNA molecule. The vector then introduces the donor DNA into the recipient plant cell, and a new plant is grown.

restriction enzyme an enzyme that cuts DNA at a particular sequence

For plants that have two seed leaves (dicots), a naturally occurring ring of DNA called a Ti plasmid is a commonly used vector. Dicots include sunflowers, tomatoes, cucumbers, squash, beans, tomatoes, potatoes, beets, and soybeans. For monocots, which have one seed leaf, Ti plasmids do not work as gene vectors. Instead, donor DNA is usually delivered as part of a disabled virus, or sent in with a jolt of electricity (electroporation) or with a "gene gun" (particle bombardment). The monocots include the major cereals (corn, wheat, rice, oats, millet, barley, and sorghum).

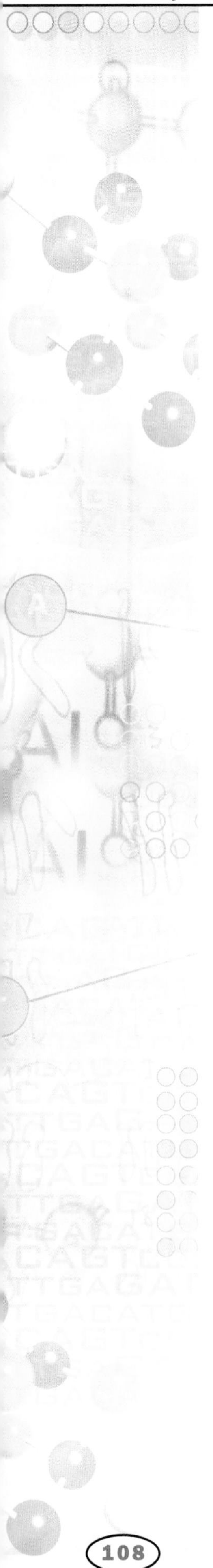

Transgenesis in plants is technically challenging because the transgene must penetrate the tough cell walls, which are not present in animal cells. Instead of modifying plant genes in the nucleus, a method called transplastomics alters genes in the chloroplast, which is a type of **organelle** called a plastid. Chloroplasts house the biochemical reactions of photosynthesis. Transplastomics can give high yields of protein products, because cells have many chloroplasts, compared to one nucleus. Another advantage is that altered chloroplast genes are not released in pollen, and therefore cannot fertilize unaltered plants. However, it is difficult to deliver genes into chloroplasts, and expression of the trait is usually limited to leaves. This is obviously not very helpful in a plant whose fruits or tubers are eaten. The technique may be more valuable for introducing resistances than enhancing food qualities. Someday, transplastomics may be used to create "medicinal fruits" or edible vaccines.

organelle membrane-bound cell compartment

GM beyond the Laboratory

After genetic modification, the valuable trait must be bred into an agricultural variety. Consider "golden rice," a grain that was given genes from daffodils and a bacterium to confer on it the ability to manufacture beta-carotene, a precursor to vitamin A. The first golden rice plants were created solely to show that the manipulation worked, and the modification of an entire biochemical pathway took a decade. The plant varieties were not edible, and the production of beta-carotene was low. In early 2002, however, researchers at the International Rice Research Institute in the Philippines began using conventional breeding to transfer the ability to produce beta-carotene from the inedible golden rice into edible varieties.

Genetic manipulation of plants can also focus on a particular species' own genes. This is the case for the potato, which has traditionally been difficult to cultivate because edible varieties must have an acceptable taste and texture, yet lack the alkaloid toxins that many natural strains produce. Breeding for so many characteristics is very time-consuming, and this is where genetic manipulation might speed the process. Researchers have identified a group of disease resistance genes on a region of one potato chromosome. The genes provide resistances to various insects, nematode worms, viruses, and *Phytophthora infestans*, which caused the blight infection that resulted in the nineteenth-century Irish potato famine. Being able to manipulate and transfer these genes will help researchers quickly breed safe and tasty new potato varieties, and perhaps transfer the potato's valuable resistance genes to related plants, such as tomatoes, peppers, and eggplants.

GM crops are widely grown in some countries, but are boycotted in others where many people object to genetic manipulation. As of 2001, 75 percent of all food crops grown in the United States were genetically modified, including 80 percent of soybeans, 68 percent of cotton, and 26 percent of corn crops. Farmers find that GM crops are cheaper to grow because their reliance on pesticides and fertilizer is less and a uniform crop is easier to harvest. Heavy reliance on the same varieties may be dangerous, however, if an environmental condition or disease should arise that targets the variety, but this dilemma also arises in traditional agriculture.

Because GM crop use is so pervasive in the United States, and because regulatory agencies evaluate the chemical composition and biological effects

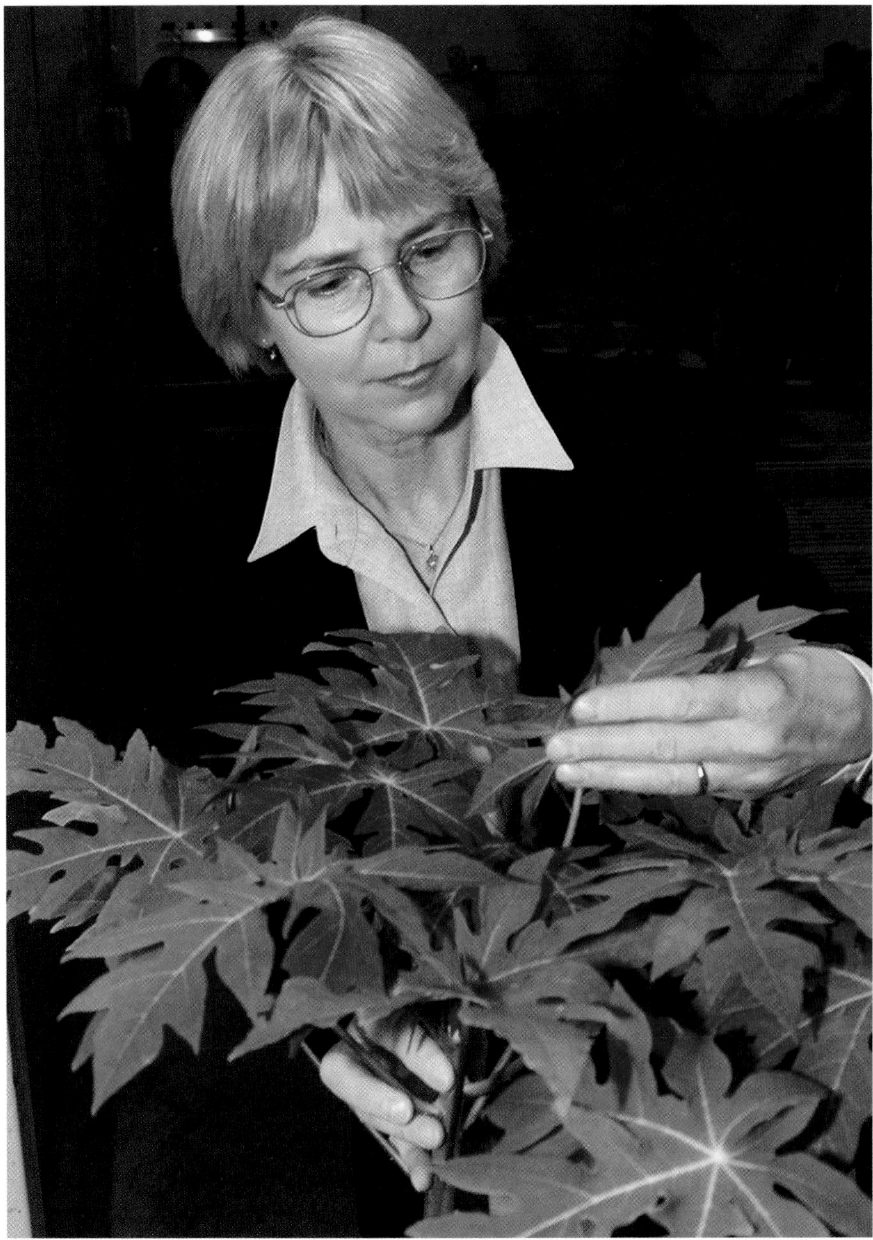

This researcher inspects a papaya plant that scientists have attempted to modify to make it resistant to a destructive virus. The benefits and drawbacks of developing, growing, and selling genetically modified crops have been a matter for public debate for decades.

of crops rather than their origin, a consumer would not know that a fruit or vegetable has been genetically modified unless it is so labeled. Some people argue that these practices prevent consumers from having a choice of whether or not to use a genetically modified food. SEE ALSO AGRICULTURAL BIOTECHNOLOGY; ANTISENSE NUCLEOTIDES; PLANT GENETIC ENGINEER; PRION; RESTRICTION ENZYMES; TRANSGENIC ORGANISMS: ETHICAL ISSUES; TRANSGENIC PLANTS.

Ricki Lewis

Bibliography

Fletcher, Liz. "GM Crops Are No Panacea for Poverty." *Nature Biotechnology* 19, no. 9 (September 2001): 797–798.

Hileman, Bette. "Engineered Corn Poses Small Risk." *Chemical and Engineering News* 79, no. 38 (September 17, 2001): 11.

Maliga, Pat. "Plastid Engineering Bears Fruit." *Nature Biotechnology* 19, no. 9 (September 2001): 826–927.

Potrykus, I. "Golden Rice and Beyond." *Plant Physiology* 123 (March 2001): 1157–1161.

Geneticist

Genetics intersects almost every other field of biology. For this reason, professionals with a genetics education have a broad range of career opportunities. The recent success of the Human Genome Project has created a great demand for genetics professionals with a variety of expertise in all areas of genetics, not only those applicable to human disease.

Geneticists are involved in identifying genes responsible for basic biological traits, including genes that cause disease or mediate a response to a medication. Geneticists also determine how those genes function in organisms, including plants, animals, and humans. They may treat individuals with a particular genetic disorder or counsel families regarding the genetics of human disease. Geneticists are also involved in identifying genetic mechanisms to improve agricultural processes, such as breeding pest-resistant fruits and vegetables. Geneticists are also interested in the distribution of genetic variations in populations and how those variations arise. Additional job responsibilities may include educating students and the general public, lobbying Congress to pass bills to support genetic research, and testifying in legal cases on the probability that a suspect committed a crime, based on evidence from DNA forensics.

Because the responsibilities of geneticists are highly varied, the educational requirements of geneticists are also quite varied. Individuals with a bachelor's degree are qualified to perform many laboratory procedures, but most often individuals in the field of genetics will obtain a graduate degree. Those geneticists with master's degrees include highly skilled laboratory geneticists and genetic counselors. Geneticists who direct their own research projects must have a Ph.D. in genetics or an M.D. with specialty training in genetics.

Geneticists with a Ph.D. are highly trained professionals who perform genetic research in the areas of expertise mentioned above. Individuals with an M.D. are uniquely skilled in the treatment of patients with genetic disorders and are also often involved in research projects. Many individuals combine a genetics education with learning in other subjects, such as business, law, or computer science. Once having obtained the proper education, geneticists will find job opportunities at academic centers (such as colleges and universities), in government agencies (such as the National Institutes of Health, the Food and Drug Administration, and the Federal Bureau of Investigation), and in industry (for example, pharmacogenetics companies). The salaries paid to genetics professionals will range from \$25,000 to well over \$100,000, depending on the level of education, area of educational expertise, and job environment (academic, federal, or industrial).

A career in genetics can be exciting; and this is true for all areas of education, expertise, and job environment. It is exhilarating to have the opportunity to participate in research that impacts the treatment of a disease, or improves the crop yield for farmers, or helps correctly identify individuals who commit crimes. SEE ALSO CLINICAL GENETICIST; DNA PROFILING;

Genetic Counselor; Genetics; Molecular Biologist; Pharmacogenetics and Pharmacogenomics; Plant Genetic Engineer; Population Genetics; Statistical Geneticist.

Allison Ashley-Koch

Bibliography

Internet Resources

Careers in Genetics and the Biosciences. Human Genome Project. <http://www.ornl.gov/hgmis/education/careers.html>.

Genetics: Educational Information. Federation of American Societies for Experimental Biology. <http://www.faseb.org/genetics/careers.htm>.

Genetics

Genetics is the scientific study of the structure, function, and transmission of genes in living things. The field of genetics includes many disciplines and uses many different techniques. Historically, genetic scientists (geneticists) investigated patterns of inheritance in whole organisms by observing the distribution and segregation of physical characteristics across several generations of breeding. This type of genetic research, called classical or Mendelian genetics, is still conducted today and remains invaluable for identifying and describing inheritance patterns and traits. The development of modern molecular biology and biochemistry has facilitated the growth of a different but complementary branch of genetic research, known as molecular genetics. This branch focuses on understanding the detailed molecular mechanisms that govern the transmission of genetic material from one generation to the next.

The observable characteristics that describe any organism (for example, height or eye color in humans, or flower size and color in plants) can be broadly grouped into two categories: those that are acquired because of environmental effects, and those that are inherited. Genetics is concerned with this second category—that is, inherited characteristics. Gregor Mendel, an Augustinian monk of the mid-nineteenth century, observed that characteristics such as shape and color in peas were passed from parent to offspring regardless of the environment the plants grew in. He meticulously counted and documented thousands of crosses between different pea varieties to deduce the principles that governed this inheritance. In the end, his work was so influential and vital to the development of genetics that the term "Mendelian" genetics is now a synonym for classical genetics.

Several key concepts put forward by Mendel have been expanded, as the science of genetics has grown. It is now known that genetic information is passed on as a series of discrete units known as genes, each of which is associated with specific traits. Furthermore, most organisms (including humans) get two copies of their genetic information, one from each parent. This means that most living things have two copies of each gene, and that these two copies are not necessarily the same, since they came from different parents. When an organism reproduces, it passes only one of its two copies to an offspring. Importantly, copies of different genes separate (segregate)

randomly into the next generation, which means that an offspring can receive either of the two copies that the parent has, and that the set of copies that is passed is different from offspring to offspring.

A growing interest in the biochemical basis of life led geneticists of the twentieth century to attempt to identify the substance that carries genetic information from one generation to the next. Intense research led to the discovery of DNA (deoxyribonucleic acid), the molecule that encodes the genetic information of almost all organisms (a few viruses use RNA—ribonucleic acid—to encode their genes). Understanding the relationship between observable, inherited traits and the structure and organization of the genetic material (DNA) is the primary focus of molecular genetic research. The powerful combination of classical genetics and molecular genetics has led to many important advances in biological and health-care research and continues to be a major force in the advancement of science in the twenty-first century. SEE ALSO Gene; Geneticist; Genome; Inheritance Patterns; Mendelian Genetics; Molecular Biologist.

Daniel J. Tomso

Genome

A genome is the complete collection of hereditary information for an individual organism. In cellular life forms, the hereditary information exists as DNA. There are two fundamentally distinct types of cells in the living world, prokaryotic and eukaryotic, and the organization of genomes differs in these two types of cells.

salinity of, or relating to, salt

Prokaryotes comprise the bacteria and archaea. The latter were originally designated "extremophiles" because they favor such extreme environments as high acidity, **salinity**, or temperature. Prokaryotic cells tend to be very small, have few or no cytoplasmic organelles, and have the cellular DNA arranged in a "nucleoid region" that is not separated from the remainder of the cell by any membrane. Eukaryotes exist as unicellular or multicellular organisms. Among the unicellular eukaryotes are the protozoa, some types of algae, and a few forms of fungi, while the multicellular organisms include animals, plants, and most fungi.

chloroplasts energy-producing cell organelle

mitochondria the photosynthetic organelles of plants and algae

Eukaryotic cells are larger than prokaryotic cells, have a complex array of cytoplasmic structures, and have a prominent nucleus that communicates with components in the cytoplasm through an elaborate nuclear envelope. The hereditary information occurs principally in the nucleus of eukaryotic cells; in addition, minuscule (but essential) amounts of hereditary information occur in some cytoplasmic organelles (specifically, in **chloroplasts** for plants and algae, and in **mitochondria** for all eukaryotic groups).

chromatin complex of DNA, histones, and other proteins, making up chromosomes

Eukaryotic cells pass through a "cycle," progressing from a newly formed cell to a cell that is dividing to produce the next generation of progeny cells. Prior to division, the cell is in an "interphase"; during division, the cell is in a "division phase." During interphase, the nuclear DNA is organized in a dispersed network of **chromatin**, which is a complex consisting of nucleic acid and basic proteins. Immediately prior to and during division, the chromatin condenses to a series of discrete, compact structures called chromo-

An achievement celebrated in international headlines, Dr. J. Craig Venter of Celera Genomics published the human genome map, pictured here, in 2000. A genome is the entirety of an organism's DNA, and can be written as a long sequence of nucleotides (A, T, C, and G).

somes. Thus, the physical organization of the genome varies from interphase to division phase. Finally, viruses (which are noncellular, parasitic "life forms") have genomes of double-stranded DNA, single-stranded DNA, double-stranded RNA, or single-stranded RNA.

Eukaryotes

In sexually reproducing eukaryotes, progeny organisms receive a portion of their genetic information from each parent, receiving half the information from each. These parental contributions are designated **haploid** complements. The haploid complement can be represented as a "C value," which expresses the haploid complement as an amount of DNA measured in **base pairs**. Alternatively, the haploid complement can be expressed as the number of chromosomes contributed by each parent: This number of chromosomes is characteristic of each species. Finally, the haploid complement can be expressed as the number of genes on the haploid set of chromosomes.

haploid possessing only one copy of each chromosome

base pairs two nucleotides (either DNA or RNA) linked by weak bonds

Chromosome Number

Each species has a characteristic number of chromosomes. For species with genetically determined sexes, the haploid set is composed of **autosomes** plus a sex chromosome. *Homo sapiens*, for example, have 22 autosomes plus an X chromosome or Y chromosome. The haploid DNA content of chimpanzees is nearly identical, but is organized into 23 autosomes plus a sex chromosome.

autosomes chromosomes that are not sex determining (not X or Y)

The record for minimum number of chromosomes belongs to a subspecies of the ant, *Myrmecia pilosula*. The females have a single pair of chromosomes, while males have only a single chromosome. Like some other members of the insect class, these ants reproduce by a process called haplodiploidy, in which diploid fertilized eggs develop into females, while haploid unfertilized eggs develop into males.

The record for maximum number of chromosomes is found in the plant kingdom, due to a condition known as polyploidy. In polyploidy, many extra sets of chromosomes beyond the normal diploid number may accumulate

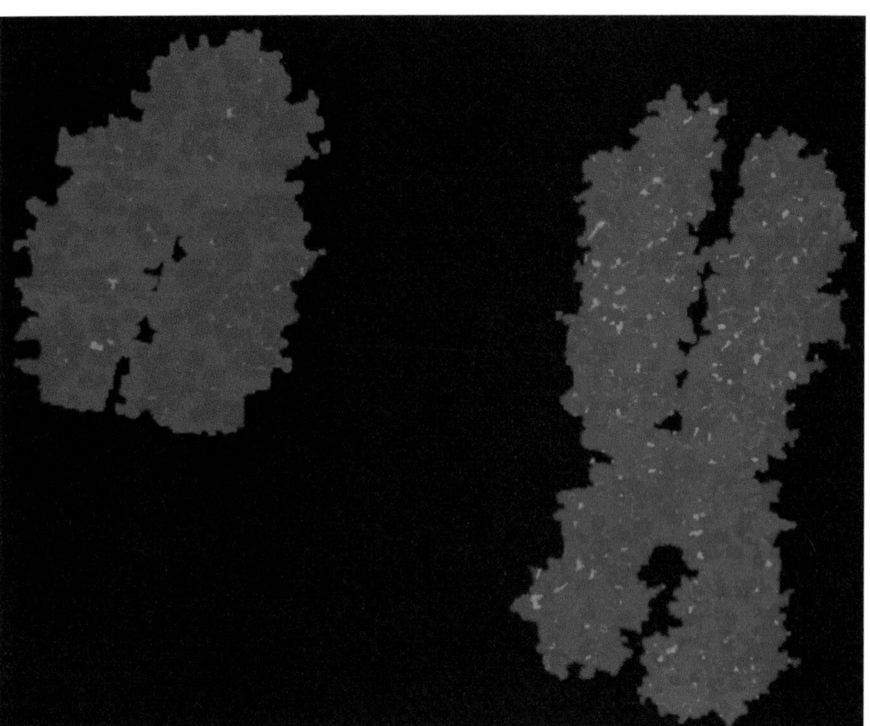

Human Y and X chromosomes. These determine human gender.

over time. Cultivars of wheat exist with diploid numbers of chromosomes equaling 14, 28, or 42 (multiples of the haploid number, which is 7). Polyploids exist for many cultivated plants, including potatoes, strawberries, and cotton, as well as in wild plants such as dandelions. Polyploidy has led to striking numbers, and the known record is held by the fern *Ophioglossum reticulatum*, which has approximately 630 pairs.

Genome Size or C Value

The C value is the amount of DNA in a haploid complement. Currently, the amount is reported as the total number of base pairs. Generally, more complex organisms have more DNA. For example, the haploid complement of *Homo sapiens* DNA contains between 3.12 and 3.2 gigabases (the prefix "giga" denotes billions), while the haploid complement of yeast (*Saccharomyces cerevisiae*) DNA contains 12,057,500 base pairs.

Unexpected genomic sizes occur, however, in a condition called the C value paradox. Two closely related species can have widely divergent amounts of DNA. For example, *Paramecium caudatum* has a C value of 8,600,000 kilobases (where the prefix "kilo" denotes thousands) while its near relative *P. aurelia* has a C value of just 190,000 kilobases. Another paradoxical circumstance occurs when a simpler organism has a C value higher than a more complex organism. For example, *Amphiuma means* (a newt) and *Amoeba dubia* (an amoeba) have, respectively, C values that are 26 and 209 times the C value of humans.

Number of Nuclear Genes, "Gene Density," and Intergenic Sequences

An important trend in genome evolution has been the accumulation, both within the genes (intragenic) and between genes (intergenic), of DNA that does not code for any gene products. *Homo sapiens* have between 31,000 and 70,000 genes; mice have 24,780; *Caenorhabditis elegans* (a roundworm) has more than 19,099; fruit flies have 13,601; and yeast approximately 6,000. A ratio of gene number to C value indicates that lower organisms have both smaller genes and lower numbers of nongene base pairs between adjacent genes. Higher eukaryotes have a larger number of intragenic inserts (introns), greater intergenic distances, and more abundant repeated sequences.

In higher eukaryotes, only a small portion of the genome is organized into genes. For example, in humans less than 2 percent of the genome specifies protein products. Another portion (about 20 percent in humans) is present as gene fragments, pseudogenes (sequences that resemble genes but are not expressed as proteins), and surrounding stretches of nucleotides. The vast majority of nucleotides (approximately 75 percent in humans) constitute extragenic sequences. Two forms of extragenic sequences are prominent: unique sequences and repetitive sequences.

For repetitive sequences, two types of organization occur: short tandem repeats (called satellite sequences) and widely distributed, interspersed repeats. Satellites are recurrent short sequences present in essential chromosomal structures such as **centromeres** and **telomeres**. Interspersed repeats are generated from transposons, which are nucleotide sequences that can replicate themselves and become distributed throughout the genome. An example of interspersed repeats that occurs in humans is a sequence of a few hundred nucleotides called Alu, which occurs approximately a million times. In higher plants, satellites and interspersed sequences constitute the bulk of the genome.

centromeres regions of the chromosome linking chromatids

telomeres chromosome tips

Ploidy

Ploidy reflects the reproductive mechanisms of an organism. Animals commonly have both a maternal and a paternal parent. Through meiosis, the former forms a haploid **gamete** called an ovum (or egg); the latter forms a haploid gamete called a sperm. During fertilization, the egg and sperm unite to form a **diploid** zygote that matures to an adult organism. Thus, the genome of adult animals is diploid, while the genome of their gametes is haploid.

gamete reproductive cell, such as sperm or egg

diploid possessing pairs of chromosomes, one member of each pair derived from each parent

Plants exhibit an alternation of generations; sporophytes (the mature, visible plant) are diploid; through meiosis, they produce spores that germinate into gametophytes; the gametophytes are haploid and produce gametes that fuse to reestablish the diploid state. Fungi also exhibit an alternation of generations. They commonly exist as **multinucleate** tubes of cytoplasm called hyphae. The individual nuclei are most often haploid (though may be diploid in the lower fungi).

multinucleate having many nuclei within a single cell membrane

Hyphae of different members of a fungal species sometimes fuse; in this circumstance (called heterokaryosis) the genome becomes the sum of the two (dikaryotic) haploid complements. Unicellular protistan organisms, a group that includes protozoans and most algae, exhibit many variations. For example, the ciliates (such as paramecia) have diploid micronuclei and polyploid macronuclei; the former are the basis of inheritance; the latter establish the genetic character of an existing organism.

Mitochondrial and Chloroplast Genomes

Two cytoplasmic organelles responsible for the production of energy are the mitochondria (present in nearly all eukaryotic cells) and chloroplasts (present only in photosynthetic organisms). Both contain small, circular DNA molecules that constitute the nonnuclear portion of a eukaryotic genome. These organelles are descended from formerly free-living bacteria that took up residence in the first eukaryotes.

The human mitochondrial genome contains 16,569 base pairs specifying 13 protein products and 24 RNA products. In both lower eukaryotes and especially plants, larger mitochondrial genomes are present. In extreme cases, mitochondrial genomes may be several hundred thousand or millions of base pairs. Chloroplast genomes contain between 100 and 200 kilobases. It is thought that each was once larger, but over time their genes have been moved to the nucleus.

Prokaryotes

Prokaryotic genomes are composed of a chromosome plus various accessory elements. The former is most commonly a circular double-stranded DNA molecule but may be a linear molecule in some major groups, such as *Streptomyces* and *Borrelia* (the causative agent of Lyme disease). Accessory elements most prominently include plasmids (commonly circular but linear in *Actinomycetes* and some *Proteobacteria*) as well as insertion sequence (IS) elements, transposons, and prophages (derived from viruses). Other variations in chromosomal geometry exist: multiple circular chromosomes are found in some organisms; combinations of circular and linear chromosomes occur in others; and, in the extreme (observed in *Streptomyces*), circular and linear chromosomes can convert between those two topologies.

The smallest bacterial chromosome, with only 580 kilobase pairs (kbp) occurs in *Mycoplasma genitalium*, and the largest, with 9,200 kbp, occurs in *Myxococcus xanthus*. Representative sizes cluster between 2,000 and 5,000 kbp (e.g., *Escherichia coli* MG1655 has 4,649,221 bp). A typical bacterial gene contains approximately a thousand base pairs. *M. genitalium* has approximately 470 genes, while *M. xanthus* has more than 10,000, and *E. coli* has approximately 4,288.

allelic variation presence of different gene forms (alleles) in a population

By 2002 the nucleotide sequences of more than seventy-five prokaryotic chromosomes had been mapped. One goal of these sequencing projects is gene annotation: establishing the location, function, and **allelic variation** for each gene. In *E. coli* MG1655, for example, the positions of the 4,288 protein-coding genes have been identified; the average distance between genes is 118 base pairs; and the noncoding sequences (some of which may function as regulatory sites) constitute less than 11 percent of the genome. The function of approximately 40 percent of the genes, however, remains unknown. Notably, the chromosomal size and gene content of another isolate of *E. coli*, the pathogenic H157:O7 strain, are quite different. The H157:O7 chromosome is 20 percent larger, while MG1655 and H157:O7 share 4.1 million base pairs (mbp) in common. H157:O7 has 1.34 mbp that are not found in MG1655 and MG1655 has 0.53 mbp that are not found in H157:O7.

The genomes of closely related prokaryotes often have different organizations. These differences arise from rearrangements (such as inversions) between repeated elements, IS elements, and transposons and from the "horizontal transfer" of nucleotide sequences between cells. The latter phenomenon is mediated most commonly by conjugative plasmids, which are nonessential, autonomous accessory genetic elements that can acquire genes (such as antibiotic resistance genes) and then move them from a donor organism to a recipient. The dynamic character of genomic organization in prokaryotes is often designated as "genomic plasticity."

A series of repeated elements exist in the chromosomes of prokaryotes. In some instances the repeats are redundant copies of essential, long nucleotide sequences, as is seen in ribosomal RNA loci. Other repeats are small and have known functions (as in the Chi sequences in *E. coli* that facilitate genetic crossing over) or unknown functions (as in the REP [repeated extragenic palindromic] sequences in *E. coli*).

Viruses

Viral genomes are composed of single-stranded or double-stranded DNA or RNA. Single-stranded RNAs are either positive (capable of being immediately translated into protein) or negative. Double-stranded RNA genomes are most often segmented, with each segment being a single gene, while the other genomes are single circular or linear molecules. The *Retroviridae* have single-stranded RNA genomes that are converted by an enzyme (reverse transcriptase) into double-stranded DNA that becomes incorporated into the **genome** of the host.

genome the total genetic material in a cell or organism

The smallest known virus, containing 5,386 bases, is a member of the *Microviridae*, which infects bacteria and is designated fX174. The largest viral genomes occur in *Poxviridae*, which can possess as many as 309 kbp.

Viruses are extraordinarily efficient in using the coding capacity of their genomes. The virus known as fX174 contains ten genes, and the end of one gene commonly overlaps with the beginning of the following gene. In addition, two smaller genes are nested within larger genes (this compaction being achieved by having the two genes expressed in alternate "reading frames"). As a consequence of this efficiency, only 36 bases are not translated into an amino acid sequence. At the opposite extreme, the various pox viruses share more than 100 similar genes and may have an equal number of unique genes. SEE ALSO ARCHAEA; CELL, EUKARYOTIC; CELL CYCLE; CONJUGATION; EUBACTERIA; EVOLUTION OF GENES; GENE; GENOMICS; HUMAN GENOME PROJECT; POLYMORPHISMS; POLYPLOIDY; READING FRAME; REVERSE TRANSCRIPTASE; TRANSPOSABLE GENETIC ELEMENTS; VIRUS.

Steven Krawiec

Bibliography

Brown, T. A. *Genomes.* New York: Wiley-Liss, 1999.

Casjens, Sherwood. "The Diverse and Dynamic Structure of Bacterial Genomes." *Annual Review of Genetics* 32 (1998): 339–377.

Gould, Stephen J. "The Ant and the Plant." In *Bully for Brontosaurus*. New York: W. W. Norton, 1991.

Genomic Medicine

A quote commonly heard these days is that in the history of medicine, the greatest advancements in the treatment of patients have occurred within the past 50 years. But what if a doctor could prevent a disease from occurring, treating the cause rather than the symptoms? We all agree this would be wonderful, but how could a doctor predict a patient's medical future? This dream is now within the realm of possibility. In fact, the greatest change in the history of medicine since the discovery of antibiotics is anticipated to occur over the next several generations. It is the movement of medicine from a discipline that reacts to and treats a disease to one that is focused on preventing disease from occurring.

This feat will be accomplished using genetic and genomic information gained from the Human Genome Project to tailor treatments and medicine to each individual patient. The individual's risk of getting a disease, or even infections, is almost always at least partly due to the combination of genes he or she was born with. Therefore, by using genetic information doctors will be able to predict, with different degrees of certainty, what diseases their patients are at risk of developing. They will then be able to use this information to reduce the chance of the disease from occurring in an individual, or prevent it entirely. This new approach of using genetics in the regular practice of medicine has been given the name "genomic medicine."

The Importance of SNPs

polymorphisms DNA sequence variant

exon coding region of genes

mutations changes in DNA sequences

Genomic medicine will be applied to patients through several mechanisms. The primary tool will be the detection of single nucleotide **polymorphisms** (SNPs). These are small variations in DNA sequences that are found in every person. These normal polymorphisms are very frequent, occurring in approximately one out of every 1,000 base pairs. The majority of these SNPs occurs in regions of the genome that do not code for proteins, and usually do not contribute to a person's disease susceptibility. However, they may serve as "markers" for a disease, lying close to an important susceptibility gene. When they occur directly in a protein-coding area of a gene (**exon**), they can cause the protein variation that helps make each of us unique. This variation in proteins is one of the primary reasons that each of us differs in our risk for disease, infections, and drug tolerance.

It is important to realize that these SNPs are not **mutations**. The term "mutation" suggests a nucleotide change that prevents a protein from performing its normal function. These mutation-caused changes are rare, occurring in less than 1 percent of the population. This type of genetic mutation is almost always severe enough that its presence alone is enough to cause the disease. In contrast, SNPs are much more common, certainly occurring in more than 1 percent of the population. Their presence may make it easier for a person to get a specific disease, but they do not cause the disease by themselves. Instead, disease can occur only in the presence of the correct environment, other gene combinations, drugs, and other such factors. This is very important, for it suggests that by knowing that an individual is at risk for a disease, a doctor can take actions to prevent it. These actions may be as simple as a dietary or lifestyle change, or taking a medicine or vaccine.

Doctors will be able to use this same identification of SNPs to screen patients prior to prescribing drugs. Drug side effects are usually due to a

This DNA chip created by Toshiba was designed to store an individual's genetic profile. The electronics company expects to release this electrochemical technology in 2003.

specific, uncommon polymorphism in a gene that produces a protein that interacts with the drug. These gene interactions may not have anything to do with how the drug helps the patient. More likely, the gene is involved in how the body metabolizes the drug. This field of studying genes and their interaction with drugs is called pharmacogenetics. And pharmacogenetics does not just study drug side effects. Every doctor knows that some drugs work better for some patients than others. This difference is also likely due to normal polymorphisms, or different genetic forms of the disease.

The Future of Medicine

The idea of preventive medicine is not new, but until the completion of the Human Genome Project medicine did not have a way of accomplishing it for common medical problems. For example, for years almost every baby born in the United States has had its urine or blood screened early in life for phenylketonuria (PKU), a rare genetic disease in which affected individuals cannot metabolize the amino acid phenyalanine, a common amino acid in our food. Untreated, PKU patients develop severe mental and psychomotor retardation. However, if a child is identified early in life to have the PKU mutation, the disease can be prevented by placing the child on a special diet that lacks phenylalanine.

Genomic medicine technique will not look at just one gene or protein in an individual. It will also look at the interaction of thousands of genes at once, using DNA chips or microarrays. These microarrays are already being used in cancer chemotherapy to predict which drugs will work best on each patient's specific tumor.

Therefore, in the future an individual will be able to go to the doctor for a regular checkup and give a small sample of blood, or maybe even just let the doctor take a mouth swab. This will provide a DNA sample that would then be placed on a genetic "microchip" or other device and quickly

be genotyped to give a genetic profile of the patient. The doctor will use the information to tailor the medical treatment for that patient. Lifestyle changes or medicine will be suggested to prevent the occurrence of diseases to which the patient is genetically susceptible, or at least to reduce the risk or severity of such diseases. If medication must be prescribed, the doctor will also use this genetic profile before choosing the medicine, to make sure that "the right medicine for the right patient" is chosen: one that works and will not harm the patient or cause side effects.

These changes in medicine are likely to take place over the first half of the twenty-first century. They will be exciting, but they will require that both patient and doctor have a strong understanding of genetics, the most powerful future tool of medicine. SEE ALSO DNA MICROARRAYS; GENETIC TESTING; PHARMACOGENETICS AND PHARMACOGENOMICS; POLYMORPHISMS.

Jeffery M. Vance

Bibliography

Guttmacher, A., et al. "Genomic Medicine: Who Will Practice It? A Call to Open Arms." *American Journal of Medical Geneticists* 106 (2001): 216–222.

Internet Resources

"Making the Vision of Genomic Medicine a Reality." Centers for Disease Control. (March 2001). <http:www.cdc.gov/genomics/info/perspectives/vision.htm>.

"Medicine and the New Genetics." Human Genome Project Information. <http://www.ornl.gov/hgmis/medicine/medicine.html>.

Pistoi, Sergio. "Facing Your Genetic Destiny." Scientific American (February 18, 2002). <http: www.sciam.com/article.cfm?articleID=00016A09-BE5F-1CDA-B4A8809EC588EEDF>.

Genomics

Genomics is a recent scientific discipline that strives to define and characterize the complete genetic makeup of an organism. Its primary approaches are to determine the entire sequence and structure of an organism's DNA (its **genome**) and then to determine how that DNA is arranged into genes. This second goal is accomplished by determining the structure and relative abundance of all messenger RNAs (mRNAs), the middlemen in genetics that encode individual proteins.

genome the total genetic material in a cell or organism

From Microorganisms to Human DNA

For many years, genomics has been focused on microorganisms, which have relatively small genomes. However, more recently the field has been energized by the advent of more industrialized, higher-throughput sequencing technologies. By 2001 more than seventy organisms had been completely sequenced, and a working draft of the human genome had been produced. Vigorous efforts have now been initiated to map the mouse genome, and one company already claims to have completed the sequence. From the description of the structure of the genetic material by James Watson and Francis Crick in 1953, it will have taken only about fifty years to determine the complete genetic codes of humans and most of the model organisms that are important in biological research.

Latin Name	Common Name	Genome Size
Eukaryotes (haploid genome)		
Oryza sativa	Rice	420,000 Kb
Homo sapiens	Human	3,200,000 Kb
Arabidopsis thaliana	Mustard cress	115,428 Kb
Drosophila melanogaster	Fruit fly	137,000 Kb
Caenorhabditis elegans	Roundworm	97,000 Kb
Saccharomyces cerevisiae	Yeast	12,069 Kb
Eubacteria		
Haemophilus influenzae	–	1,830 Kb
Escherichia coli	Human colon bacterium	4,639 Kb
Helicobacter pylori	Stomach ulcer bacterium	1,667 Kb
Mycobacterium	Tuberculosis	4,411 Kb
Yersinia pestis	Plague	4,653 Kb
Archaea		
Halobacterium	Salt-tolerant archaean	2,014 Kb
Methanobacterium thermoautotrophicum	Methane-producing archaean	1,751 Kb

Kb=one thousand base pairs

Size comparison of selected completed genomes. Most of these organisms are of economic, medical, or scientific importance.

Of what value is the knowledge of these genomes? How are they being used within the scientific community? The first fully sequenced genomes included the fruit fly, a worm, and a number of bacteria and yeast. One of the first analyses performed was to simply compare the sequences between organisms, in order to identify what is shared in common and what is different. This allows the very specific comparison of organisms that will enable the refining of phylogenic relationships. This kind of information is also very valuable for asking questions about how organisms have evolved, how they adapt to different circumstances, and what gene products contribute to their survival in various environmental conditions.

Applications

Genomics has brought us to the threshold of a new era in controlling infectious diseases. These studies will likely lead to the development of new disease prevention and treatment strategies for plants, animals, and humans alike. For instance, understanding pathogen genes, their expression, and their interaction will lead to new antibiotics, antiviral agents, and "designer" immunizations. These new DNA-based immunizations are by-products of genomic research and will undoubtedly eventually replace the traditional vaccines made from whole, inactivated microorganisms. This is highly relevant to domesticated animals, where viruses still kill billions of dollars worth of livestock every year.

Understanding the genomes of plants and animals has additional benefits. Gene mapping should allow us to understand the basis for disease resistance, disease susceptibility, weight gain, and determinants of nutritional value. The use of genomic information provides the opportunity to select optimal environments for the healthy growth of plants and animals, to develop disease-resistant strains, and to achieve improved nutritional value such as with the "golden" rice. Success in these species may well provide important insights needed to improve the health of humans.

The Human Genome Project and Future Research

The Human Genome Project reached a major milestone in 2001, with two separate publications of working drafts of the human genome. Although

much knowledge has been generated, the sequence is not complete. Neither the actual number of genes nor all their structures have been determined. However, several major lessons have been learned. First the number of genes is estimated to be between 30,000 and 70,000, fewer than previously thought. In addition, it is clear that a very large proportion of our genes are highly similar to those in other organisms, such as the fruit fly and the microscopic worm, *C. elegans*. The observation that we can build humans with between 30,000 and 70,000 genes and a fruit fly with 15,000 genes suggests that we owe much of the complexity of humans to the fine regulation of genes and not their absolute number.

Genomics has also forced biologists to begin to look at the function of genes in an industrialized mode. This new field of functional genomics takes advantage of a number of new technologies. Since many fly and worm genes are so similar to human genes (homologs), these animals can be used as model systems to study gene function. In these model systems it is possible to mutate (or alter) the structure of every single gene, enabling researchers to determine each gene's function and how several of the genes interact in complex metabolic pathways. Similar efforts using systematic gene mutations are also underway to create DNA "libraries" of two vertebrates, mice and zebrafish, whose genes are surprisingly similar to humans. Once these genomes are fully sequenced and characterized, it will be possible to create animals with disorders that are more precisely like those of humans, allowing for a better understanding of complex diseases and determination of novel and effective therapies.

nucleotide the building block of RNA or DNA

Genomics allows for the comparison of sequences between individuals, too. These studies can be used as a basis for the understanding and diagnosis of disease, especially of the complex disorders not governed by single genes. Knowledge of the entire human sequence is also the basis of the fields of pharmacogenetics and pharmacogenomics. Pharmacogenomics seeks a broader understanding of how genes influence drug response and toxicity, and the discovery of new disease pathways that can be targeted with tailor-made drugs. Pharmacogenetics is the study of the genetic factors involved in the differential response between patients to the same medicine. Polymorphisms, **nucleotide** changes that occur in more than 1 percent of the population, are the basis for our individuality but also account for our differential susceptibility to disease and the variable outcome of treatments. Through a variety of research efforts, more than one million polymorphisms have been identified in the human genome. The study of these variants, that occur once every 500 to 1,000 nucleotides in the human genome, should enable pharmacogenetics to define the optimal treatment regimens for subsets of the population, allowing a wider range of patients to be treated and more effective outcomes to be produced with any given drug. SEE ALSO AGRICULTURAL BIOTECHNOLOGY; DNA LIBRARIES; GENOMIC MEDICINE; GENOMICS INDUSTRY; HIGH-THROUGHPUT SCREENING; HUMAN GENOME PROJECT; MODEL ORGANISMS; PHARMACOGENETICS AND PHARMACOGENOMICS.

Kenneth W. Culver and Mark A. Labow

Bibliography

Bloom, Mark V., Greg A. Freyer, and David A. Micklos. *Laboratory DNA Science: An Introduction to Recombinant DNA Techniques and Methods of Genome Analysis.* Menlo Park, CA: Addison-Wesley, 1996.

Koonin, Eugene V., L. Aravind, and Alexy S. Kondrashov. "The Impact of Comparative Genomics on Our Understanding of Evolution." *Cell* 101 (2000): 573–576.

O'Brien, Stephen J., et al. "The Promise of Comparative Genomics in Mammals." *Science* 286 (1999): 458–462, 479–481.

Ye, Xudong, et al. "Engineering the Provitamin A (-Carotene) Biosynthetic Pathway into (Carotenoid-Free) Rice Endosperm." *Science* 287 (2000): 303-305.

Internet Resources

Celera, Inc. <http://www.celera.com>.

"Entrez Genomes." National Center for Biotechnology Information. <http://www.ncbi.nlm.nih.gov/Entrez/Genome/main_genomes.html>.

Genomics Industry

Fundamental to the myriad of genomic research efforts in operation around the world is the mapping and sequencing of whole **genomes**. The entire genomes of more than seventy organisms had been completed by early 2002, including the working drafts of the human genome, first published in 2001. The successful completion of the sequencing of these genomes was made possible in part by companies developing and utilizing new technologies and processes to increase the speed and accuracy of mapping and sequencing. This effort has also spawned entire new fields that aim, in various ways, to capitalize on genomics and identify disease-causing genes and new therapeutic strategies.

genomes the total genetic material in cells or organisms

The advent of the global Human Genome Project in 1989 provided significant, additional incentive for the development of a variety of new genomics-based companies. These companies can be loosely divided into seven major types: large-scale sequencing companies, gene mining companies, functional genomics companies, population-based genomics companies, **bioinformatics** companies, established pharmaceutical companies, and new biopharmaceutical companies. These are general distinctions and are not absolute, since many companies are blending and using a variety of these technologies with overlapping applications.

bioinformatics use of information technology to analyze biological data

Large-Scale Sequencing Companies

Early stages of the human genome project involved the physical mapping and subsequent sequencing of genes and intervening segments. A number of companies were founded specifically to identify genes, sequence them, and determine their function as a means to lead to new diagnostics and pharmaceuticals. These companies include Incyte, Human Genome Sciences, and Celera Genomics. This group of companies has contributed to public and private databases millions of bases of sequence data not only for the human genome, but for many microorganisms as well.

Gene Mining Companies

Now that the sequence data is available and placed in the public domain, companies have been created to "mine" the data, that is, to analyze the genomic sequences to identify genes, their function, and their relationships to health and disease processes. Companies pioneering in this area included Sequana and Millennium Pharmaceuticals (although Sequana did not survive

as an independent company). Once genes were identified as relevant to specific disease processes, some of the companies in this sector focused on discovering and developing small molecules, **antibodies**, proteins, or a combination of the three, in search of drugs that will target the consequences of the defective gene or dysregulated pathways.

antibodies immune-system proteins that bind to foreign molecules

Functional Genomics Companies

The challenge of the postgenomics era has been to identify those genes that are of clinical value for drug development. Inherent to that is a basic understanding of the function of all genes. Using industrialized biology approaches, these companies develop technologies for their own use or for sale to other companies. These new technologies are designed to pinpoint specific genes associated with disease and to determine causal relationships with **pathology**. Companies specializing in this area include Affymetrix, Excelesis, and Lexicon. In addition, most major pharmaceutical companies have established functional genomics departments.

pathology disease process

Population-Based Genomics Companies

In order to determine which genes are relevant and underlie complex human diseases such as diabetes mellitus, Alzheimer's disease, or hypertension, companies have been created to collect patient materials. These firms collect relevant clinical information and DNA on people suffering from defined disease as well as people who have no known disorders. They then look at DNA sequence variations in an attempt to identify the genetic factors that may predispose an individual to develop these types of disorders, or factors that directly lead to the development of disease. One company working in this area is deCODE, which has access to the genetic and health data of the entire population of Iceland.

Bioinformatics Companies

In order to extract the maximum amount of information from genomic sequence data, a new discipline has emerged, called bioinformatics. Bioinformaticians are involved in the capture and interpretation of biological data. The need for bioinformatics support throughout the genomics industry is significant and, as a result, a number of companies have been started that provide these services. Companies engaged in this type of research include DoubleTwist, Genomica, and Spotfire. Because genomics has created terabytes (a trillion bytes) of data, virtually all genomics efforts now require a substantial investment in in-house bioinformatics.

Established Pharmaceutical Companies

The pharmaceutical industry has been working to integrate genetic and genomic information into the drug discovery process. Essentially all of the major pharmaceutical companies have such programs, including GlaxoSmithKline, Merck, and Novartis. From the beginning of compound synthesis through the identification of drug targets, genomics is changing the traditional drug development process. Understanding the influence of **polymorphisms** on drug target binding is critical to optimizing drug development. This is especially important now, because successes in combinatorial

polymorphisms occurring in several forms

chemistry have given rise to a tremendous increase in the number compounds to be analyzed.

New Biopharmaceutical Companies

As a result of molecular biology and genomics, a number of new companies have been formed that focus on creating drugs from specific genes. As genes are identified and the function of the protein is learned, these companies manufacture the proteins made by specific genes for use in the treatment of human disease. This includes a variety of drugs made through recombinant DNA technology, such as insulin and growth hormone. These new drugs provide patients with a safer therapeutic alternative to other protein sources. Companies working in this sector include Genentech and Amgen. In addition, some companies focus on making protein therapeutics in novel ways. One such company is Genzyme Transgenics, which has genetically altered livestock to produce therapeutic protein in their milk. Other companies focus on producing artificial **antibodies** that are engineered to look just like normal human antibodies. These so called humanized antibodies, produced by companies such as Protein Design Laboratories, are already on the market for use in treating diseases such as cancer and arthritis.

antibodies immune system proteins that bind to foreign molecules

From Present to Future

Technological advances clearly drive innovation in the pharmaceutical and health-care industries. Advances in the speed and accuracy of DNA sequencing and the growing understanding of the genome have created hundreds of companies, with billions of dollars in assets, that are using this information to contribute to the creation of new diagnostic and therapeutic agents. This has been accomplished even before science has achieved full knowledge of the human DNA sequence or a complete understanding of the function of most human genes. It is expected that the next decade will provide the remaining human genomic information, which will enable a breathtaking array of new therapeutic and preventive options for the treatment of human disease. SEE ALSO BIOINFORMATICS; BIOTECHNOLOGY; COMBINATORIAL CHEMISTRY; GENOME; HIGH-THROUGHPUT SCREENING; HUMAN GENOME PROJECT; PHARMACOGENETICS AND PHARMACOGENOMICS.

Kenneth W. Culver and Mark A. Labow

Bibliography

Internet Resource

Genomic Industry Web Guide. <http://www.business2business.com/webguide/0,1660,49834,ff.html>.

Genotype and Phenotype

An individual's genotype is the composition, in the individual's **genome**, of a specific region of DNA that varies within a population. (The genome of the individual is the total collection of the DNA in a cell's chromosomes. It includes all of the individual's genes, as well as the DNA sequences that lie between them.)

genome the total genetic material in a cell or organism

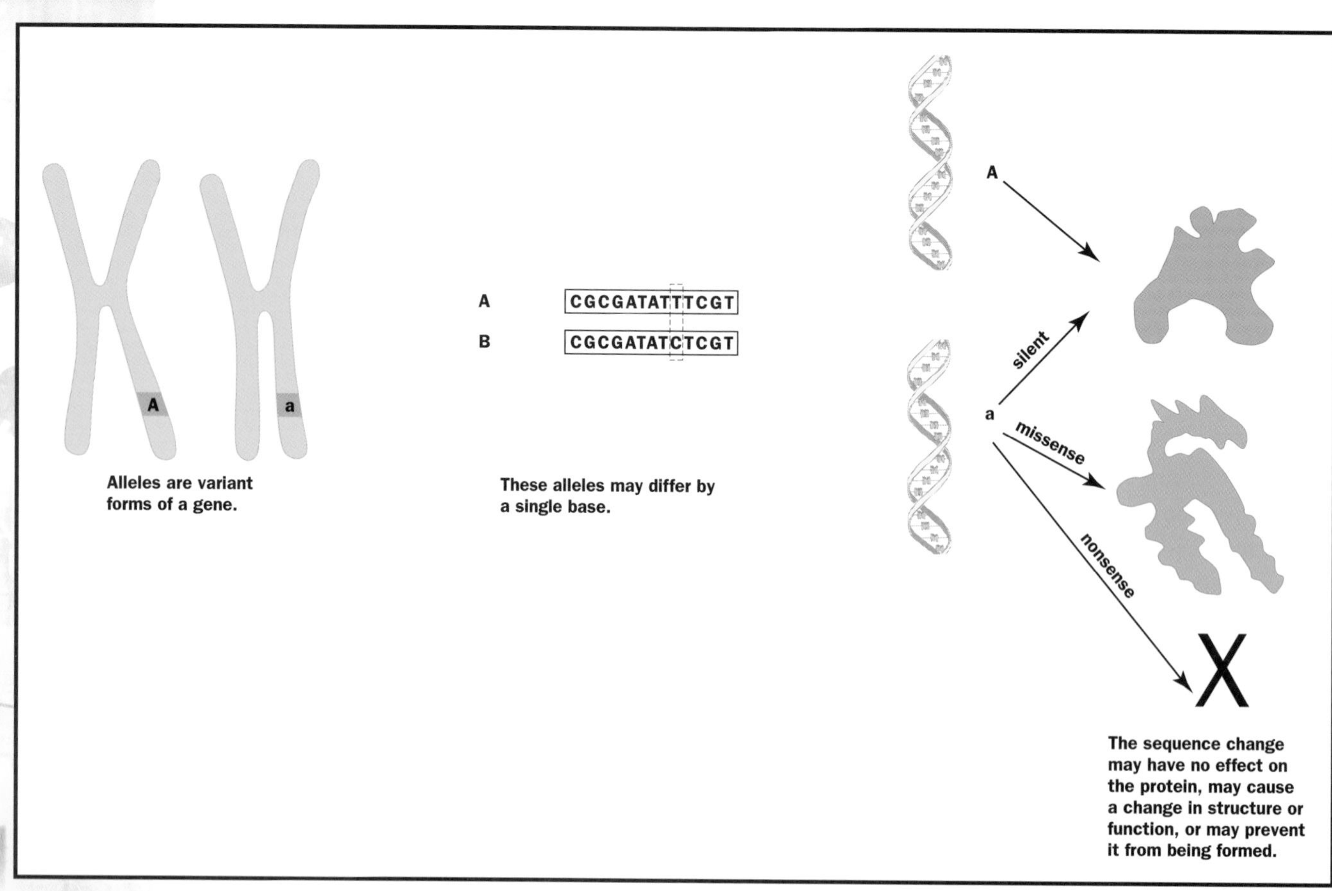

Alleles are variant forms of a gene.

These alleles may differ by a single base.

The sequence change may have no effect on the protein, may cause a change in structure or function, or may prevent it from being formed.

Alleles differ in their DNA sequence, and may lead to production of alternative forms of the protein they encode.

A genotype could represent a single DNA nucleotide, at a specific location on a chromosome. It could also be a sequence repeated multiple times, a large duplication, or a deletion. Most variation in genotypes does not cause any difference in the proteins being produced by the cell, because genes, which code for proteins, occupy only about 2 percent of the total genome. However, when a specific genotype does affect the composition or expression of a protein, disease or changes in physical appearance can result. The physical effect of a particular genotype is known as its phenotype, or trait.

Alleles, Polymorphisms, and Mutations

An individual genotype is composed of two distinct parts, the inherited sequences from the maternal and paternal genomes. Therefore, for every genotype, there are two copies of the sequence. For genes, we refer to these individual sequence copies on each chromosome that make up a genotype as the two alleles. Alleles may be identical or different, and various combinations of alleles can create a range of phenotypes.

polymorphisms occurring in several forms

mutations changes in DNA sequences

A DNA change within a gene may or may not alter the protein that is encoded by the gene. If the change contributes to normal genetic variability and is found in more than 1 percent of the population being studied, the variation is called a **polymorphism**. Variations in a gene that modify, seriously disrupt, or prevent the functioning the encoded protein are called **mutations**. The effect of a mutation can range from harmless to harmful depending on the type of mutation.

One type of sequence variation is the substitution of one **nucleotide** for another, which can thus affect the three base-pair unit (codon) that codes for a specific amino acid. In some cases, where the substitution does not change the amino acid, the result is a neutral or silent mutation. A substitution that alters the coding sequence, and thus substitutes a different amino acid, is called a missense mutation. This type of mutation can have a beneficial effect whereby there is a positive gain of function for the protein, it can also have a drastically negative effect and result in loss of function or the gain of a new function that is deleterious to the organism. By far the most severe change is a nucleotide substitution that creates a stop codon where one should not be. This phenomenon, called a nonsense mutation, will prevent the full formation of the protein. A nonsense mutation can completely abolish the function of the protein, which can be lethal to the organism if the protein is essential for sustaining a key biological process.

nucleotide the building block of RNA or DNA

Dominant, Codominant, and Recessive Genes

Differences in phenotype can result from differences in genotype. Two individuals may have different bases at a particular location in the coding region of a gene. While this will result in different codons in that part of the sequence, the two different codons may code for the same amino acid, which will in turn result in the same protein. Thus it can be said that the two individuals have different genotypes in their DNA, but that their phenotypes are identical. This example illustrates the first factor one must look for in predicting phenotypes: how the genotype affects the translation of DNA.

The second factor is the inheritance pattern of the genotype. If each parent transmits an identical gene sequence, A, to a child, then the genotype of the offspring will be AA. This is called a **homozygous** genotype. However, if the child inherits a different allele, a, from one of the parents, then its resulting genotype, Aa, contains two different sequences. Such genotypes are called **heterozygous**.

homozygous containing two identical copies of a particular gene

heterozygous characterized by possession of two different forms (alleles) of a particular gene

An Aa genotype can result in the same phenotype as either an AA or aa genotype, if one of the alleles acts in a dominant fashion. If the A allele is dominant over the a allele, then the phenotype of a heterozygous (Aa) individual will be the same as the phenotype of a homozygous dominant (AA) individual.

Huntington's disease, a nervous system disorder, follows a dominant inheritance pattern. The presence of one mutated Huntington gene will result in the conditions of the disease. Even if a patient has one normal Huntington gene on another chromosome, one mutated copy is enough to produce Huntington's disease. Thus, the Huntington allele is said to be dominant over the normal allele.

Another type of inheritance pattern is called recessive. Here, two copies of the mutant allele, a, must be inherited (resulting in genotype aa) to cause a change in phenotype. This is exemplified by cystic fibrosis, a recessive disorder marked by digestive and respiratory problems. In this case a heterozygous individual (genotype Aa) who carries one copy of the faulty gene will not display clinical symptoms, because the normal gene is dominant over the *CFTR* gene. The normal gene is able to express the protein that **epithelial cells** need for transporting chloride. If two parents are

epithelial cells one of four tissue types found in the body, characterized by thin sheets and usually serving a protective or secretory function

heterozygous for the mutated cystic fibrosis gene, there is a 25 percent chance that their child will inherit a mutated copy from each parent and will have the disease. Heterozygous carriers, who live normal lives, pass on the mutant gene to half of their children, enabling it to stay within the population for generations and to persist at relatively high frequencies.

Although it is convenient to illustrate inheritance concepts by talking about diseases, if we consider the diversity of phenotypes expressed by all organisms in the living world it is obvious that not all variation is bad and most genetic changes do not lead to disease.

Multiple Alleles and Pleiotropy

loci sites on a chromosome (singular, locus)

Some genetic **loci** are multiallelic, having more than one allele that will manifest in a variety of phenotypes. In most mammal species, for example, the immune response genes of the major histocompatibility complex are extremely polymorphic—meaning that there are many different alleles at each gene locus. The combination of alleles in each individual may result in either susceptibility or resistance to specific disease-causing agents.

The human blood group system is another example of multiple alleles resulting in many different phenotypes. The genes that determine ABO blood type encode enzymes that add particular sugar groups to proteins in blood cells. A person's specific blood type is due to the presence or absence of A and B sugar-protein complexes on the surface of red blood cells.

There are three alleles involved, A, B, and O, and six possible genotypes: AA, BB, OO, AB, AO, and BO. The various genotypes result in four different phenotypes or blood types: A, B, O, and AB. Individuals have blood type A if their genotypes are AA or AO. Individuals have blood type B if their genotypes are BB or BO. Individuals have blood type O if their genotype is OO, and they have blood type AB if their genotype is AB.

Many of these examples describe the concept of genotype and phenotype in terms of proteins or diseases that have been thoroughly analyzed by scientists. However, genotypic differences are abundant in nature and are evident in the most extraordinary ways.

A small mutation in a viral gene may make an otherwise harmless strain of the influenza virus capable of causing disease or even death. Other genetic variations in mammals, including humans, may influence aggression or other social interactions.

Some genes affect more than one unrelated characteristic. These genes are said to be pleiotropic. One gene, for example, produces melanin, which is responsible for skin pigmentation and is also involved in nerve pathways. If there is a certain mutation in the melanin gene, no melanin will be produced. In humans, this condition, called albinism, causes white skin and, usually, vision problems. In domestic cats, it causes a white-hair, blue-eye phenotype along with hearing loss.

In the past, scientists thought that pleiotropic genes were unusual. However, the Human Genome Project has shown that humans have only about one-third of the number of genes that was predicted. It is now believed that most genes are pleiotropic, serving many functions. SEE ALSO ALTERNATIVE SPLICING; BLOOD TYPE; DISEASE, GENETICS OF; GENE; GENETIC CODE;

Immune System Genetics; Inheritance Patterns; Mutation; Pleiotropy; Polymorphisms.

Joelle van der Walt and Jeffery M. Vance

Growth Disorders

Growth, which usually refers to skeletal growth since it determines final adult height, is an extremely complex process. As such, it is susceptible to a wide range of genetic and physiologic disturbances. Indeed, growth is adversely affected by many if not most chronic diseases of childhood, through many different mechanisms.

Skeletal growth depends on hormonal signals for regulation. It also requires the production of adequate amounts of cartilage, because most bone forms within a model or template made from cartilage. Primary disorders of growth, that is, disorders in which growth is intrinsically affected, therefore fall into two major categories: disorders of the endocrine (hormone) system and disorders of the growing skeleton itself (skeletal dysplasias). Many of the former and most of the latter are genetic disorders.

Endocrine Disorders

Growth hormone (GH) is produced by the pituitary gland at the base of the brain and is a major regulator of growth. Deficiency of the hormone is the prototype of the inherited endocrine disorders of growth. Although normal in size at birth, infants with GH deficiency exhibit severe postnatal growth deficiency while maintaining normal body proportions. If untreated, children typically have a "baby-doll" facial appearance and a high-pitched voice that persists after puberty.

Isolated GH deficiency most often results from deletion of all or part of the GH gene. Humans carry two copies of the GH gene, but having just one good copy is usually sufficient to prevent GH deficiency. Thus, this disorder is inherited as an autosomal recessive trait. Rarely, **point mutations** of this gene can lead to a dominantly inherited form of GH deficiency, in which the product of the mutant GH **allele** is thought to interact with and prevent secretion of the product of the normal GH allele.

point mutations gains, losses, or changes of one to several nucleotides in DNA

allele a particular form of a gene

transcription factors proteins that increase the rate of gene transcription

GH deficiency also results from mutations of genes that encode **transcription factors**, such as PIT1, PROP1, and POU2F1, which are necessary for development of the pituitary gland and of the cells that produce pituitary hormones. Patients usually have small pituitary glands and exhibit deficiencies of several pituitary hormones, including gonadotropins (FSH, LH), prolactin, and thyroid-stimulating hormone (TSH) in addition to GH. Multiple pituitary hormone deficiency of this type is inherited in an autosomal recessive fashion.

At their target cells, hormones exert their efforts by binding to receptors. The clinical manifestations of GH deficiency can also result from mutations of the GH receptor, in the autosomal recessive Laron syndrome. There are also a number of birth-defect syndromes in which hypopituitarism (reduced pituitary output) results in the abnormal development of craniofacial structures. Examples include anencephaly, holoprosencephaly, Palister-Hall syndrome, and some cases of severe cleft lip and cleft palate.

Deficiencies of other hormones relevant to growth and their receptors also occur on a genetic basis. For instance, thyroid hormone deficiency can be due to reduced TSH, as discussed above, but it can also result from loss-of-function mutations of enzymes that are involved in thyroid hormone biosynthesis. There are also several forms of thyroid hormone resistance due to mutations of thyroid hormone nuclear receptors. The biosynthetic defects are inherited as recessive traits, whereas thyroid resistance is usually inherited in a dominant fashion. Mental retardation, growth deficiency, and delayed skeletal development are the main clinical manifestations of thyroid hormone deficiency.

Skeletal Dysplasias

In contrast to endocrine growth disorders, the hallmark of the skeletal dysplasias ("-plasia" means "growth") is disproportionate short stature. In other words, the limbs are disproportionately shorter than the trunk or vice versa. These disorders result from mutations of genes whose products are required for normal skeletal development. In most cases they are involved in endochondral **ossification**, the process by which the skeleton grows through the production of the cartilage template that is converted into bone. The mutated genes encode cartilage and bone extracellular matrix proteins, growth factors, growth factor receptors, intracellular signaling molecules, transcription factors, and other molecules whose functions are needed for bone growth.

ossification bone formation

Growth Factor Receptor Mutations. The prototype of the skeletal dysplasias is achondroplasia, which is one of a graded series of dwarfing disorders that result from activating mutations of **fibroblast** growth factor receptor 3 (FGFR3). Achondroplasia is the most common form of dwarfism that is compatible with a normal life span, while thanatophoric dysplasia, which lies at the severe end of the spectrum of FGFR3 disorders, is the most common lethal dwarfing condition in humans. Both are characterized by the shortening of limbs, especially proximal limb bones, and a large head with a prominent forehead and hypoplasia (reduction of growth) of the middle face. The mildest disorder in this group is hypochondroplasia, in which patients exhibit mild short stature and few other features.

fibroblast undifferentiated cell normally giving rise to connective tissue cells

All of the disorders in this group result from **heterozygous** mutations of FGFR3. Except for the lethal thanatophoric dysplasia, they are inherited as autosomal dominant traits. The vast majority of mutations arise anew, during sperm formation (spermatogenesis), and especially in older fathers. FGFR3 is a very mutable (easily mutated) gene and there are certain extremely mutable regions within the gene where disease-causing mutations cluster.

heterozygous characterized by possession of two different forms (alleles) of a particular gene

There is a very strong correlation between clinical **phenotypes** and specific mutations. In fact, essentially all patients with classic features of achondroplasia have the same amino acid substitution in the receptor. The mutations that cause these disorders enhance the transduction of signals through FGFR3 receptors in **chondrocytes** in growing bones. This inhibits the proliferation of these cells that is necessary for linear growth to occur.

phenotypes observable characteristics of an organism

chondrocytes cells that form cartilage

A six-year-old girl with achondroplasia (right) stands beside her younger, not quite four-year-old brother. Features of achondroplasia evident in the girl include small stature, short arms and legs, relatively large head size, and subtle differences of facial features.

Cartilage Matrix Protein Mutations. Another major class of skeletal dysplasias result from mutations of genes that encode cartilage matrix proteins such as collagen types II, IX, X, and XI, and cartilage oligomeric matrix protein (COMP). The type II collagen mutations cause a spectrum of autosomal dominant disorders called spondyloepiphyseal dysplasias because they primarily affect the spine (spondylo) and the ends of growing limb bones (epiphyses). They range in severity from lethal before birth to extremely mild. In addition to dwarfism that affects the trunk more than the limbs, patients with these disorders develop precocious **osteoarthritis** of weight-bearing joints such as the hips and knees. Many patients have eye problems that reflect disturbances of type II collagen in the vitreous portion of the eye.

osteoarthritis a degenerative disease causing inflammation of the joints

Mutations of COMP cause two clinically distinct disorders: pseudoachondroplasia and multiple epiphyseal dysplasia. Both are inherited as autosomal dominant disorders, have onset after birth, and are dominated by osteoarthritis of hips and knees. Dwarfism is severe and skeletal deformities are common in pseudoachondroplasia.

Cartilage collagens and COMP are multimeric molecules, that is, they are composed of multiple subunits, three for collagens and five for COMP. Like a square wheel on a car, the products of mutant alleles interfere functionally with the products of normal alleles when they combine during molecular assembly, a so-called dominant negative effect. Most collagen mutations are thought to act through this mechanism to reduce the number of collagen molecules in cartilage matrix, which in turn alters the ability of cartilage to function as a template for bone growth.

Similar types of mutations occur in genes encoding type I collagen, which is the principal matrix protein of bone. These mutations lead to osteogenesis imperfecta (OI), which is a spectrum of disorders of varying severity. The hallmark of OI is bone fractures, although patients often have blue sclerae (the "whites" of the eye), fragile skin, and dental problems that reflect the widespread distribution of type I collagen in many connective tissues.

Excessive Growth

Genetic growth disorders also include conditions with excessive growth. Beckwith-Wiedemann syndrome is characterized by an enlarged tongue, abdominal wall defects (omphalocele), and generalized overgrowth during the fetal and neonatal period. Most of the findings can be attributed to the excess availability of insulin-like growth factor II (IGF2) that results from duplication, loss of heterozygosity, or disturbed imprinting of the IGF2 gene. The syndrome behaves as an autosomal dominant trait in many families. The excessive growth slows with age, but patients are predisposed to childhood tumors, especially **Wilms tumor**.

Wilms tumor a cancerous cell mass of the kidney

Simpson-Golabi-Behmel syndrome is an X-linked overgrowth syndrome with many of the features of Beckwith-Wiedemann syndrome. It results from mutations of glypican 3, which is a cell surface proteoglycan that binds and may sequester growth factors such as IGF2. Glypican 3 mutations appear to enhance IGF2 signaling through its receptor, explaining the clinical similarities between the two syndromes. SEE ALSO BIRTH DEFECTS; DISEASE, GENETICS OF; GENETIC COUNSELING; HORMONAL REGULATION; IMPRINTING; INHERITANCE PATTERNS; SIGNAL TRANSDUCTION.

William Horton

Bibliography

Karsenty, G., and E. F. Wagner. "Reaching a Genetic and Molecular Understanding of Skeletal Development." *Developmental Cell* 2, no. 4 (2002): 389–406.

MacGillivray, M. H. "The Basics for the Diagnosis and Management of Short Stature: A Pediatric Endocrinologist's Approach." *Pediatric Annual* 29 (Sept., 2000): 570–575.

Wagner, E. F., and G. Karsenty. "Genetic Control of Skeletal Development." *Current Opinion in Genetic Development* 5 (Oct., 2001): 527–532.

Hardy-Weinberg Equilibrium

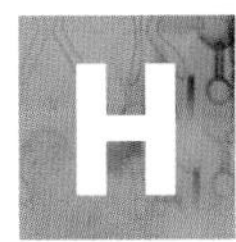

The Hardy-Weinberg equilibrium is the statement that **allele** frequencies in a population remain constant over time, in the absence of forces to change them. Its name derives from Godfrey Hardy, an English mathematician, and Wilhelm Weinberg, a German physician, who independently formulated it in the early twentieth century. The statement and the set of assumptions and mathematical tools that accompany it are used by population geneticists to analyze the occurrence of, and reasons for, changes in allele frequency. Evolution in a population is often defined as a change in allele frequency over time. The Hardy-Weinberg equilibrium, therefore, can be used to test whether evolution is occurring in populations.

allele a particular form of a gene

Basic Concepts

A population is a set of interbreeding individuals all belonging to the same species. In most sexually reproducing species, including humans, each organism contains two copies of virtually every gene—one inherited from each parent. Any particular gene may occur in slightly different forms, called alleles. An organism with two identical alleles is called homozygous for that gene, and one with two different alleles is called heterozygous. During the formation of **gametes**, the two alleles separate into different gametes. Mating unites egg and sperm, so that the offspring obtains two alleles for each gene.

The two alleles for a gene typically have different effects on the phenotype, or characteristics, of the organism. For many genes, one allele will control the phenotype if it is present in either one or two copies; this allele, which is often represented by a single, uppercase letter—*B*, for example—is said to be dominant. The other allele will only exert a visible effect if the dominant allele is not present; it is said to be recessive and is often represented by a lowercase letter—*b*, for example. The genotype of an organism specifies both alleles for a particular gene and is often symbolized by pairs of letters, such as *BB*, *Bb*, or *bb*, with each letter representing an allele.

gametes reproductive cells, such as sperm or eggs

It is important to understand that "dominant" does not mean an allele is more common in the population—lethal dominant alleles are very rare, for instance. Nor does dominant necessarily mean an allele will spread through the population. Likewise, "recessive" does not necessarily mean an allele will become less common. Indeed, the Hardy-Weinberg equilibrium shows conditions under which allele frequencies remain unaltered over generations.

Assumptions of the Hardy-Weinberg Model

Before examining the mathematical model underlying the Hardy-Weinberg equilibrium, let us look at the assumptions under which it operates:

1. Organisms reproduce sexually.
2. Mating is random.
3. Population size is very large.
4. Migration in or out is negligible.

5. Mutation does not occur.
6. Natural selection does not act on the alleles under consideration.

While the list appears to be so restrictive that no population can meet its requirements, in fact many do, to a very good first approximation. Even more to the point, variation from the Hardy-Weinberg equilibrium tells a population geneticist that one or more of these assumptions is not being met, thereby providing a clue about the forces at work within the population. Perhaps surprisingly, populations need not be very big to meet the conditions above—populations with as few as one thousand to two thousand individuals can do so.

Allele Frequencies Remain the Same Between Generations

Suppose we want to study the allele frequencies of the gene for coloration in a population of moths. The allele for the dark color pattern, *B*, is dominant to the allele for the light color pattern, *b*. In a certain population, the frequency of *B* is found to be 0.9, and that of b is 0.1 (we will see, below, how to determine these frequencies by studying the moths themselves). This means that 90 percent of all the alleles are *B*, and 10 percent are *b*.

The Hardy-Weinberg equilibrium states that, given the above conditions, allele frequencies will not change from one generation to the next. To show this is true, we need some algebra.

Random mating means each allele has an equal chance of being paired with each other allele. During random mating, the likelihood that a *B* allele from a mother will unite with a *B* allele from a father is given by

$$B \times B = 0.9 \times 0.9 = 0.81.$$

The genotype of this offspring will be *BB*.

Similarly, the likelihoods of other combinations:

$$B \times b = 0.9 \times 0.1 = 0.09 \text{ for genotype } Bb;$$

$$b \times B = 0.1 \times 0.9 = 0.09 \text{ for genotype } Bb;$$

and

$$b \times b = 0.1 \times 0.1 = 0.01 \text{ for genotype } bb.$$

Note that the two *Bb* genotypes are the same. Therefore the frequency of *BB* is 0.81, the frequency of *Bb* is 0.18, and the frequency of *bb* is 0.01. These add to 1, just as we would expect, since they represent all the members of the next generation.

Are the allele frequencies still 0.9 and 0.1? For simplicity, imagine we're looking at one hundred individuals, so that eighty-one are *BB*, eighteen are *Bb*, and one is *bb*. Since each individual has two alleles, there are 200 alleles in all.

The number of *B* alleles is given by $(81 \times 2) + (18 \times 1) = 180$.

The number of *b* alleles is given by $(1 \times 2) + (18 \times 1) = 20$.

By comparing 180 to 20, you can see the frequency of *B* is still 0.9 and that of *b* is still 0.1.

Allele Frequencies Can Be Calculated from Phenotypes

This is all very interesting, but it requires knowing the allele frequencies in a population. The power of the Hardy-Weinberg equilibrium formulas is in their ability to allow us to determine these frequencies by simple observation of the population.

When we see an organism with the dominant phenotype, we do not know whether we are looking at a homozygote (*BB*) or a heterozygote (*Bb*). When we see one with the recessive phenotype, though, we know it has the genotype *bb*. If we determine the proportion of individuals in the population with the *bb* genotype, it is a simple step to calculate the frequency of *b*. Let's work backward through our example above. If we determine that 1 percent of the population is homozygous recessive,

$$bb = 0.01.$$

Since the frequency of *bb* organisms is the product $b \times b$, we can take the square root of this number to get the frequency of *b*:

$$b = (bb)^{1/2} = (0.01)^{1/2}$$

and

$$b = 0.1.$$

Since $B + b$ must equal 1 (assuming there are only two alleles), *B* must be 0.9. With these calculations in hand, we can predict what the frequency for each of the other genotypes should be: *BB* is 0.81, and *Bb* is 0.18.

CALCULATING ALLELE FREQUENCIES AND GENOTYPES FROM THE OBSERVED FREQUENCY OF HOMOZYGOUS RECESSIVES

B: dominant allele frequency.
b: recessive allele frequency.
b = (observed homozygote recessive frequency)$^{1/2}$.
$B = 1 - b$.
$B \times b$ = expected frequency of heterozygotes in the population.
B^2 = expected frequency of homozygous dominants in the population.

Departures from Equilibrium Indicate Evolutionary Forces at Work

With these simple tools, we can look at populations to see if they conform to these numerical patterns. If they differ, we seek the reasons for the difference in some violation of the Hardy-Weinberg assumptions. Two processes, natural selection and genetic drift, are the most common and important factors at work in most populations that are not at equilibrium.

For example, suppose we find a population in which the recessive allele frequency is declining over time. We might then investigate whether homozygous recessives are dying earlier. (Many genetic diseases, such as cystic fibrosis, are due to recessive alleles.) This could be due to natural selection, in which those that are better adapted to the environment survive longer and reproduce more frequently.

Or suppose we find a population in which there is a smaller-than-expected number of homozygotes of both types, and a larger number of heterozygotes. This could be due to heterozygote superiority—where the heterozygote is more fit than either homozygote. In humans, this is the case for the allele causing sickle cell disease, a type of hemoglobinopathy.

Nonrandom mating is another potential source of departure from the Hardy-Weinberg equilibrium. Imagine that two alleles give rise to two very different appearances. Individuals may choose to mate with those whose appearance is closest to theirs. This may lead to divergence of the two groups over time into separate populations and perhaps ultimately separation into two species.

In very small populations, allele frequencies may change dramatically from one generation to the next, due to the vagaries of mate choice or other random events. For instance, half a dozen individuals with the dominant allele may, by chance, have fewer offspring than half a dozen with the recessive allele. This would have little effect in a population of one thousand, but it could have a dramatic effect in a population of twenty. Such changes are known as genetic drift. SEE ALSO GENE FLOW; GENETIC DRIFT; INHERITANCE PATTERNS; MUTATION; POPULATION BOTTLENECK; POPULATION GENETICS.

Richard Robinson

Bibliography

Hartl, D. L., and A. G. Clark. *Principles of Population Genetics, 3rd ed.* Sunderland, MA: Sinauer, 1997.

Heatshock Protein *See Chaperones*

Hemoglobinopathies

Hemoglobinopathies are diseases caused by the production of abnormal hemoglobin or by a deficiency of hemoglobin synthesis. Hemoglobin is the protein in red blood cells (erythrocytes) that binds to oxygen, to distribute it throughout the body. The major hemoglobinopathies are sickle cell disease and several forms of thalassemia.

Hemoglobin Structure and Function

In the lungs, where oxygen concentration is high, each hemoglobin molecule can bind with one molecule of oxygen. The erythrocyte containing the hemoglobin then travels through the bloodstream to the body's cells, where oxygen concentration is low, and the hemoglobin releases the oxygen for use by local tissue. It also picks up carbon dioxide, and this waste product is transported back to the lungs, where it can be released and exhaled.

Hemoglobin is made up of heme and globin. Heme is an iron-containing pigment that binds to oxygen. Globin, which holds the heme and influences how easily it stores and releases oxygen, is a protein consisting of two pairs of polypeptide chains. Globin can contain several different types of polypeptide chains, termed alpha, beta, and gamma. Each is coded for by a separate gene. The genes are evolutionarily related, and their differences are the result of ancient mutation events in an ancestral form that gave rise to each modern type.

The type of hemoglobin found in healthy adults contains two alpha chains and two beta chains. This form of hemoglobin is called HbA (hemoglobin A). As discussed below, sickle cell disease is due to mutations in the beta chains in HbA. A fetus or newborn baby does not produce HbA. Instead, it produces fetal hemoglobin, or HbF. Like HbA, fetal hemoglobin contains a pair of alpha chains. But in place of the beta chains, it contains a pair of gamma chains. As infants grow older, their bodies produce less and less HbF and more and more HbA.

The Genetics of Hemoglobinopathies

Each person possesses two copies of the beta globin gene, on separate **homologous** chromosomes. In most people, the two copies are identical. A person with two identical gene copies is said to be homozygous.

In some people, the two beta copies are not identical. These people, who have two different **alleles** of the beta globin gene, are said to be heterozygous. The beta globin allele that leads to sickle cell disease is called the hemoglobin S (HbS) allele.

People who have inherited one HbA and one HbS allele are heterozygous for the beta chain gene. They are said to have the sickle cell trait, but not sickle cell disease. As long as they have one HbA allele, these individuals produce sufficient HbA to remain healthy, and they usually do not have any medical problems, or they experience only very mild symptoms. When both alleles must be abnormal to cause a disease, the condition is said to be recessive. Sickle cell disease is a recessive condition.

Sickle cell disease can occur when two individuals who have the sickle cell trait (they are called carriers) have children. Recall that the two beta chain alleles occur on different chromosomes. These homologous chromosomes separate during **gamete** formation, so that each gamete has a fifty-fifty chance of possessing an HbS allele. There is a one-in-four chance that a child conceived by two carriers will inherit a **recessive**, abnormal allele from each parent, and therefore be **homozygous** for the abnormal allele and develop sickle cell disease.

Homozygous forms of hemoglobinopathy can be very serious. Some cause so much damage that the fetus dies before birth, while others require lifelong treatment.

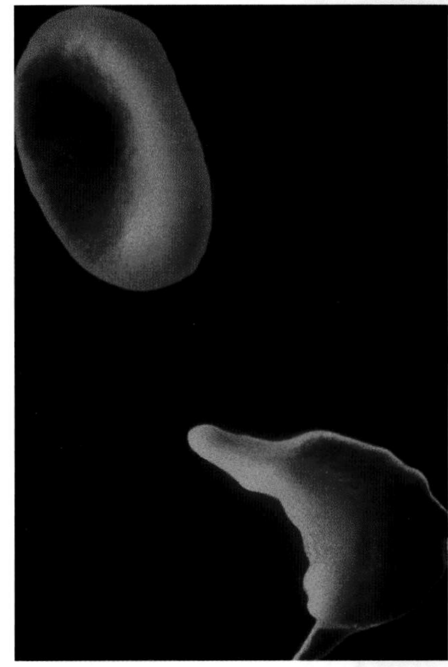

Hbs (sickling hemoglobin) causes red blood cells to take on a sickle shape under conditions of low oxygen.

homologous similar in structure

alleles particular forms of genes

gamete reproductive cell, such as sperm or egg

recessive requiring the presence of two alleles to control the phenotype

homozygous containing two identical copies of a particular gene

Sickle Cell Disease

Sickle cell disease is the most prevalent genetically based disease in the United States. Approximately 1 in 12 Americans of African descent are carriers, having one allele coding for HbS and one gene for HbA. About 1 in 375 Americans of African descent are homozygous for HbS and have the active disease. High occurrence of the HbS allele also occurs in people who live, or whose ancestors lived, in certain parts of Asia, the Mediterranean, and the Middle East.

The alpha chain gene is found on chromosome 11. Each gene is made up of a very long strand of nucleotides. In sickle cell disease, there is a change in only one **nucleotide** in the sequence that codes for the beta chain: A thymine is substituted for an adenine.

nucleotide the building block of RNA or DNA

Genes code for proteins. Because of that change in one nucleotide, a slightly different protein is produced. HbS differs from HbA by only one amino acid: Glutamic acid in HbA is replaced by valine in the sixth position on the beta chain. The substitution does not affect the hemoglobin molecule's ability to bind with oxygen. HbS can carry oxygen just as effectively as HbA. However, glutamic acid is a hydrophilic ("water-loving") amino acid, whereas valine is hydrophobic ("water-hating"). The valine occurs on the outside of the beta chain. The hydrophobic portions of HbS molecules are attracted to each other. When the concentration of oxygen is

low, as it is deep in the body's tissues, HbS molecules will attach to each other. Since a single red blood cell contains about 250 million hemoglobin molecules, this can result in very long chains, or **polymers**.

polymers molecules composed of many similar parts

polymerization linking together of similar parts to form a polymer

The **polymerization** that occurs distorts the red blood cell into a curved, sickle shape. Whereas normal erythrocytes travel smoothly through the blood vessels, these unusually elongated and pointed erythrocytes move much more slowly and can block smaller blood vessels. Both the slow movement and the blockages further reduce the amount of oxygen in the blood, promoting even more polymerization and sickling.

The decreased amount of oxygen in the blood also damages local tissues and will cause permanent damage if it lasts long enough. The lack of oxygen is very painful. This progressive cycle of worsening symptoms, called a vaso-occlusive crisis, can last for more than a week.

People with sickle cell disease often develop other health problems. For example, the crescent shaped erythrocytes have shorter life spans than normally shaped cells do. A healthy red blood cell lives about 120 days, while a sickle cell lives only for 10 to 30 days. The body is unable to replace the red blood cells quickly enough, resulting in anemia.

Situations that cause the body to use up oxygen, such as exercise, can precipitate a vaso-occlusive crisis. Also, because dehydration causes the hemoglobin molecules to be packed more tightly together within the erythrocyte, insufficient fluid intake can also cause red blood cells to sickle.

Treatment Options and Continuing Research

Therapy for sickle cell disease used to focus on easing symptoms and treating infections, which are the most common cause of death in children who have this disease. Newer therapies actually treat the disease.

Hydroxyurea and erythropoietin, for example, are two medications that stimulate the bone marrow to produce more fetal hemoglobin, HbF. Production of both red blood cells and hemoglobin occurs in this spongy tissue, which is located in certain bones.

Fetal hemoglobin can transport oxygen but does not polymerize, so the red blood cells cannot sickle. Thus these drugs can prevent vaso-occlusive crises. However, they do have side effects that can limit their usefulness. Hydroxyurea, for example, can suppress bone marrow function.

Normally, the production of HbF is turned off shortly after birth. Scientists are trying to determine how to reactivate the gene for HbF so that the bone marrow of people with sickle cell disease can continually produce fetal hemoglobin without the use of medications. Other research focuses on learning how to insert normal beta chains and regulatory genes into **stem cells**, which are cells that develop into erythrocytes.

stem cells cells capable of differentiating into multiple other cell types

Bone marrow transplants are a new treatment and have largely been conducted in Europe. The donor bone marrow will produce normal hemoglobin and normal red blood cells. However, the tissue must come from an immunologically compatible donor. Also, a bone marrow transplant is a complicated process, and some people have died during the procedure.

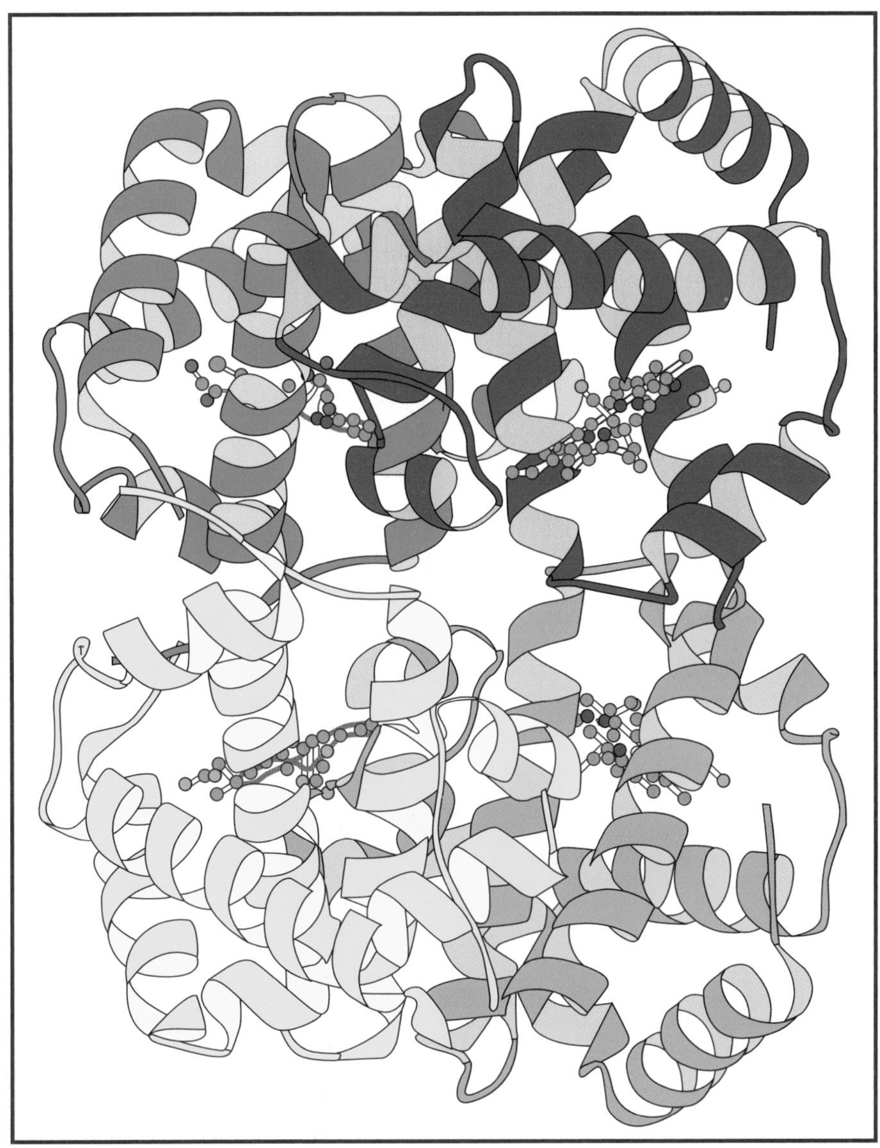

A ribbon diagram of hemoglobin. The four ball-and-stick structures are the heme groups. The helices and loops are the two alpha and two beta globin chains. Adapted from <http://www.ee.seikei.ac.jp/user/seiichi/lecture/Biomedical/02/graphics/hemoglobin.gif>.

Hemoglobin C Disease

Hemoglobin C (Hbc), which is also found in people of African or Mediterranean descent, is very similar in structure to HbS. Both are caused by a change in the sixth residue of the beta chain. While valine replaces glutamic acid to form HbS, the amino acid lysine is found in this position, in HbC.

The substitution of lysine does not cause **pathological** changes in the hemoglobin molecule. People who are homozygous for HbC typically have red blood cells that appear unusual, but they do not sickle. These individuals have no symptoms, and they do not require treatment.

pathological disease-causing

Some people have one gene for HbC and another for HbS. They have hemoglobin SC disease, which usually is much less severe than sickle cell disease.

The Thalassemias

The thalassemias are a group of hemoglobinopathies that, like sickle cell disease, are caused by a genetic change. Unlike sickle cell disease, however,

the genetic change does not result in the production of an abnormal form of the globin molecule. Instead, the bone marrow synthesizes insufficient amounts of a hemoglobin chain. This, in turn, reduces the production of red blood cells and causes anemia.

Either the alpha or beta chain may be affected, but beta thalassemias are more common. Individuals who are heterozygous for this disorder have one allele for this disease and one normal allele and are said to have thalassemia minor. They usually produce sufficient beta globin so that they have only mild anemia. They may not have any symptoms at all. Thalassemia minor is sometimes misdiagnosed as iron deficiency anemia.

If two individuals with thalassemia minor have children, there is a one-in-four chance that each child will inherit an abnormal gene from both parents and will be homozygous for the disorder.

Individuals who are homozygous for this condition may develop either thalassemia intermedia or thalassemia major. Newborn babies are healthy because their bodies are still producing HbF, which does not have beta chains. During the first few months of life, the bone marrow switches to producing HbA, and symptoms start to appear.

In thalassemia major, also called Cooley's anemia, the bone marrow does not synthesize beta globin at all. Children affected by thalassemia major become very anemic and require frequent blood transfusions. They are so ill that they often die by early adulthood.

In thalassemia intermedia, the production of beta globin is decreased, but not completely. People with this disease have anemia, but they do not require chronic blood transfusions to stay alive.

Alpha thalassemia is more complicated, because an individual inherits two alpha globin genes from each parent for a total of four alpha globin genes. Thus a person can inherit anywhere from zero to four normal genes.

The more abnormal alpha genes that are inherited, the greater the symptoms. If an individual does not have any functional alpha genes, the body cannot produce any alpha globin. Since HbF requires alpha chains, the developing fetus does not produce healthy hemoglobin and shows severe symptoms even before birth. This condition is almost always fatal, with affected infants dying either before or shortly after delivery.

The loss of three functional alpha genes produces severe anemia, the loss of two functional genes typically causes mild anemia, and the loss of only one gene usually does not produce any symptoms. The thalassemias most commonly occur in people from Italy, Greece, the Middle East, Africa, and Southeast Asia; and in their descendants.

Treatment Options and Continuing Research

Blood transfusions have been a common therapy for severe thalassemia, but transfusions do not cure the disease, and frequent transfusions can cause iron overload, an illness caused by excessively high levels of iron. A drug, called an iron chelator, may be given to bind with the excess iron. Iron chelators can produce additional side effects, such as hearing loss and reduced growth.

As with sickle cell disease, gene therapy and bone marrow transplants are very promising therapies for severe thalassemias. While transplants are

risky procedures and can cause death, they are more likely to be successful when performed on young and relatively healthy children.

The Heterozygous Advantage

Being homozygous for either sickle cell disease or thalassemia can result in serious illness, but being heterozygous for either condition may actually be beneficial under certain circumstances. Both diseases occur primarily in people who live, or whose ancestors lived, in parts of the world where malaria occurs.

Malaria is spread by a mosquito, but it is caused by plasmodia, single-celled organisms that, during an infection, reproduce inside red blood cells. Before the development of modern sanitation and medicine, malaria was a common cause of death. But people who had either the sickle cell trait or thalassemia minor—people who were heterozygous for either condition—were much more likely to survive an infection than were people homozygous for HbA.

This "heterozygote advantage" meant that these individuals tended to live longer, have children and pass their genes on to the next generation. While some of their children died from thalassemia or sickle cell disease, about half of them were heterozygous and benefited from the heterozygote advantage. This survival advantage explains the high prevalence of these alleles in these populations.

Today, especially in developed countries, there are effective methods for preventing and treating malaria. Nevertheless, the genes for sickle cell disease and thalassemia still exist and are passed down to children who will never be exposed to malaria. It is likely that these genes will very slowly be lost from the gene pool. SEE ALSO GENOTYPE AND PHENOTYPE; HETEROZYGOTE ADVANTAGE; INHERITANCE PATTERNS; MUTATION; POPULATION SCREENING; PROTEINS; TRANSPLANTATION.

Sue Wallace

Bibliography

Weatherall, D. J. "ABC of Clinical Haematology: The Hereditary Anaemias." *British Medical Journal* 314 (1997): 492–496.

Internet Resources

"Bioelectronics Laboratory Index." Seikei University. <http://www.ee.seikei.ac.jp/user/seiichi/lecture/Biomedical/02/graphics/hemoglobin.gif>.

Facts about Sickle Cell Disease. National Heart, Lung, and Blood Institute. <http://www.nhlbi.nih.gov/health/public/blood/sickle/sca_fact.txt>.

Joint Center for Sickle Cell and Thalassemic Disorders. <http://sickle.bwh.harvard.edu>.

Hemophilia

Hemophilia A and hemophilia B are genetic disorders in the blood-clotting system, characterized by bleeding into joints and soft tissues, and by excessive bleeding into any site experiencing trauma or undergoing surgery. Hemophilia A and B are clinically indistinguishable. Both have the same type of bleeding manifestations, and both affect males almost exclusively.

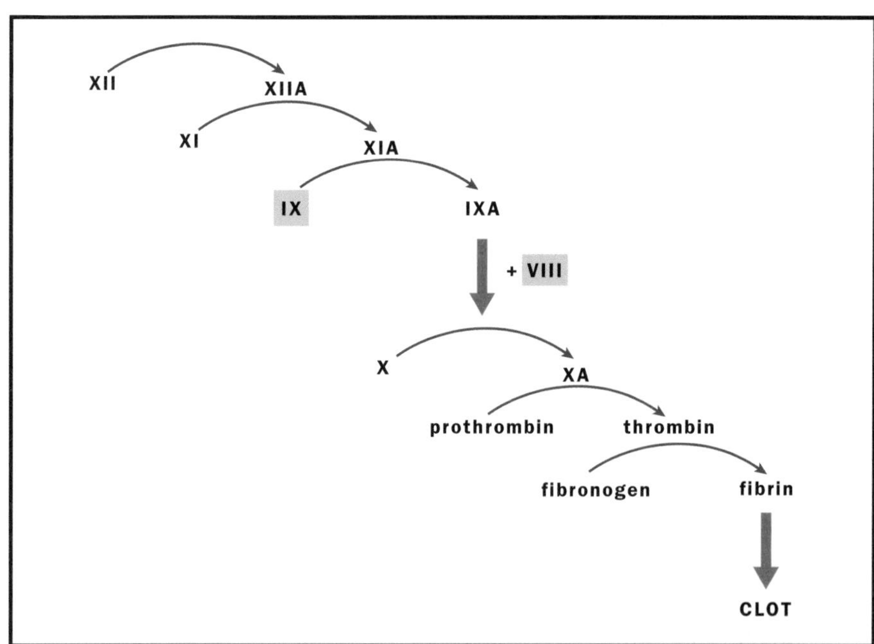

The coagulation cascade involves multistep conversion of several factors into their active (A) forms. In hemophilia, factors VIII or IX are missing or present in insufficient amounts.

The two conditions can be distinguished by detecting the responsible defective proteins.

Contrary to popular belief, individuals with hemophilia rarely bleed excessively from minor cuts or scratches. Hemophilia A and B are both worldwide in distribution, affecting all racial and ethnic groups. The prevalence of hemophilia A is approximately 1 in 10,000 males, and that of hemophilia B is 1 in 30,000 males.

The disorder was recognized (although not named) in Babylonian times. A second-century Jewish rabbi gave permission for a woman not to have her third son circumcised after her first two sons died from the procedure. No doubt the most famous carrier of hemophilia was the nineteenth-century Queen Victoria, whose son, Prince Leopold, had the disorder, and whose two daughters inherited and passed on the gene. Ultimately several of the European royal families were affected by Queen Victoria's gene.

There are a number of clotting factors that interact to form a stable blood clot following injury or surgery. Each factor has a name and is produced in certain cells (usually in the liver), encoded by a certain gene. Hemophilia is caused by a defect in one of these genes. In the case of hemophilia A, factor VIII (FVIII) is deficient or absent. In hemophilia B, FIX is deficient in amount or absent, or it does not function properly.

The genes encoding FVIII and FIX are on the long arm of the X chromosome. Because males have only one X chromosome, hemophilia A and hemophilia B affect males almost exclusively. If a boy's X chromosome has the defective gene that causes hemophilia A or B, the boy will have hemophilia.

Since boys with hemophilia have only an X chromosome carrying the defective gene that causes hemophilia A or B, and no gene for the production of normal FVIII (or FIX), and since fathers pass an X chromosome on to their daughters, all of their daughters will be "obligate carriers" for hemophilia. Obligate carriers—individuals who are definitely carriers—include daughters of men with hemophilia and women who have a maternal family history of hemophilia and one or more affected sons or grandsons.

In general, such carrier females do not bleed excessively, as their other X chromosome, with a gene for normal FVIII (or FIX) production, results in intermediate levels of FVIII (or FIX). However, carrier females have a fifty-fifty chance of passing their X chromosome that bears the hemophilia gene to each child they might have. A son would have an equal chance of being normal or having hemophilia. A girl would have an equal chance of being normal or being a carrier, like her mother.

Gene Defects Causing Hemophilia

Hundreds of defects in the FVIII gene have been shown to cause hemophilia A. These include deletions of varying sizes in the gene, **stop codons**, frameshift mutations, and point mutations. Inversion of the gene is the most common mutation. The same types of defects are found in the FIX gene. This makes population screening for hemophilia impractical: There are too many possible mutations to screen for. However, affected members of a given family will all have the same defect.

stop codons RNA triplet that halts protein synthesis

It is useful to determine (by gene analysis) which defect is present in the FVIII or FIX gene of a particular family with hemophilia, so that one can look for this defect in possible carrier females. Identification of carrier females permits genetic counseling and decision making, on the part of parents, regarding childbearing.

One form of hemophilia is due to the insertion of a transposable genetic element. (DNA sequence that can be copied and moved in the genome).

Detection of FVIII or FIX Gene Defect in Family: Carrier Detection

A number of different techniques are available for carrier detection. Linkage analysis using DNA **polymorphisms** to track defective FVIII or FIX genes is possible in large families. Use of intragenic polymorphisms in both the FVIII and FIX genes allows precise detection of carrier females in most families studied. In persons with hemophilia A, the FVIII gene inversion can be tested for by **Southern blotting**.

polymorphisms occurring in several forms

Southern blotting separating DNA fragments by electrophoresis and then identifying a target with a DNA probe

If the gene defect in the family is not known, the inversion mutation in an affected male is generally sought first, as this mutation accounts for at least 20 percent of all cases of hemophilia A, and the test is relatively simple to perform. Although more time-consuming, point-mutation screening also can be done, using a variety of methods. For researchers working on the FVIII or FIX genes, there are sites on the Internet that are valuable resources, as they contain regularly updated listings of all reported mutations in each of these genes, and other useful information.

Because of a high mutation rate, approximately one-third of infants found to have hemophilia A or B have no family history of the disorder, the condition having occurred spontaneously. However, hemophilia is genetically transmitted to future generations.

Differences in Severity of Hemophilia

There are different degrees of severity of hemophilia A and B. Clinical severity usually correlates with the individual's circulating FVIII or FIX level

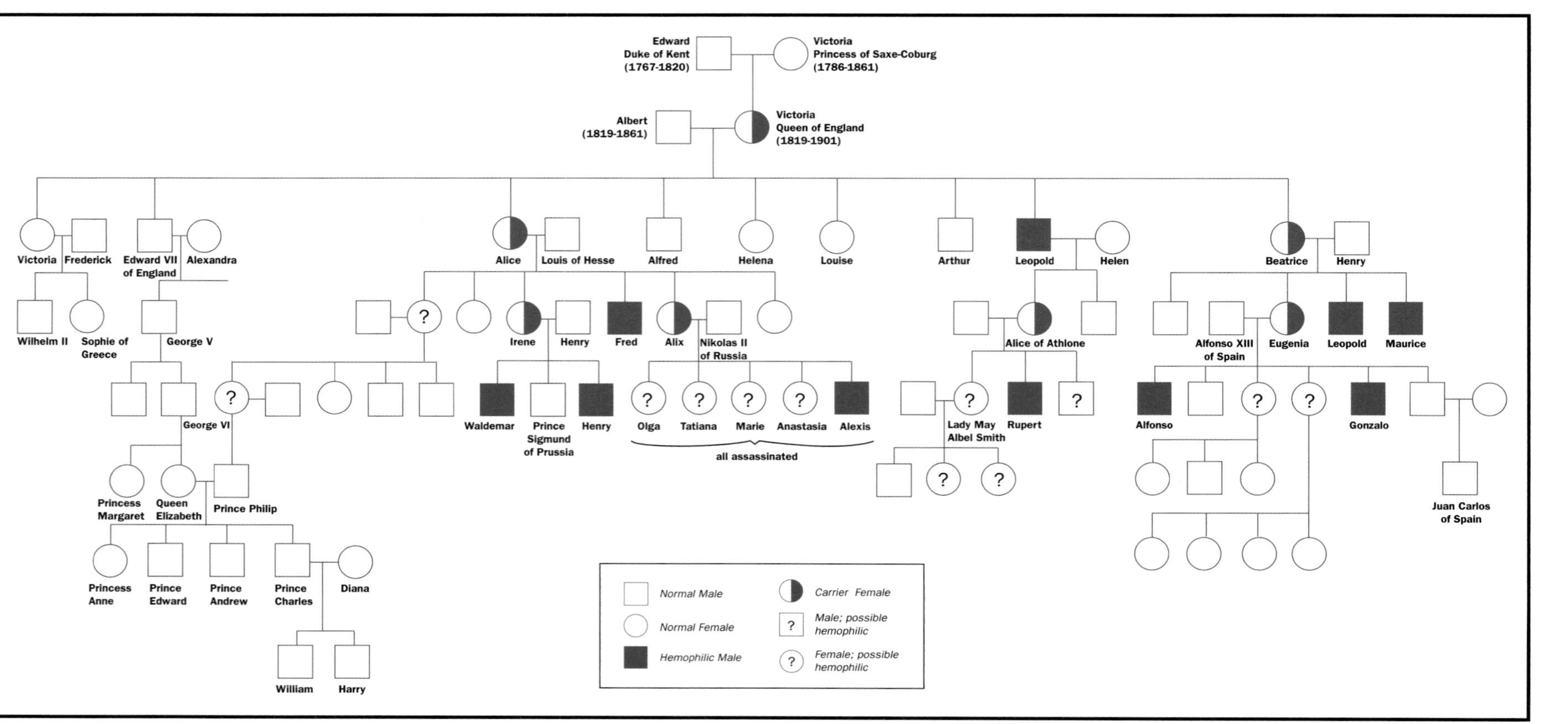

Pedigree of hemophilia in the royal families of Europe. Queen Victoria was a carrier of this X-linked trait. Her son Leopold inherited the X chromosome that carried it, while her son Edward VII did not. Adapted from http://ublib.buffalo.edu/libraries/projects/cases/hemo.htm.

(determined by doing an FVIII or FIX assay on a venous blood sample). The severity is generally the same in all affected members of a family.

Normal values for FVIII and FIX can range between 50 and 150 percent of the mean value, while severely affected individuals generally have levels of less than or equal to 1 percent, moderately affected persons 1 to 5 percent, and mildly affected persons 5 to 35 percent. Severely affected individuals often have spontaneous joint and muscle hemorrhages, whereas mildly affected persons bleed only with trauma or surgery.

Treatment

Treatment for bleeding episodes consists of replacing the missing clotting factor by intravenous infusion of FVIII (or FIX). There are both human plasma-derived FVIII and FIX concentrates and recombinant DNA-derived FVIII and FIX concentrates. The useful life for both proteins (FVIII and FIX) once infused is relatively short (on average, half is degraded within twelve hours for FVIII and within eighteen hours for FIX). Thus for serious bleeding episodes or surgery, frequent repeat dosing (or continuous infusion) is often necessary.

Prophylaxis is also used, particularly in persons with severe hemophilia A or B. This consists of giving FVIII three times weekly, and FIX twice weekly (in view of its longer half-life, once infused). The aim of prophylaxis (which is often begun between age one and three) is to prevent joint bleeding (and the resulting increase in joint destruction and disability).

Persons with mild hemophilia A can often be treated with the synthetic agent DDAVP (1-deamino-8-D-arginine vasopressin). This analogue of the naturally occurring **antidiuretic** hormone vasopressin results in a rapid release of whatever FVIII (and another large plasma glycoprotein, von Willebrand factor) is in the individual's body storage sites. Thus, following intravenously administered DDAVP, FVIII (and von Willebrand factor) increase (two- to tenfold), but then fall back to baseline within approximately twelve to fifteen hours. This drug comes in several formulations, for **intravenous**, **subcutaneous**, and intranasal use.

antidiuretic a substance that prevents water loss

intravenous into a vein

subcutaneous under the skin

It is hoped that gene therapy for persons with severe hemophilia may eventually become a realistic option. In early 2002 there were several ongoing phase-one trials (very early research studies) in human subjects with severe hemophilia, using different **vectors** and different techniques. However, formidable challenges remain. SEE ALSO BLOTTING; DISEASE, GENETICS OF; GENETIC COUNSELING; INHERITANCE PATTERNS; MUTATION; TRANSPOSABLE GENETIC ELEMENTS; X CHROMOSOME.

vectors carriers

Jeanne M. Lusher

Bibliography

Lakich, Delis, et al. "Inversions Disrupting the Factor VIII Gene as a Common Cause of Severe Haemophilia A." *Nature Genetics* 5 (1993): 236–241.

Lillicrap, David. "The Molecular Basis of Haemophilia B." *Haemophilia* 4 (1998): 350–357.

Potts, D. M., and W. T. W. Potts. *Queen Victoria's Gene: Haemophilia and the Royal Family.* Gloucestershire, U.K.: Sutton Publishing, 1995.

Internet Resources

Haemophilia B Mutation Database. King's College London. <http://www.kcl.ac.uk/ip/petergreen/haemBdatabase.html>.

Hemophilia: The Royal Disease. University at Buffalo, SUNY. <http://ublib.buffalo.edu/libraries/projects/cases/hemo.htm>.

National Hemophilia Foundation. <http://www.hemophilia.org>.

Heterozygote Advantage

alleles particular forms of genes

Heterozygote advantage is the superior fitness often seen in hybrids, the cross between two dissimilar parents. A heterozygote is an organism with two different **alleles**, one donated from each parent. Fitness means the ability to survive and have offspring. Heterozygote advantage also refers more narrowly to superior fitness of an organism that is heterozygous for a particular gene, usually one governing a disease.

Inbreeding is the practice of repeatedly crossing a single variety of organism with itself, in order to develop a more uniform variety. During this process, the organism becomes homozygous for many genes, meaning that its two gene copies are identical. This is often accompanied by loss of vigor: slower growth, less resistance to disease, and other signs of decreased fitness. This is known as inbreeding depression. Breeding with another variety ("outcrossing") produces offspring that are heterozygous for many genes, and is often accompanied by an increase in size and vigor. This phenomenon had been known to farmers and plant breeders for many years, and was given the name "heterosis" by the U.S. geneticist George Shull in 1916. It is also known as hybrid vigor or, more commonly, as heterozygote advantage.

Agricultural Significance

Hybrid vigor has had a profound impact on agriculture. Yields of hybrid corn are much greater than those of nonhybridized, "open-pollinated" varieties. This of course means that the hybrid "seed corn" must be produced each generation by crossing two distinct, inbred lines. This in turn has given employment to several generations of teenagers who, in summer, detassel corn plants, that is, remove the pollen from one of the parental plants to prevent self-pollination and assure cross-hybridization with the desired variety.

In domesticated animals used for meat or milk production, selection of the best producers to produce high-yielding offspring has been in effect for centuries and has produced dramatic results. In this case, however, essentially all crossing is performed within the same breed and usually involves inbreeding. No serious attempts are made to outcross among different breeds, which would take advantage of heterozygote advantage. Early experiments suggested that outcrossing does not yield favorable results and thus is avoided to this day.

Hypotheses of Heterozygote Advantage

There are two plausible explanations for heterozygote advantage and inbreeding depression, which may both act in a single organism on different genes. No single hypothesis is likely to explain all cases of heterozygote superiority.

Poppy hybrid flowers are larger and more elaborately colored than the parental types. (*Eschschoizia* crossed with *Hailigain*).

The first hypothesis is known as the favorable dominance hypothesis. It is based on the fact that recessive alleles are very often **deleterious** in the homozygous condition, often because recessives code for a defective form of the protein. Thus, possessing at least one dominant allele is favored. Under this hypothesis, the two inbred parents are each homozygous recessive for one or more (different) traits, and each has decreased fitness. Hybridization creates heterozygote offspring that have a dominant (functional) allele for each trait, thus increasing their fitness or vigor. In this hypothesis, heterozygotes are superior to the homozygous recessive condition, but not the homozygous dominant condition.

deleterious harmful

The other explanation for heterosis is that the heterozygote is superior to both homozygotes. This is usually referred to as the overdominance hypothesis. Hybrid vigor in corn is due to overdominance rather than favorable dominance. Overdominance may help explain why harmful recessive alleles remain in the gene pool. Despite the disadvantage of possessing two copies, possessing one copy is advantageous. The molecular explanation for overdominance is less simple, and may be different for different genes (see discussion below).

Heterozygote Superiority in Humans

Is hybrid vigor a common phenomenon in humans? Do traits such as birth weight, height, and other factors improve with outcrossing? In a large, well-

designed study of interracial crosses in Hawaii, Newton Morton and his colleagues found no significant beneficial effects among the offspring of an interracial cross when compared to offspring whose parents were from the same racial group. Other studies have found some beneficial effects of outbreeding but, in general, these studies tended to be flawed and are thus unreliable.

In a small number of cases, humans do show a heterozygote advantage, in which the fitness of the heterozygote is superior to either homozygote. The best known example is the β-hemoglobin locus and its relationship to sickle cell disease. Adult hemoglobin is composed of four polypeptides: two α chains and two β chains, coded for by different genes. The β chain is a sequence of 146 amino acids, with glutamic acid in position 6. This normal hemoglobin is referred to as type A. In sickle cell disease, a mutation causes glutamic acid to be replaced by valine at position 6 and is referred to as hemoglobin S. Individuals who are homozygous for the S allele (SS) have sickle cell disease. Untreated, this condition is lethal, and affected individuals do not survive to be old enough to have offspring.

If there were no selective advantage to the recessive allele, we would expect it to slowly be removed from the gene pool, and the frequency of individuals with sickle cell disease should approximate the mutation rate of the normal A allele to the S allele, which is extremely rare. The disease, however, is relatively common in western and central Africa, where the frequency of S allele can be over 15 percent. The reason for this high frequency is due to the heterozygote AS being resistant to the malarial parasite *Plasmodium falciparum.* Thus in areas where malaria is common, AS individuals, who possess one sickling allele, have an advantage over AA individuals, who possess none. Copies of the S allele are lost from the gene pool when they occur in individuals affected with sickle cell disease since they do not reproduce, but more copies are created in the offspring of AS individuals since they are resistant to a severe parasitic disease, namely malaria. This advantage compensates for the loss of individuals with sickle cell disease, and therefore keeps the S gene at a relatively high frequency in western and central African populations. Other hemoglobin abnormalities, including hemoglobin C and E, also seem to have the same effect as sickle cell disease, though the effect is not as pronounced.

Thalassemia is another hemoglobin disorder that causes severe anemia. There are two types, α and β, which are due to mutations at the α and β hemoglobin loci respectively. The disease has its highest frequency in areas bordered by the Mediterranean Sea, especially Sardinia, Greece, Cyprus, and Israel. Similar to sickle cell disease, the disorder is fatal at an early age and the heterozygotes are resistant to malaria.

The same is true for a deficiency of the enzyme glucose-6-phosphate dehydrogenase (G6PD) deficiency, which has a similar geographic distribution as thalassemia. The disease is X-linked, thus affecting boys. Its persistence in this case is due to heterozygous females being resistant to malaria. Other proposed cases of heterozygote superiority in humans are more speculative, and have not been confirmed by repeated studies. SEE ALSO HARDY-WEINBERG EQUILIBRIUM; HEMOGLOBINOPATHIES; INHERITANCE PATTERNS; POPULATION GENETICS.

P. Michael Conneally

Bibliography

Cavalli-Sforza, L. L., and W. F. Bodmer. *The Genetics of Human Populations.* Mineola, NY: Dover Publications, 1999.

Hartl, D. L., and A. G. Clark. *Principles of Population Genetics*, 3rd ed. Stamford, CT: Sinauer, 1997.

Li, Ching Chun. *First Course in Population Genetics.* Pacific Grove, CA: Boxwood Press, 1976.

Morton, Newton E., Chin S. Chung, and Ming-Pi Mi. *Genetics of Interracial Crosses in Hawaii.* New York: S. Karger, 1967.

High-Pressure Liquid Chromatography *See HPLC: High-Performance Liquid Chromatography*

High-Throughput Screening

High-throughput screening (HTS) is an automated method for rapidly analyzing the activity of thousands of chemical compounds. It has become a key tool in modern drug discovery. Paired with combinatorial chemistry and **bioinformatics**, HTS allows potential drugs to be quickly and efficiently screened to find candidates that should be explored in more detail.

bioinformatics use of information technology to analyze biological data

How the Process Works

Most drugs work by binding to a protein target on or in a living cell. One of the first steps in drug discovery and development is finding molecules that will bind to the target. Imagine, for instance, you want to develop an anticancer drug that binds to and inactivates a particular mutant protein known to promote aberrant cell growth. You have a couple of compounds that bind very weakly to your protein, and these serve as the starting point for generating a large number of related compounds, through combinatorial chemistry. In this method, many thousands of related compounds can be quickly and automatically synthesized. Those that bind best can be modified and tested further, and ultimately may go on to be tested in animals and people as candidate drug therapies.

Initial screening of these compounds for their binding ability is the job for HTS. The key to HTS is to develop a test, or assay, in which binding between a compound and a protein causes some visible change that can be automatically read by a sensor. Typically the change is emission of light by a **fluorophore** in the reaction mixture. One way to make this occur is to attach the fluorophore to the target protein in such a way that its ability to fluoresce is diminished (quenched) when the protein binds to another molecule. A different system measures the difference in a particular property of light (polarization) emitted by bound versus unbound fluorophores. Bound fluorophores are more highly polarized, and this can be detected by sensors. Other detection methods are possible as well.

fluorophore fluorescent molecule

The details of HTS differ with different systems, but all depend on automated or robotic systems to combine the chemicals and read the outputs. Reactions between the target protein and the compound usually occur in

microplates, which are plastic trays with multiple indentations, or wells. Systems currently in use can handle plates with 96, 384, 1,536, or even higher numbers of wells at once. HTS typically uses extremely small volumes in each well, often 10 microliters or less. Small volumes have numerous advantages, including keeping to a minimum the amount of each compound used. This is especially important for many proteins targets, which may be difficult and costly to isolate and purify.

The time required for reactions varies with the substances involved, and may range from several minutes to several hours. Fast robotic systems combined with rapid reactions can screen 10,000 or more compounds in a single day. This is an enormous increase over traditional chemical assays, in which a chemist may be able to handle fewer than 100 tests in the same amount of time.

The Uses of HTS Assays

Storing, processing, analyzing, and accessing the wealth of data generated in an HTS assay poses special problems, simply because there is so much of it. Bioinformatics strategies are used to develop databases relating chemical structure, target characteristics, and assay results, allowing researchers to learn more from their results than just whether or not a particular compound was successful. Analyzing the common features of successful compounds may lead to rational development of better drug candidates.

genomics study of gene sequences

High-throughput technology can also be put to use in other areas besides drug development. Indeed, any system in which there are many similar candidates to be screened, and in which a visible output can be designed, is amenable to high-throughput methods. **Genomics** applications are a principal area for applying HTS technology, in DNA sequencing, protein analysis, and other fields. HTS methods can be combined with DNA microarray technology, for instance, to analyze the expression of hundreds of different genes under varying conditions. SEE ALSO BIOINFORMATICS; BIOTECHNOLOGY; COMBINATORIAL CHEMISTRY; DNA MICROARRAYS; PROTEOMICS.

Richard Robinson

Bibliography

Brush, Michael. "High-Throughput Technology Picks Up Steam." *Scientist* 13, no. 4 (February 15, 1999): 11.

Internet Resource

High Throughput Screening. <http://www.htscreening.net/>.

HIV

HIV, the human immunodeficiency virus, is the virus that causes AIDS, a debilitating and deadly disease of the human immune system. HIV is one of the world's most serious health problems: at the end of 2001, more than 40 million people worldwide were infected with HIV and living with the virus or AIDS. The World Health Organization estimates that about 20 million people have died from AIDS since the infection was first described in 1981. Nearly 500,000 of those deaths have occurred in the United States. Although there is no cure for the disease, therapies exist that reduce the

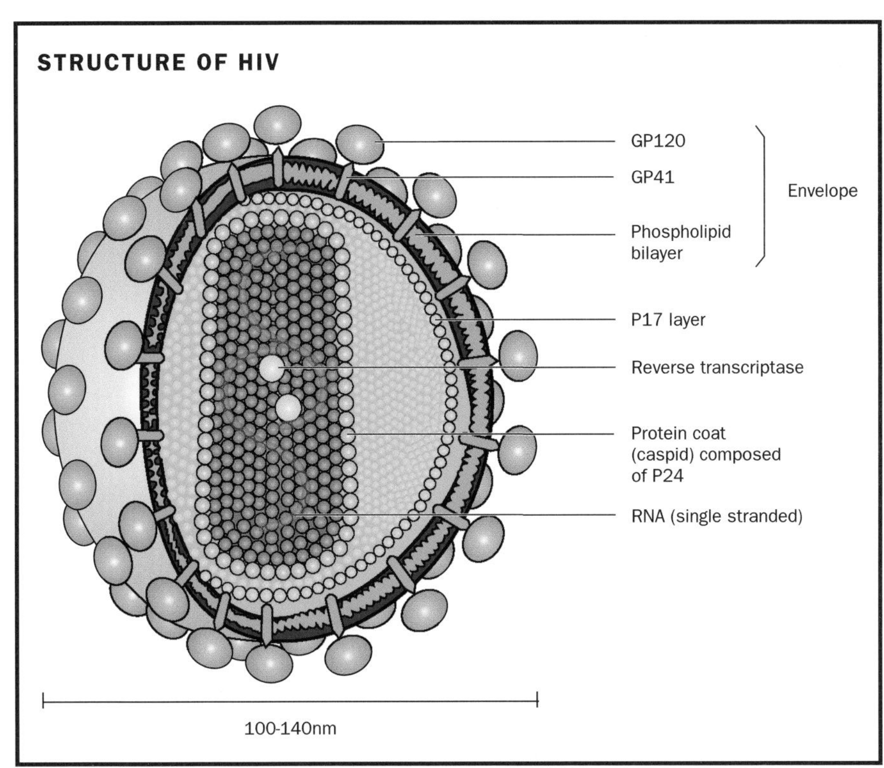

The HIV virion (virus particle) has two glycoproteins on the outside, which aid its passage into human cells. It contains two copies of its RNA genome, plus the reverse transcriptase enzyme used to copy the RNA into DNA once it is inside the host cell.

symptoms of AIDS and can extend the life spans of HIV-infected individuals. Researchers are also pursuing protective vaccines, but a reliable vaccine might still require years to develop.

HIV and AIDS

HIV infects certain cells and tissues of the human immune system and takes them out of commission, rendering a person susceptible to a variety of infections and cancers. These infections are caused by so-called opportunistic agents, **pathogens** that take advantage of the compromised immune system but that would be unable to cause infection in people with a healthy immune system. Rare cancers such as Kaposi's sarcoma also take hold in HIV-infected individuals. The collection of diseases that arise because of HIV infection is called acquired immune deficiency syndrome, or AIDS. HIV is classified as a lentivirus ("lenti" means "slow") because the virus takes a long time to produce symptoms in an infected individual.

pathogens disease-causing organisms

HIV Life Cycle: Entering Cells

Like a typical virus, HIV infects a cell and appropriates the host's cellular components and machinery to make many copies of itself. The new viruses then break out of the cell and infect other cells. HIV stores its genetic information on an RNA molecule rather than a DNA chromosome. This is a distinguishing characteristic of **retroviruses**, which are viruses that must first convert their RNA genomes into DNA before they can reproduce.

retroviruses RNA-containing viruses whose genomes are copied into DNA by the enzyme reverse transcriptase

Each HIV virion (viral particle) is a small sphere composed of several layers. The external layer is a membrane coat, or envelope, obtained from the host cell in which the particle was made. Underneath this membrane lies

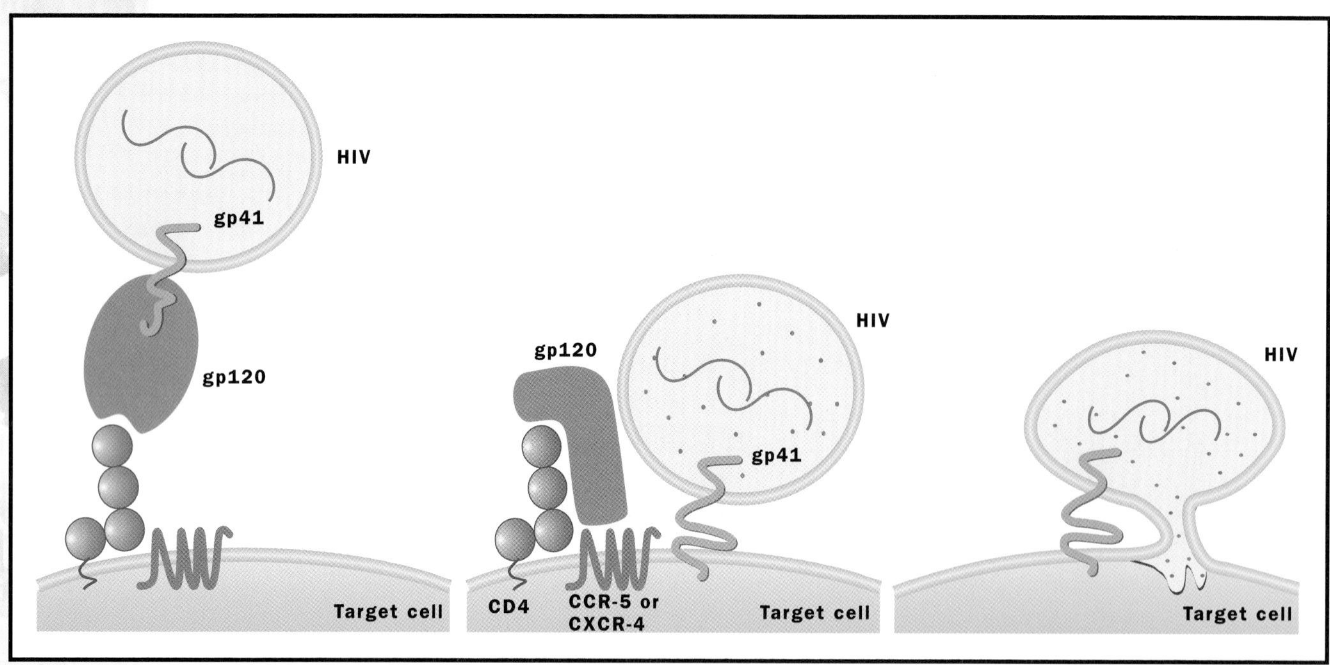

HIV's gp120 protein interacts with two proteins on the surface of the target cell. Gp41 then promotes fusion of the virus membrane with the cell membrane, allowing the viral contents to enter the cell. Adapted from an article in *Molecular Medicine Today,* 1998.

antibodies immune-system proteins that bind to foreign molecules

a shell made from proteins, called a nucleocapsid. Inside the protein shell are two copies of the virion's RNA genome and three kinds of proteins, which are used by the virion to establish itself once inside the cell that it infects.

Two proteins, called gp120 and gp41, enable the virion to recognize the type of cell to enter. These proteins project from the HIV membrane coat. Gp120 binds to two specific proteins found on the target cell's surface (these target-cell proteins are called receptors). The first receptor, CD4, is found on immune system cells known as CD4 T cells, and also sometimes on two cell types known as macrophages and dendritic cells. The immune system uses CD4 T cells in the initial step in making **antibodies** against infectious agents. After binding to CD4, the HIV protein called gp120 binds with a second cell membrane protein, commonly referred to as the co-receptor. The co-receptor can be one of many different proteins, depending on the cell type. The two most common are CXCR4, which is normally found on CD4 T cells, and CCR5, a receptor found on CD4 T cells as well as on certain macrophages and dendritic cells. In the absence of HIV, CXCR4 and CCR5 allow these immune system cells to respond to chemical signals, but when HIV infects the cells, the HIV commandeers their usage. In some cases, individuals have a mutation in their co-receptor that prevents HIV from entering their cells.

Once gp120 has bound to both the CD4 receptor and co-receptor, the gp41 protein fuses HIV's membrane envelope with the cellular membrane, injecting the virus into the target cell. Once in the cytoplasm, the viral protein shell opens up and releases the viral proteins—a reverse transcriptase, a viral integrase, and a protease—along with the viral RNA strands. The reverse transcriptase copies the RNA strands into DNA. The viral integrase then helps insert the DNA copies into the cell's chromosome. At this point, the virus is called a provirus, and the life cycle halts. The provirus may remain dormant in the cell's chromosome for months or years, waiting for the T cell to become activated by the immune system.

HIV Life Cycle: Reproduction

When the immune system recruits T cells to fight an infection, the T cells start producing many proteins. Along with the normal cellular protein products, a T cell carrying an HIV provirus also produces HIV proteins. The first HIV proteins made are called Tat and Rev. Tat encourages the cellular machinery to copy HIV's proviral DNA into RNA molecules. These RNA molecules are then processed in the nucleus to become templates for several of the HIV proteins, some of whose functions are not well understood.

Rev, on the other hand, ushers the HIV's RNA molecules from the nucleus, where they are being reproduced, into the host cell's **cytoplasm**. Early in HIV reproduction, with only a few RNA molecules from which to make protein, a small quantity of Rev is made. Therefore, most of the RNA molecules remain in the nucleus long enough to get processed. As time passes, however, and Tat continues to instigate RNA production, more Rev is made. A higher amount of Rev protein increases the speed with which RNA molecules are ejected from the nucleus. These RNA molecules, which have undergone little or no processing, become templates to make different HIV proteins. The newer proteins are made in long chains that require trimming before they become functional. One of the proteins in the chain is the protease, the protein that trims. Other proteins include those that make up the protein shell, the reverse transcriptase, and integrase.

cytoplasm the material in a cell, excluding the nucleus

After the newly created proteins are processed to the right size, they form new virions by first assembling into a shell, then drawing in two unprocessed RNA molecules and filling up the remaining space with integrase, protease, and replicase. The new virions bud from the host-cell membrane, appropriating some of that membrane to form an outer coat in the process. The mature virus particles are now ready to infect other cells.

HIV's Immune-System Impairment Mechanism

One of the most disastrous effects of HIV infection is the loss of the immune system's CD4 T cells. These cells are responsible for recognizing foreign invaders to a person's body and initiating antibody production to ward off the infection. Without them, people are susceptible to a variety of diseases. HIV destroys the T cells slowly, sometimes taking a decade to destroy a person's immunity. However, in all the time before an HIV-infected individual shows any symptoms, the virus has been reproducing rapidly. The lymph tissue, the resting place for CD4 T cells, macrophages, and dendritic cells, becomes increasingly full of HIV, and viral particles are also released into the bloodstream.

HIV's main target is the population of CD4 T cells within a host's body. HIV kills them in one of three ways. It kills them directly by reproducing within them, then breaking them upon exit; it kills them indirectly by causing the cells to "commit suicide" by inducing **apoptosis**; or it kills them indirectly by triggering other immune cells to recognize the infected T cell and kill it as part of the immune system's normal function.

apoptosis programmed cell death

As infected T cells die, the immune system generates more to take their place. As new T cells become infected, they are either actively killed or induced to commit suicide. Meanwhile, the HIV virus is not completely

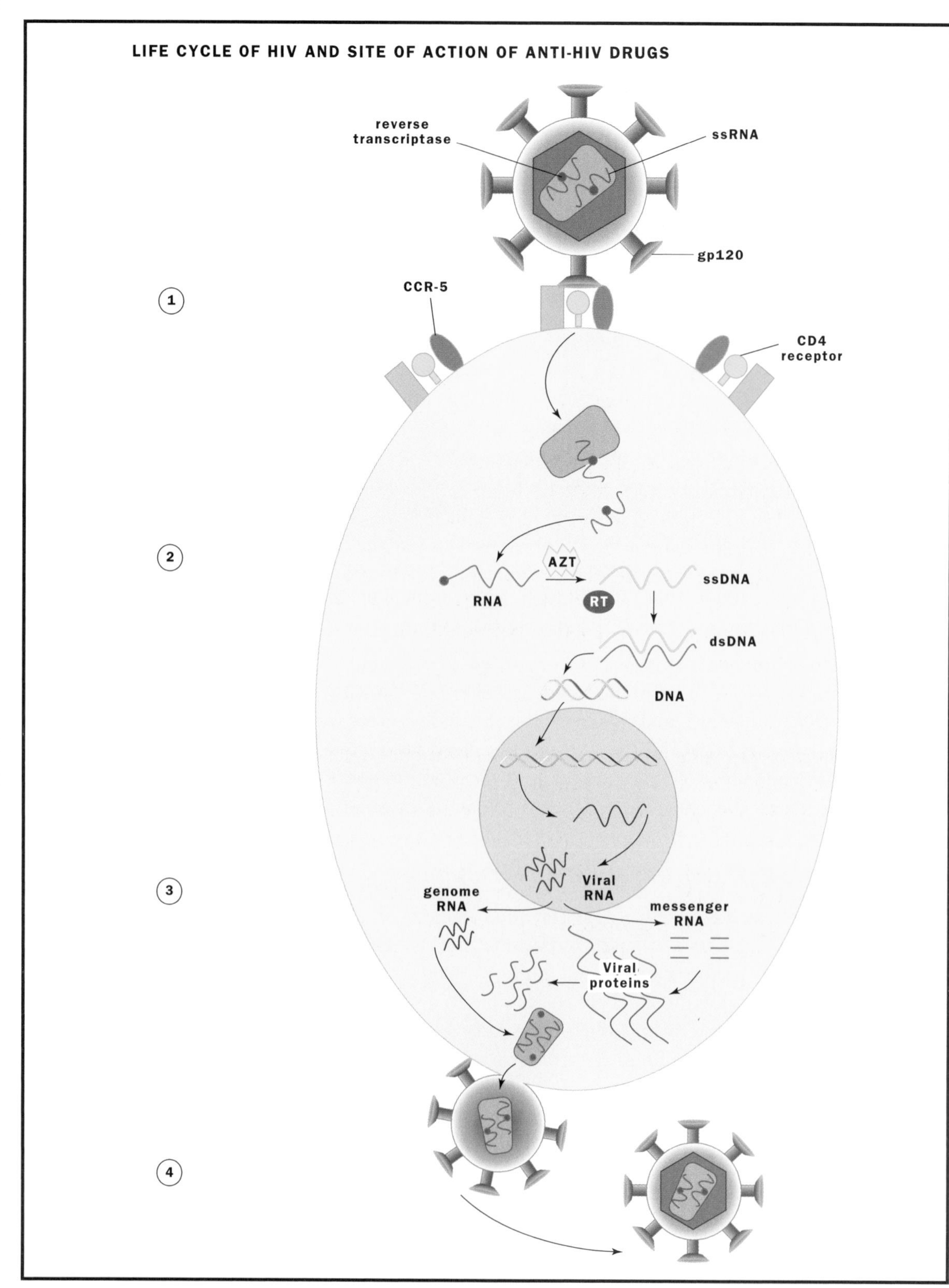

HIV binds to CD4 and CCR-5 or CXCR4 on the surface of the target cell, allowing the virus coat to fuse with the cell membrane and virus contents to enter the cell. (2) Reverse transcriptase (RT) enzyme transcribes viral RNA into single-strand viral DNA, a step blocked by RT inhibitors (AZT, ddi, d4t, ddc). Single-strand DNA is converted to double-strand DNA which then enters the nucleus and integrates into the host chromosome. (3) Upon activation of the cell, viral RNA triggers manufacture of viral proteins. In the absence of protease inhibitors, proteins are processed into shorter lengths. Protease inhibitors (saquinavir, ritonavir, indinavir, nelfinavir) prevent this step, interrupting the viral life cycle. (4) After all components are present in the cell, HIV particles assemble and bud from the cell. Adapted from Stine, 1997.

hidden from the immune system. As with any infectious agent, HIV presents its proteins to the immune system, which develops antibodies against it. This antibody production, however, is hampered by the fact that HIV mutates rapidly, changing the proteins it displays to the immune system. With each new protein, the immune system must generate new antibodies to fight the infection. Thus, an HIV infection is a dramatic balance between a replicating, ever changing virus and the replenishing stores of T cells that are fighting it. Unfortunately, the immune system, without therapeutic intervention, eventually loses the battle.

Once the CD4 T cells are depleted, the immune system can no longer ward off the daily bombardment of pathogens that all human organisms experience. Common infectious agents thus overwhelm the system, and HIV patients become susceptible to a variety of "opportunistic" diseases that take advantage of the body's reduced ability to fight them off. AIDS doctors report at least twenty-six different opportunistic diseases specific to HIV infection. These include unusual fungal infections such as thrush. The chickenpox virus may come out of dormancy, manifesting itself as the painful disease known as shingles. An obscure form of pneumonia, called pneumocystis pneumonia, is also common in AIDS patients. In addition, patients can acquire cancers such as B-cell lymphoma, which is a cancer of the immune system. Doctors generally consider patients with fewer than 200 CD4 T cells per cubic milliliter of blood as having AIDS. (In contrast, a healthy person counts more than 1,000.)

Anti-HIV Drug Therapy

Drugs that interfere with viral replication can slow down HIV disease. Early trials relied on the administration of one drug at a time. While patients' health improved and their T cell count rose, in time HIV mutated enough to render the drugs ineffective. Since 1995, however, doctors have found that rotating patients through three different drugs in very high doses significantly improves the health of AIDS patients. Known as "highly active antiretroviral therapy" (HAART), this therapeutic approach also reduces the amount of HIV circulating in the bloodstream to nearly undetectable levels. People infected with HIV who are treated by HAART are now living longer, healthier lives than ever before.

Targeting Life-Cycle Points

Drugs meant to knock out HIV target the activities of two HIV proteins, the reverse transcriptase and the protease. HAART requires drugs of both types. Drugs called protease inhibitors prevent the viral protease from trimming down the large proteins made late during infection. Without those proteins, the viral shell cannot be assembled. In addition, the proteins that reproduce HIV's genetic information, the reverse transcriptase and the integrase, are not functional.

Drugs that inhibit the reverse transcriptase prevent it from copying the RNA into DNA. These drugs work early in the life cycle of HIV. Reverse transcriptase inhibitors include azidothymidine (AZT), whose structure resembles the DNA nucleotide thymine. When reverse transcriptase builds DNA with AZT instead of thymine, the AZT caps the growing DNA molecule and halts DNA production, due to AZT's slight difference in structure

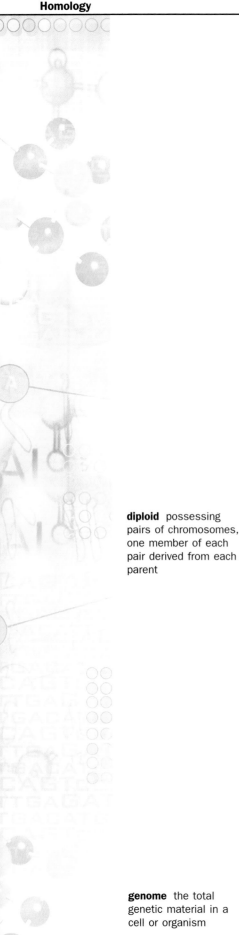

from the thymine that DNA production requires. SEE ALSO IMMUNE SYSTEM GENETICS; RETROVIRUS; REVERSE TRANSCRIPTASE.

Mary Beckman

Bibliography

Janeway, Charles A., et al. "Failures of Host Defense Mechanisms." In *Immunobiology: The Immune System in Health and Disease*, 4th ed. New York: Current Biology Publications, 1999.

———. *HIV Infection and AIDS: An Overview.* Washington DC: National Institute of Allergy and Infectious Diseases and U.S. Department of Health and Human Services, 2001.

Shilts, Randy. *And the Band Played On: Politics, People, and the AIDS Epidemic.* New York: St. Martin's Press, 2000.

Stine, Gerald. *Acquired Immune Deficiency Syndrome: Biological, Medical, Social and Legal Issues*, 3rd ed. New York: Prentice-Hall, 1997.

Homology

Homology is used to describe two things that share a common evolutionary origin. In genetics and molecular biology, homology means that the sequences of two different genes or two different proteins are so similar that they must have been derived from the same ancestral gene or protein.

The word "homology" has several meanings in biology, each related to the word's origin, meaning "same knowledge." At a molecular level, the term "homology" describes sequences, either DNA or protein, that share a common evolutionary origin. On a larger scale, a pair of chromosomes from a **diploid** organism that have the same size and shape, are considered homologous chromosomes. Regions of each member of a chromosome pair, which carry the same set of genes, are homologous regions. Finally, physical features with a common evolutionary origin, such as the wing of bat and the hand of a human, are homologous structures.

diploid possessing pairs of chromosomes, one member of each pair derived from each parent

Diversity and Natural Selection

Biologists have long been fascinated by the diversity of life. The amazing variety of living things makes it natural to wonder how so many different life-forms came to be. Physical characteristics that could be easily observed, such as the shape of wing, the structure of a shell, or the size of a beak, provided the first means to search for an answer. Recognition of the variation within a species (imagine a Chihuahua and a Great Dane) led Charles Darwin to propose that new species emerge when selection favors certain traits within a population.

Today's biologists continue to study the effects of natural selection on the evolution of species, but they are no longer limited to beak size and wing shape. Now they can compare the positions of genes on chromosomes, the amino acid sequences of proteins, and the nucleotide sequences of genes. With DNA or protein sequences from over 133,000 species represented in the taxonomy database at the National Center for Biotechnology Information (NCBI) and over 800 **genome** sequences either published or in progress, researchers have an unprecedented opportunity to study evolution at a molecular level.

genome the total genetic material in a cell or organism

CAAAGCTCTTGCTTTGACAATTTTGGTCTTT---CAGAA-TACTATAAAT	human
CAAAGTTGTAGTCTTGACAATTCTGCTCTTT---ACATA-AAATTGAAGC	rabbit
CAAAATTGTAGCCCTGACAATTTGACTCCTTTTACAAGA-AAATATAGGA	cow

Sequence comparison of the globin gene for human, rabbit, and cow. Dashes represent bases that are not present in the gene. Note the large amount of homology, as well as the highlighted differences.

Homology and Computer Analysis

To study homologous sequences, researchers use computer programs, such as BLAST (Basic Local Alignment Search Tool), to compare a DNA or protein sequence with a collection of other sequences. One such collection is GenBank, the genetic sequence database operated by the NCBI that contains all publicly available DNA and protein sequences. Biologists use databases such as GenBank to find out if a test sequence matches any known sequences, how well it matches, and which portions of the sequence match.

Computer programs identify matching sequences by similarity. However, similar sequences are not always homologous, because they may not have a common origin. Although many sequences that show similarity did evolve from a common ancestor, the appearance of similar sequences can also result from independent events. For example, mutations frequently occur in the gene for the envelope protein of the AIDS virus, HIV-1, changing the amino acid sequence of the protein. The human immune system recognizes and destroys unmutated viruses, while leaving unharmed (selecting for) those viruses that contain mutations that make them unrecognizable. As a result, viruses from different patients can show identical mutations in the envelope protein, even though the patients were infected by different strains of the virus.

Exploring the Mechanisms of Mutation

The ability to compare protein and DNA sequences not only shows us where evolution has occurred but provides insight into its mechanisms. By comparing genomes, we find that mutations can occur on a small scale: Even a single nucleotide change is a mutation. They can also occur on a large scale, as happens when sequences are inserted, deleted, duplicated, or moved between chromosomes.

Many mutations that replace single nucleotides have no effect because of the "degeneracy" or redundancy of the genetic code. The genetic code has more **codons** (sixty-four) than amino acids (twenty). As a consequence, most amino acids are specified by two to four different codons. Because of this, some mutations can be "silent," with one nucleotide replacing another but without changing the specified amino acid. Other mutations are said to be "conservative." This occurs when a mutation replaces one amino acid with another that has similar properties: They may be chemically similar sharing the same charge, shape, or polarity.

codons sequences of three mRNA nucleotides coding for one amino acid

If, however, the mutation affects the function of an important protein, that mutation may result in an evolutionary dead end, because it is less likely to be passed on to a future generation. As a result, important sequences show fewer mutations, whereas less important sequences show more change. Such properties can be deduced by comparing sequences from different organisms. Proteins that interact with other molecules, such as DNA or RNA, tolerate fewer changes in structure, and show little change

through evolution. The histone proteins that form the backbone of the eukaryotic chromosome are important examples.

Evolutionary Relatedness

molecular systematics the analysis of DNA and other molecules to determine evolutionary relationships

The number and types of differences that accumulate between genes or proteins of two different species can be used to assess their evolutionary relatedness and the amount of time since they diverged from a common ancestor. Such studies, termed "**molecular systematics**," can be used to show that humans are more closely related to chimps than to gorillas, for instance, and how long ago the split in these lineages occurred.

Homologous proteins that perform the same function in different species are called orthologs. For example, hemoglobin, a protein that transports oxygen, has a similar amino acid sequence in both horses and dogs. If the predicted amino acid sequence of a newly discovered protein is similar to a known protein in another species, researchers can make guesses about the function of the newly discovered gene. If the sequence of a newly discovered protein was similar to hemoglobin, one might guess that the new protein is able to bind to oxygen and function in transporting oxygen. In the way, orthologs help researchers about the functions of newly discovered genes.

Natural selection acts against harmful mutations in critical genes. Gene duplication, however, makes extra copies of less critical genes, which are more free to acquire mutations. Members of these gene families are known as paralogs. Researchers look for paralogs in order to find proteins with new abilities. Cytokine genes, for example, are all derived from the same ancestral gene and share common sequence motifs, yet they fill a variety of roles in the immune system. New members of the cytokine family might be valuable tools for fighting disease. Just as species diverge and fill new biological niches, genes become duplicated and acquire new functions. On a molecular scale, the evolution of the genome reflects the evolution of all living things. SEE ALSO BIOINFORMATICS; CHROMOSOME, EUKARYOTIC; EVOLUTION OF GENES; MOLECULAR ANTHROPOLOGY; MUTATION.

Sandra G. Porter

Bibliography

Lander, Eric, et al. "Intitial Sequencing and Analysis of the Human Genome." *Nature* 409 (2001): 860–921.

Strachan, Tom, and Andrew P. Read. *Human Molecular Genetics*, 2nd ed. New York: John Wiley & Sons, 1999.

Venter, J. C., et al. "The Sequence of the Human Genome." *Science* 291 (2001): 1304–1351.

Internet Resource

National Center for Biotechnology Information. <http://www.ncbi.nlm.nih.gov>.

Hormonal Regulation

chemotaxis movement of a cell stimulated by a chemical attractant or repellent

pheromone molecule released by one organism to influence another organism's behavior

All types of cells are capable of receiving signals from their environment and mounting an appropriate response to the signal, such as **chemotaxis** toward a nutrient source or toward other cells emitting a **pheromone**. The

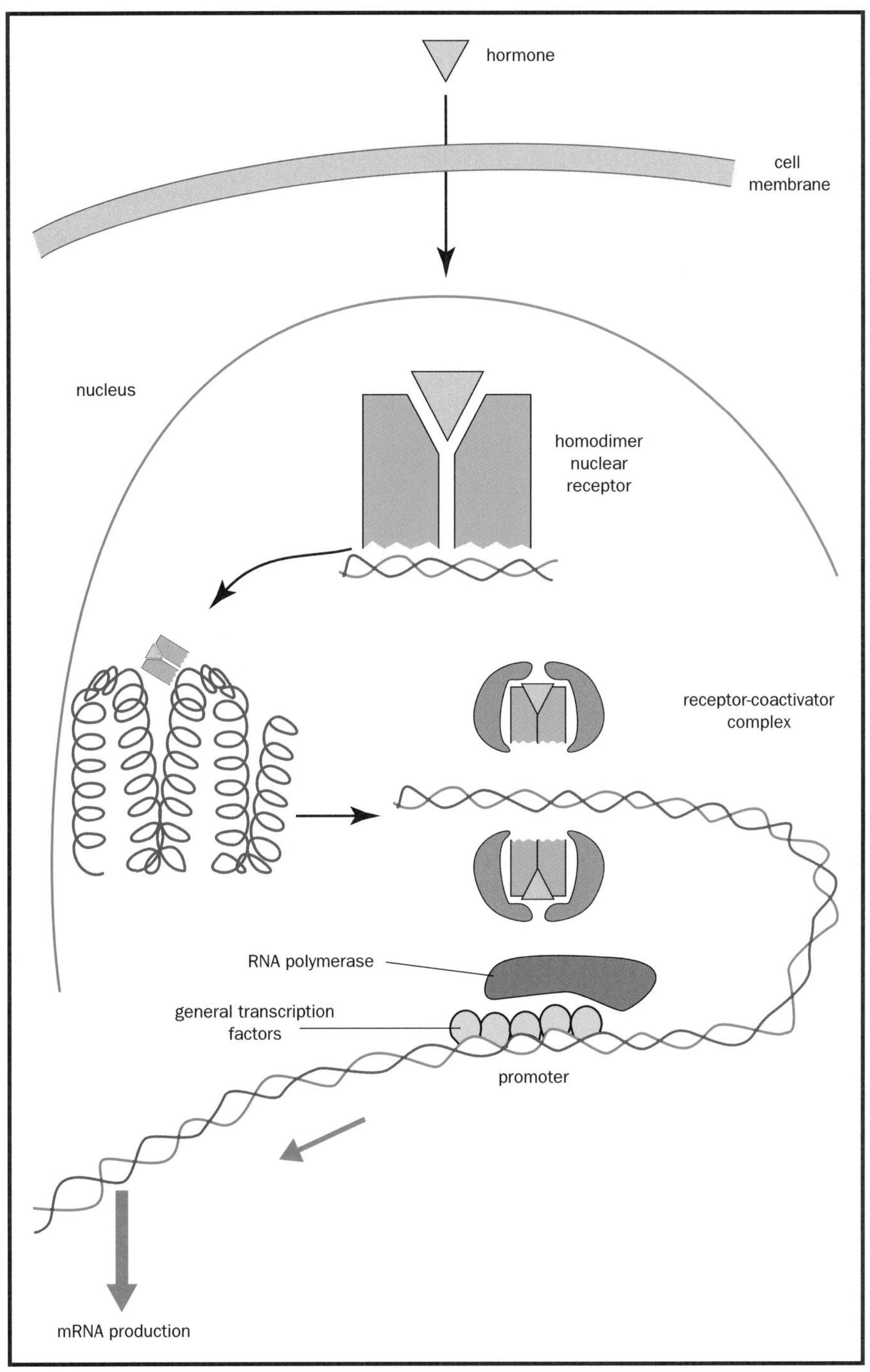

A steroid hormone binds to a two-part receptor within the cell. This links with a coactivator, making a complex that binds to DNA. This triggers transcription of the target gene, through interaction with general transcription factors.

key difference between microorganisms and more-complex plants and animals is that the former are largely independent, with each cell in contact with the environment. In contrast, more complex plants and animals are self-contained entities whose interior is mostly insulated from the environment. Animals have complex organ systems, with each organ specialized for a particular function. Therefore, the survival of the organism depends on the precise regulation of growth, differentiation, and metabolism in different groups of cells throughout the animal.

The endocrine system is the set of glands and other tissues responsible for coordinating cellular growth and differentiation, many aspects of

NUCLEAR RECEPTOR PARTNERS FOR RXR

Receptor	Hormone	Function
retinoic acid receptor (RAR)	all-trans retinoic acid (vitamin A derivative)	regulates important aspects of early embryonic development
thyroid hormone receptor (TR)	triiodothyronine	controls the body's basal metabolic rate
vitamin D3 receptor (VDR)	vitamin D3	regulates calcium homeostasis, other functions
fatty acid receptor (PPAR)	fatty acid and eicosanoid ligands	regulates fat metabolism and the the body's ability to utilize insulin
bile acid receptor (FXR)	bile acid	regulates the transport of bile acids and cholesterol out of the cell
oxysterol receptor (LXR)	oxysterol	regulates the formation of bile acids from cholesterol
benzoate receptor (BXR)	unknown	unknown
steroid and xenobiotic receptor (SXR)	multiple compounds	regulates degradative and detoxification enzymes
constitutive androstane receptor (CAR)	multiple compounds	regulates degradative and detoxification enzymes

homeostasis maintenance of steady state within a living organism

reproduction and embryological development, the maintenance of **homeostasis**, and a variety of cyclical phenomena (e.g., reproductive cycles). Because these varied processes require coordinated gene expression, they are regulated by a large and diverse group of inter- and intracellular signaling pathways.

The primary mediators of these pathways are a large group of chemical messengers, called hormones, produced by specialized cells in response to physiological requirements. Many of these specialized cells are located in endocrine glands. Some of the most well-known endocrine glands are the pituitary (which produces many important hormones, such as ACTH), the thyroid (which produces thyroid hormone to regulate metabolic rate), the adrenals (which produce glucocorticoids to regulate blood sugar and stress, and which also produce epinephrine, or adrenaline), the testes (which produce testosterone), and the ovaries (which produce estrogens and progesterone).

Hormones may need to act at different distances from their source, depending on the requirements of the organism, and they can be broadly classified according to the distance across which they signal. Endocrine hormones (e.g., adrenocorticotropic hormone) are produced by endocrine glands (in this case the pituitary gland, located at the base of the brain) at a distance from their site of action (in this case adrenal glands, sitting atop the kidneys) and must be transported throughout the body via the circulatory system. Paracrine hormones, such as the prostaglandins that mediate local inflammatory processes, are produced near their site of action. Autocrine hormones, such as interleukins, act on the cells that produce them, in this case the white blood cells of the immune system.

Hormone Receptors

Regardless of the distance across which a hormone acts, only those cells that contain a specific receptor can respond to the corresponding hormonal signal. The expression of receptors only in the target cells ensures that these (and only these) cells respond in the appropriate way to the hormone, despite the possible presence of a large number of other hormones in the immediate surroundings.

In addition to the distance across which they act, hormones may be further divided into two large groups based on where in the target cell the hormone receptors are located. The first class consists of extracellular hormones that act via specific cell-surface receptors. Most hormones of this family are proteins (such as insulin, interferons, interleukins, and growth factors), fatty acid derivatives (such as prostaglandins and leukotrienes), or amino acid derivatives (such as serotonin and melatonin).

Extracellular hormones bind to specific receptors on the cell surface, triggering a chain of events inside the cell. These events may include the modification (e.g., **phosphorylation** or dephosphorylation) of one or more "second messengers"—small molecules that act inside the cell to continue the signaling cascade. In addition to any other short-term effects they may have, virtually all hormonal signaling processes culminate in a change in the expression of a set of target genes. Depending on the cell, the transcription of these genes may be increased or decreased, or be turned completely on or turned off, in response to the presence of the hormone.

phosphorylation addition of the phosphate group PO_4^{3-}

The other major class of hormonal signals consists of small, typically fat-soluble molecules that are able to diffuse freely into cells. Once inside the cell, the hormone binds to its receptor to directly regulate the expression of target genes. The hormone-receptor complex is a functional **transcription factor** that in most cases leads to the expression of its target genes. This modulation primarily occurs directly, via binding of the hormone-receptor complex to target DNA sequences, although additional regulation can occur indirectly, via interaction with other transcription factors.

transcription factor protein that increases the rate of transcription of a gene

Since the vast majority of these receptors are always in the nucleus, the family is referred to as the "nuclear hormone receptor superfamily." It is also often also called the "steroid receptor superfamily," because steroid receptors were the first of this family to be identified. Steroid hormones include testosterone, progesterone, and the estrogens. The discussion that follows will focus on hormones that interact with nuclear receptors to directly influence gene expression.

Nuclear Receptors and Their Hormones

The nuclear receptors are a large group of related proteins that mediate many of the effects of steroid hormones, thyroid hormone, vitamin D3, the vitamin A derivative retinoic acid, and modified forms of cholesterol, such as hydroxycholesterol and bile acids. The number of nuclear hormone receptor genes varies widely among animals. Humans and other vertebrates have about forty-nine receptor genes, whereas the nematode *Caenorhabditis elegans*, with only 959 cells in the adult worm, has more than 250 receptor genes. This was a somewhat unexpected finding, and it led to the speculation that

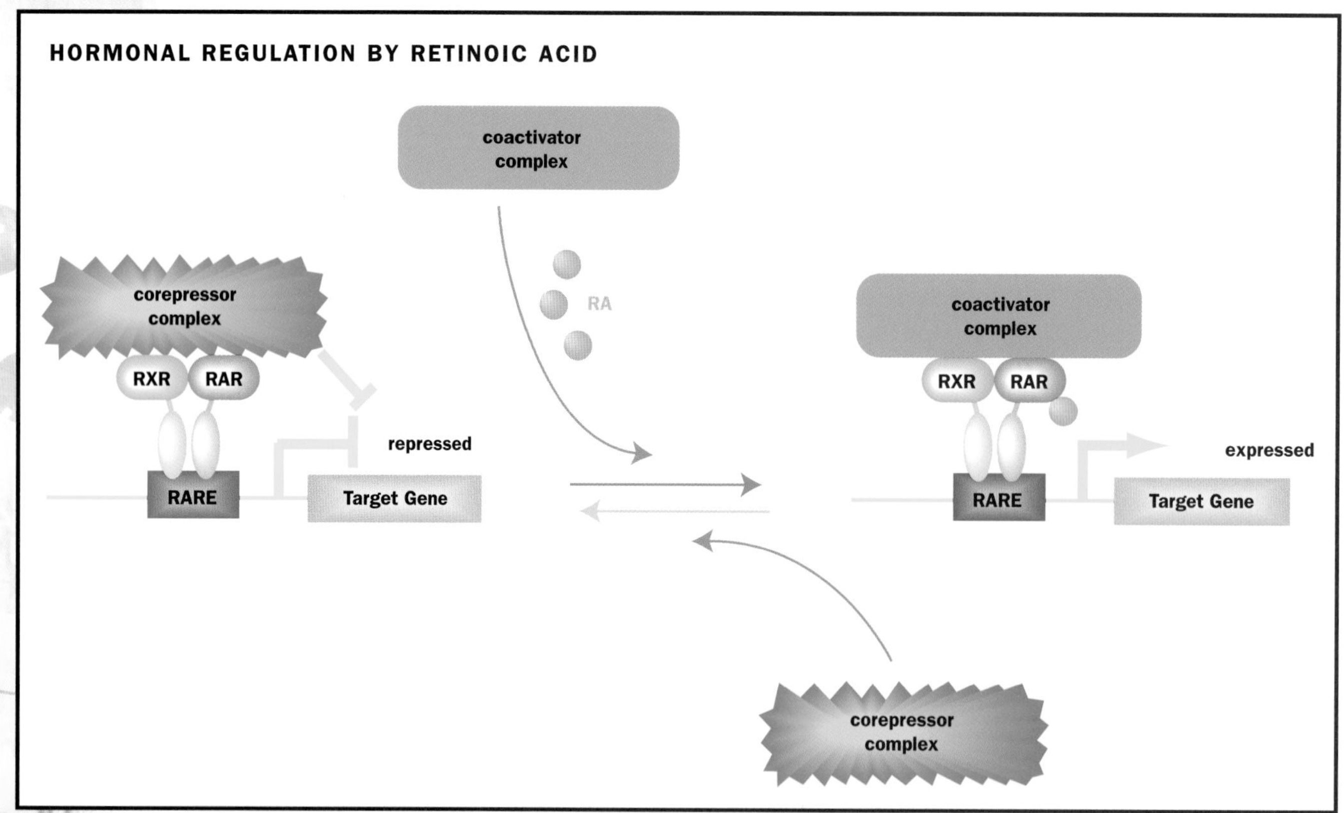

Schematic depiction of how the retinoic acid receptor activates transcription of target genes. In the absence of retinoic acid, the receptor interacts with the corepressor complex to condense chromatin and silence transcription from target genes. In the presence of retinoic acid, the corepressor complex dissociates and the receptor can then interact with a coactivator complex that leads to chromatic decondensation and subsequent activation of target genes. The active and repressed states exist in a dynamic equilibrium that is regulated by the presence of retinoic acid.

C. elegans may use nuclear hormone receptors to regulate processes that are controlled by different transcription factors in vertebrates.

The classical steroid receptors are all quite similar to each other, and they all function as homodimers—a complex of two identical proteins. With the exception of the estrogen receptor, all of these receptor homodimers bind to exactly the same target DNA sequences. For this reason, high levels of one hormone may cause inappropriate activation of another pathway and multiple consequences.

It is currently unclear how one receptor (e.g., the glucocorticoid receptor) distinguishes its correct target genes from those of other receptors (e.g., the progesterone receptor), when multiple receptors are present in the same cell. Estrogens, progesterone, and androgens are important steroid hormones that influence many aspects of later development.

The estrogen receptor is expressed in the brain, kidney, liver, and lungs, and throughout the female reproductive tract. Interestingly, the estrogen receptor is also present and required in male reproductive tissues. The major human estrogen, 17-β estradiol, activates the receptor to regulate cell proliferation in, for example, the uterus. The progesterone receptor is also important for female development, with its effects restricted to the female reproductive tract and mammary tissue.

The androgen receptor is primarily responsible for male development and secondary sexual characteristics, such as muscle mass. The major hormone acting through this receptor is dihydrotestosterone (DHT). The androgen receptor is also the major receptor targeted by so-called **anabolic steroids**, which function by mimicking the activities of DHT on muscle growth. Some predictable and unfortunate consequences of increasing the circulating levels of testosterone-like molecules include atrophy of the testes (since they sense high levels of testosterone and react by shutting down their own production) and the development of female secondary sexual characteristics, such as breasts, in men (because excess DHT is converted to estradiol).

anabolic steroids hormones used to build muscle mass

The largest and most diverse group of nuclear receptors contains those that function as heterodimers, meaning they are composed of two different parts. Each heterodimer is composed of one unique receptor protein and one protein common to the whole group, called the 9-cis-retinoic acid receptor (RXR). There are nine distinct hormone-regulated receptor-signaling pathways wherein RXR is used as a common heterodimeric partner.

One of these is the retinoic acid receptor (RAR), which binds with all-trans retinoic acid, a vitamin A derivative, to regulate many important aspects of early embryonic development, including limb formation, central nervous system patterning, growth and differentiation of many tissues, **hematopoiesis**, and eye, brain, and craniofacial development. Since retinoic acid affects so many important developmental processes, too much or too little retinoic acid has profound effects on early development.

hematopoiesis formation of the blood

Another RXR partner is the steroid and **xenobiotic** receptor (SXR). Steroid and xenobiotic **ligands** for SXR regulate the breakdown of foreign chemicals by degradative enzymes in the liver and intestines, protecting the body from toxic chemicals and bioactive dietary compounds. SXR is known to directly regulate the transcription of genes such as *CYP3A4*, which mediates the breakdown of 60 percent of clinically useful drugs, as well as the transcription of the multidrug resistance protein MDR1, which transports drugs out of the cell. Thus, SXR is a key mediator of the body's defense system against foreign chemicals, controlling both their metabolism and clearance from the cell.

xenobiotic foreign biological molecule, especially a harmful one

ligands molecules that bind to receptors or other molecules

Nuclear Hormone Receptors and Transcriptional Regulation

As noted above, nuclear receptor hormones generally act as transcription factors to increase transcription of their target genes. They do this by increasing the rate at which **RNA polymerase** binds to the target gene's **promoter**. This occurs in several steps.

RNA polymerase enzyme complex that creates RNA from DNA template

promoter DNA sequence to which RNA polymerase binds to begin transcription

The binding of a hormone to the receptor triggers the assembly of other proteins to form a "coactivator complex." The hormone-receptor-coactivator complex binds to a specific DNA sequence (called the hormone response element, a type of transcriptional enhancer). This complex then alters the local DNA structure by directly or indirectly chemically modifying the **histones**. These modifications open up the DNA, increasing access to the target genes and thereby allowing RNA polymerase and other (general) transcription factors to reach the gene promoter region. Additionally,

histones the proteins around which DNA wind in the chromosome

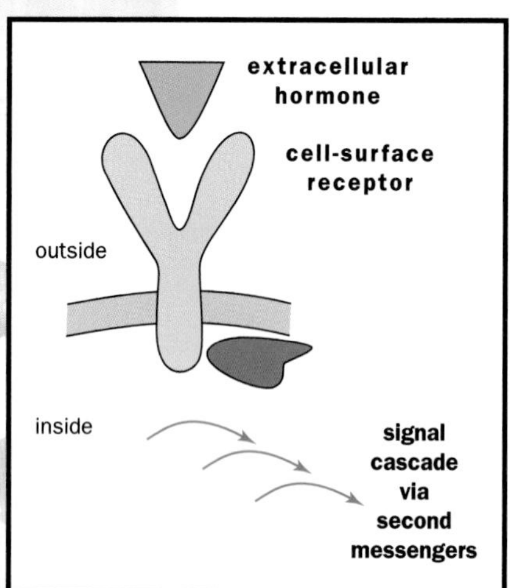

Some hormones trigger second-messenger signal cascades, quickly turning on cellular enzymes for a rapid response.

the hormone-receptor-coactivator complex can directly interact with general transcription factors to help form a "preinitiation complex" of proteins on the target gene promoter. RNA polymerase then interacts with this complex, and the transcription of the gene into **mRNA** begins.

mRNA messanger RNA

All hormone-regulated nuclear receptors activate transcription in this manner. Some, such as the steroid receptors, exist in cells as cytoplasmic complexes with "chaperone proteins," such as HSP90, and are excluded from the nucleus in the absence of the hormone. In the presence of hormone, the complexes dissociate, and the receptors **dimerize** and are transported to the nucleus, where they activate transcription.

dimerize linkage of two subunits

Other receptors, such as the retinoic acid, thyroid hormone, or vitamin D receptors, are always found in the nucleus and interact with their specific target genes in the presence or absence of the hormone. When the hormone is absent, the receptor interacts with "corepressor" proteins. The complex of receptor and corepressor interacts with histone deacetylases, leading to local chromatin condensation and silencing of the target gene. Hormone binding leads to a change in the three-dimensional structure of the receptor, causing dissociation of the corepressor complex and leading to the recruitment of the coactivator complex, which enables the target gene to be transcribed.

The Importance of Hormone Concentration

Because the hormones that act through nuclear hormone receptors are nearly all fat-soluble, they are readily absorbed into the body, freely transported, and stored and accumulated in fatty tissues. Steroids, retinoic acid, thyroid hormone, and vitamin D3 are active at extremely low concentrations, ranging from about 0.3 to 30 parts per billion, with 3 parts per billion considered a physiological concentration for retinoic acid and many steroids. Since the hormones are present and act at such low concentrations, it is critical that their levels be precisely regulated. Consequently, hormone synthesis and degradation is regulated by the activity of specific biosynthetic and catabolic **enzymes**.

enzymes proteins that control a reaction in a cell

It should also be noted that some chemicals in the environment and natural compounds found in the diet can affect the activity of hormone receptors, particularly the estrogen receptor. Such interaction can potentially lead to disturbances in hormone homeostasis and inappropriate regulation of target genes. These xenobiotic "endocrine disrupting chemicals" have the potential to impact many body systems by inappropriately activating or interfering with the activity of hormone receptors. As a result, endocrine disruption is a growing concern that is being studied intensively in many laboratories around the world. SEE ALSO CHAPERONES; ROUNDWORM: *CAENORHABDITIS ELEGANS*; SIGNAL TRANSDUCTION; TRANSCRIPTION FACTORS.

Bruce Blumberg

Bibliography

Chawla, A., et al. "Nuclear Receptors and Lipid Physiology: Opening the X Files." *Science* 294 (2001): 1866–1870.

Evans, R. M. "The Steroid and Thyroid Hormone Receptor Superfamily." *Science* 240 (1988): 889–895.

Kliewer, S. A., J. M. Lehmann, and T. M. Willson. "Orphan Nuclear Receptors: Shifting Endocrinology into Reverse." *Science* 284 (1999): 757–760.

HPLC: High-Performance Liquid Chromatography

High-performance liquid chromatography (HPLC) is an advanced form of liquid chromatography used in separating the complex mixture of molecules encountered in chemical and biological systems, in order to understand better the role of individual molecules. In liquid chromatography, a mixture of molecules dissolved in a solution (mobile phase) is separated into its constituent parts by passing through a column of tightly packed solid particles (stationary phase). The separation occurs because each component in the mixture interacts differently with the stationary phase. Molecules that interact strongly with the stationary phase will move slowly through the column, while the molecules that interact less strongly will move rapidly through the column. This differential rate of migration facilitates the separation of the molecules.

The advantages of HPLC over other forms of liquid chromatography are several. It allows analysis to be done in a shorter time and achieves a higher degree of resolution, that is, the separation of constituents is more complete. In addition, it allows stationary columns to be reused a number of times without requiring that they be regenerated, and the results of analysis are more highly reproducible. A further advantage of HPLC is that it permits both instrumentation and quantitation to be automated.

Components of HPLC Analysis

HPLC has four basic components: a solvent delivery system to provide the driving force for the mobile phase; a means by which samples can be introduced into the solvent; the column; and some type of detector. A recorder is used to display the results and an integrator performs the calculations.

The column used for a specific separation is based on the type of the molecules to be analyzed. Various types of chromatographic modes can be used for the separation of the molecules. For example, ion exchange columns separate charged molecules such as amino acids, proteins, or **nucleotides**. Size exclusion columns separate organic polymers such as polyvinyls and silicones or biopolymers such as proteins, nucleic acids, or sugars. Adsorption columns separate molecules based on their interaction with the stationary phase. This mode is useful for the separation of vitamins, dyes, lipids, phenols, and antioxidants. Partition columns are used to separate molecules based on the way that the solvent becomes partitioned into stationary and mobile layers, and is useful in analyzing steroids, aromatics, vitamins, and antibiotics. The molecules **eluting** from any one of these different types of column are then analyzed by various types of detectors, measuring absorbance, fluorescence, or electrochemical or radiochemical properties. Other types of detectors include mass spectroscopy and refractive index.

nucleotide the building block of RNA or DNA

eluting exiting

HPLC Applications

A recent advancement of HPLC has been the development of the denaturing HPLC method (DHPLC). This procedure can separate

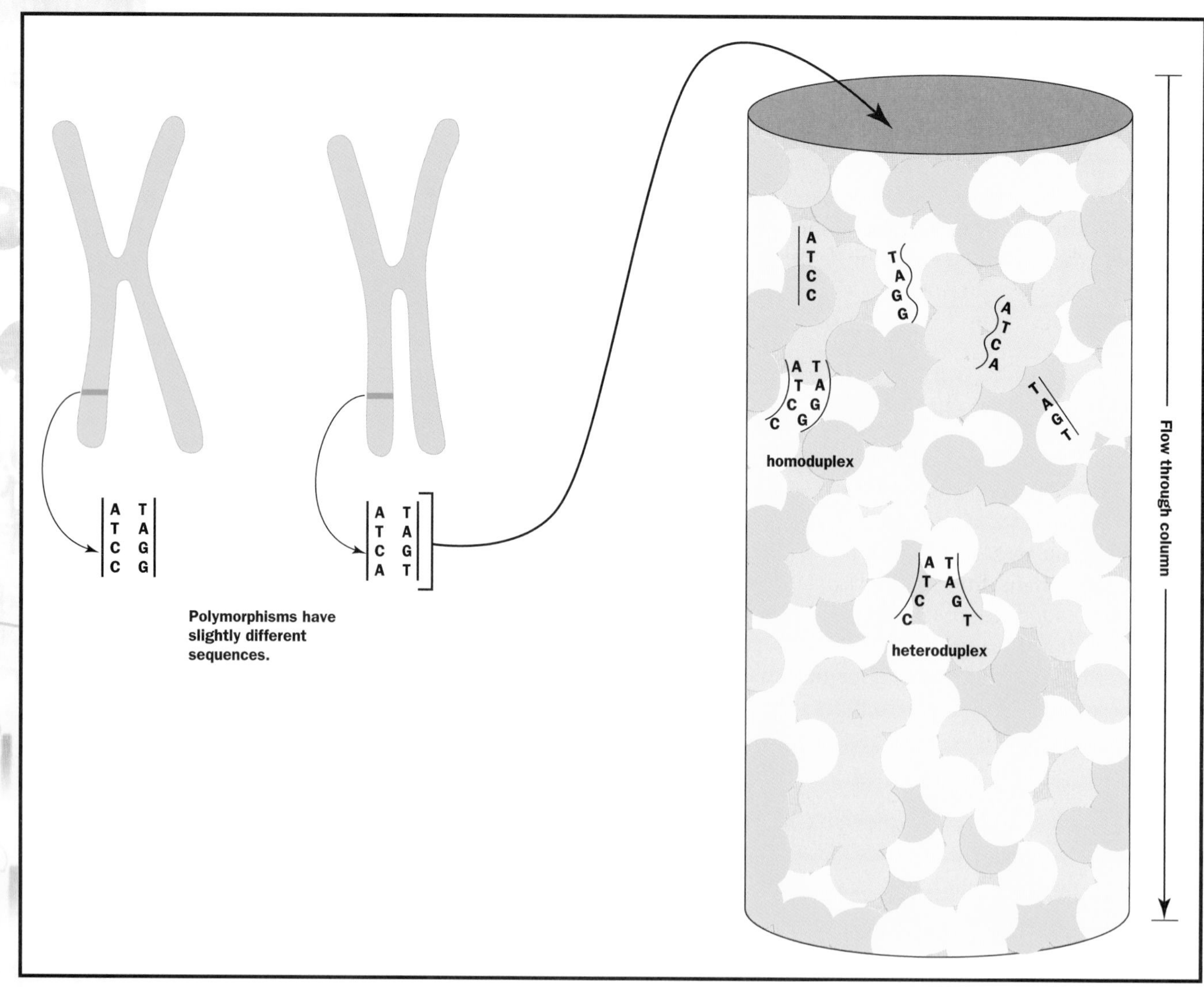

DNA from different sources is placed into the column and treated to allow it to separate and recombine. Two strands from the same source, termed a homoduplex, bind strongly and tend to remain together longer. Heteroduplexes do not attract as strongly, and remainn unattached more often.

base pair two nucleotides (either DNA or RNA) linked by weak bonds

polymorphisms occurring in several forms

double-stranded DNA molecules that differ by as little as one **base pair**. The speed of analysis (approximately 5 minutes per sample) and the size of DNA fragment that can be analyzed (up to 2.0 kilobytes) has made it a preferred method for a variety of applications in the field of molecular biology. Applications of DHPLC include the detection of single nucleotide **polymorphisms** (SNPs). These are single base-pair variations in DNA that can give valuable information on genetic variation within a population. They can also help to identify the genes that cause certain human diseases.

To determine if the two genes of interest differ, they are first amplified by the polymerase chain reaction and then injected together into a so-called reversed-phase column. In this type of column, the stationary phase is less polar than the mobile phase, which is the opposite of the arrangement found in standard columns. Once the genes are injected, their DNAs bind to the stationary phase. Increasing the temperature causes each gene to separate into its two strands (a phenomenon called denaturing). Once denatured, the strands can enter the mobile phase and move through the column.

Cooling the column causes the strands of the genes to rejoin, and the DNAs reattach to the column. The temperature is manipulated to make the strands constantly separate and rejoin, with the balance determined by the strength of the attraction that exists between the strands. If the two genes are exactly identical, they will spend more time in the stationary phase, and elute from the column more slowly. If the genes differ even by a single nucleotide, however, they will spend more time in the mobile phase, and leave the column more quickly.

Y chromosome analysis is one of the most powerful molecular tools for tracing human evolution. Polymorphisms in the human Y chromosome, detected by DHPLC, can be used as markers for tracing human evolution. This will eventually help to elucidate the patterns of human origins, migration, and mixture. The ability to rapidly and efficiently **genotype** SNPs by use of DHPLC is also useful in medicine, through the identification of mutations that result in susceptibility to certain diseases or that affect physical responses to certain drugs. SEE ALSO DNA; POLYMERASE CHAIN REACTION; POLYMORPHISMS; Y CHROMOSOME.

Prema Rapuri

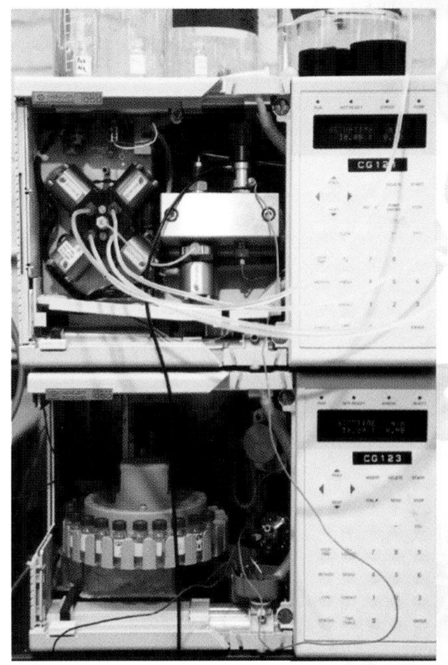

A high-performance liquid chromatography system allows geneticists to perform molecular research efficiently by automating necessary processes.

genotype set of genes present

Bibliography

Bidlingmeyer, Brian A. *Practical HPLC Methodology and Applications.* New York: John Wiley & Sons, 1992.

Underhill, Peter A., Peidong Shen, Alice A. Lin, et al. "Y Chromosome Sequence Variation and the History of Human Populations." *Nature Genetics* 26 (2000): 358–361.

Human Disease Genes, Identification of

In order to help treat human diseases, it is important to understand what causes them to occur. Understanding what causes a disease is the first step in understanding the entire abnormal course of disease. Sometimes it is fairly easy to determine what causes a disease. For example, pneumonia is caused by the *Pneumococcus* bacterium. However, in other cases it is not nearly as easy to tell what is causing a disease, so scientists look for clues from a number of different sources.

One such clue is having a disease run in families, which suggests that a disease might be caused by a gene or genes passed from parent to child, perhaps over many different generations. The process of identifying these genes, called disease gene discovery, is important because it helps scientists to understand what is going wrong as a result of such diseases, called the disease pathogenesis. By understanding the disease process, it is possible to figure out where it is easiest to either stop or correct what is going wrong. Variations or mutations within a gene may themselves act to cause the disease. Alternatively, they may modify the risk of developing the disease or modify how it is expressed (i.e., what the symptoms are). Thus genes may be important both directly and indirectly in causing disease.

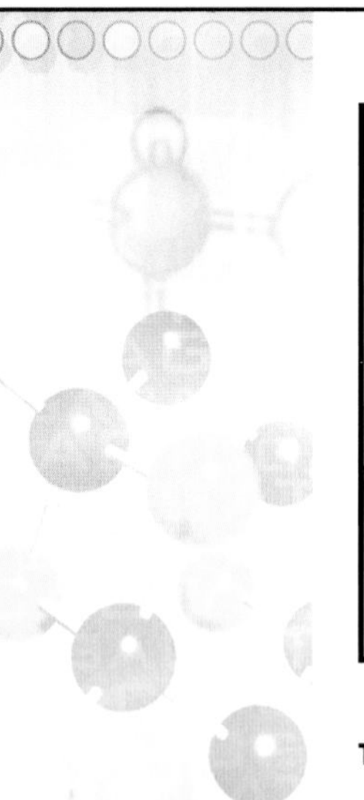

SOME EARLY DISEASE-CAUSING GENES FOUND USING GENOMIC SCREENING

Disease name	Symptoms	Year identified	Chromosome
Chronic Granulomatous Disease	Poor immune system	1986	X
Duchenne Muscular Dystrophy	Muscle weakness, muscle deterioration	1986	X
Cystic Fibrosis	Difficulty breathing, poor lung function	1989	7
Neurofibromatosis	Benign tumors	1990	17
Fragile-X Syndrome	Mental retardation	1991	X
Huntington's Disease	Uncontrollable movements, brain deterioration	1993	4
Tuberous Sclerosis	Benign tumors	1993	16
Alzheimer's Disease	Loss of memory and brain function	1993	19
Breast Cancer	Tumors in the breast	1994	17
Glaucoma	Loss of eyesight	1997	1

Table 1.

The Process of Disease-Gene Discovery

Performing disease-gene discovery requires asking and answering five important questions: What does the disease actually look like? Do genes really play an important role in this disease? Can enough families with enough affected members be found to help study the disease? What are the genes? What do these genes do?

What does the disease actually look like? It may seem obvious that to study the genetics of a disease such as epilepsy (seizures), families with seizure sufferers should be studied. However, it is possible that genes are important only in some, not all, types of epilepsy. It may also be true that different genes are important in different types of epilepsy. Thus it may be better not to study all seizures but only one type. For example, it might be better to study seizures that affect only one part of the body or that cause only a loss of consciousness with no effects on the rest of the body. Which type of seizures to study depends mostly on how important genes are for that particular type of disease.

Do genes really play an important role in this disease? There are many ways to see if genes are important in a disease (or a particular subtype of disease) without having to know what the genes are. Figuring this out requires studies that look at information from very large collections of families and individuals. These are usually called **population studies**.

population studies collection and analysis of data from large numbers of people in a population, possibly including related individuals

One important type of population study looks at a large set of twins to see how often two identical twins (who have all the same genes) both have a disease and compares that to how often two fraternal twins (who are just like brothers and sisters in that they have only half of their genes in common) both have the disease. If genes are important, the identical twins will share the disease (be concordant) much more often than fraternal twins. For example, in over 80 percent of cases where one identical twin is diagnosed with autism, the other is also diagnosed with the disorder. When one fraternal twin, on the other hand, has autism, the other will have the disorder in only about 5 percent of cases. This suggests that genes have a very strong effect in autism.

Another way to see if genes are important is to look at a large set of adopted children who have the disease and compare how often their adoptive parents have the same disease with how often their natural parents have it. If the rate of the disease is much higher in the natural parents than in the adoptive parents, genes are likely to be important. For example, the frequency of multiple sclerosis in the natural parents of people with multiple sclerosis is about 3 percent. Among the adoptive parents of people with multiple sclerosis, the frequency was much smaller and about the same as the general population.

Yet another way to see if genes are important is to look at the occurrence of the disease in the brothers and sisters of someone who has the disease. If the rate at which brothers and sisters have the disease is much higher than the rate in the overall population, then it is likely that genes are important. Autism again is a good example. The rate at which autism is found in the brothers and sisters of an affected person is about 3 in 100, which does not seem very high. But the rate in the general population is only about one in five hundred (0.2%). It is the comparison of these rates that is important, not just the actual frequencies.

Each of these kinds of studies requires looking at a lot of people and their families, and each will usually take several years to complete. However, such studies are very important, since there is no point in looking for genes that affect a disease if we know that genes are not important. It is only after this information is available and genes are known to be important that the next step can be taken.

Can enough families with enough affected members be found to help study the disease? Depending on the disease, the type of family that can be found for genetic studies may differ a lot. For diseases caused by a **mutation** in a single gene, families with many affected people can be found. Sometimes these families may have as many as twenty or thirty affected people, over three or four generations. These types of families are usually quite rare, and thus it can take a lot of work to find them. On the other hand, having even a single large family can be enough to allow a gene to be found.

mutation change in DNA sequence

For diseases where genes may have only a moderate effect, smaller families, where only two or three people (usually brothers and sisters) are affected, may be the only ones that can be found. Since such diseases tend to be more common in the general population, it is easier to find these families than to find the very large families. However, the smaller families cannot, by themselves, help much in finding the location of a disease gene. Therefore many such families (usually hundreds) are needed. How the actual process of finding the genes is done depends on the type of families that are studied.

One of the important aspects of finding the families is deciding how to ask them to be part of the study. It is important that each person who participates is told what he is going to have to do. In most cases, participants will just be answering a lot of questions, giving permission to get medical records about their disease, and giving a blood sample. Sometimes additional hospital tests might be needed. Before they can be studied, potential subjects must be asked to participate and must give "**informed consent**," which simply means that they have been told what they will need to do, what the risks (if any) are, and have agreed that this is fine with them. It is one of the overriding rules of human genetic studies that each person must volunteer for the study, that he gives informed consent, that he has the right to refuse, without the refusal affecting his medical care, and that he can withdraw from the study at any time.

informed consent knowledge of risks involved

What are the genes? Once the information and blood samples are collected from the families, the process of finding the genes can begin. There

are two ways to search for them. The first involves looking across all the chromosomes, using genetic maps, and trying to correlate the occurrence of the disease in a family with the occurrence of one or more genetic markers from the genetic maps. This approach is called genomic screening and tries to find the genes based only on their location. It does not require that anything be known about what any of the genes do. Genomic screening uses a number of special statistical techniques to look at the probability that the disease gene (whose location is unknown) and one or more of the markers (whose location is known) are located near each other. One difficulty is that the location will not be known precisely. That is, this method will point to a region that may contain as many as 500 genes. This is a lot less than the 50,000 or so thought to exist in the human genome, but it is still a lot of genes that need to be tested. Genomic screening has worked spectacularly well for hundreds of diseases where there is a single causative mutation in a gene. A list of some of these genes in given in Table 1.

The second approach is to look at one or more specific genes to see if they might be directly involved in the disease. This is called the candidate gene approach. A gene becomes a candidate when something is known about its function and when this function might have something to do with the disease. For example, if a gene was involved in the development of the cornea of the eye, it would be a good candidate for any disease that affects the development of the cornea. The success of the candidate gene approach depends on two things. The first is how much is known about the disease process, and the second is how much is known about the function of the genes.

It is also possible to combine the two approaches. The genomic screening approach may identify several regions on several chromosomes that might contain a disease gene, but these regions may contain hundreds of genes each. By looking at the functions of these genes, it may be possible to identify one or just a few that are the most likely to be involved in the disease. These genes can then be tested using the candidate gene approach.

What do these genes do? Once a gene has been identified as being involved in the disease, it is important to further study the gene to find out what it does under normal circumstances, and what it is doing when it is changed and causing disease. Sometimes a lot is known about the normal function, but many times very little is known. Studies to look at the function may include studies of the normal gene in living cells that are grown in the laboratory. Another type of study involves testing the same normal genes in other living organisms such as mice, rats, and fruit flies. Animal studies are very helpful because animals can be tested in ways not possible in humans. Studies similar to those done for the normal gene will have to be done on the changed (mutated) copy of the gene, to see how the change in the gene changes the function of the protein that the gene makes. SEE ALSO COMPLEX TRAITS; GENE DISCOVERY; MAPPING; TWINS.

Jonathan L. Haines

Bibliography

Haines, Jonathan L., and Margaret A. Pericak-Vance, eds. *Approaches to Gene Mapping in Complex Human Diseases.* New York: John Wiley & Sons, 1998.

Internet Resource

Dolan DNA Learning Center. Cold Spring Harbor Laboratory. <http://www.dnalc.org>.

Human Genome Project

The **genome** represents the entire complement of DNA in a cell. The Human Genome Project is the determination of the entire nucleotide sequence of all 3 billion+ bases of DNA within the nucleus of a human cell. It is one of the greatest scientific undertakings in the history of mankind. The first draft of the human genome sequence was completed in the year 2001 and published simultaneously in the British journal *Nature* and the American journal *Science*.

genome the total genetic material in a cell or organism

The data obtained from sequencing the human genome promise to bring unprecedented scientific rewards in the discovery of disease-causing genes, in the design of new drugs, in understanding developmental processes and cancer, and in determining the origin and evolution of the human race. The Human Genome Project has also raised many social and ethical issues with regard to the use of such information.

Origins of the Human Genome Project

One could say that the Human Genome Project really began in 1953, when James Watson and Francis Crick deduced the molecular structure of DNA, the molecule of which the genome is made. (Watson and Crick were awarded the Nobel Prize for this work in 1962.) Since that time, scientists have wanted to know the complete sequence of a gene, and even dreamed that some day it would be possible to determine the complete sequence of all of the genes in any organism, including humans.

The original impetus for the Human Genome Project came almost a decade earlier, however, from the U.S. Department of Energy (DOE) shortly after World War II. The atomic bombs that were dropped on Hiroshima and Nagasaki, Japan, left many survivors who had been exposed to high levels of radiation. The survivors of the bomb were stigmatized in Japan. They were considered poor marriage prospects, because of the potential for carrying mutations, and the rest of Japanese society often ostracized them. In 1946 the famous geneticist and Nobel laureate Hermann J. Muller wrote in the *New York Times* that "if they could foresee the results [mutations among their descendants] 1,000 years from now . . . , they might consider themselves more fortunate if the bomb had killed them."

Muller had firsthand experience with the devastating effects of radiation, having studied the biological effects of radiation on the fruit fly *Drosophila melanogaster*. He predicted similar results would follow from the human exposure to radiation. As a consequence, the Atomic Energy Commission of the DOE set up an Atomic Bomb Casualty Commission in 1947 to address the issue of potential **mutations** among the survivors. The problem they faced was how to experimentally determine such mutations. At that time there were no suitable methods to study the problem. Indeed, it would be many years before the appropriate technology was available.

mutations changes in DNA sequences

During the 1970s molecular biologists developed techniques for the isolation and cloning of individual genes. Paul Berg was the first to create a recombinant DNA molecule in 1972, and within a few years gene cloning became a standard tool of the molecular biologist. Using cloning techniques, scientists could generate large quantities of a single gene, enabling researchers

Table 1.

MODEL ORGANISMS SEQUENCED

Date sequenced[a]	Species	Total bases[b]
7/28/1995	*Haemophilis influenzae* (bacterium)	1,830,138
10/30/1995	*Mycoplasma genitalium* (bacterium)	580,073
5/29/1997	*Saccharomyces cerevisiae* (yeast)	12,069,247
9/5/1997	*Escherichia coli* (bacterium)	4,639,221
11/20/1997	*Bacillus subtillis* (bacterium)	4,214,814
12/31/1998	*Caenorhabditis elegans* (round worm)	97,283,371 99,167,964[c]
3/24/2000	*Drosophila melanogaster* (fruit fly)	~137,000,000
12/14/2000	*Arabidopsis thaliana* (mustard plant)	~115,400,000
1/26/2001	*Oryza sativa* (rice)	~430,000,000
2/15/2001	*Homo sapiens* (human)	~3,200,000,000

[a]First publication date.
[b]Data excludes organelles or plasmids. These numbers should not be taken as absolute. Scientists are confirming the sequences; several laboratories were involved in the sequencing of a particular organism and have slightly different numbers; and there are some strain variations. Data were obtained from the (NCBI) Web site.
[c]The first number was originally published, and the second is a correction as of June 2000.

to study its structure and function. In 1977 Drs. Walter Gilbert and Fred Sanger independently developed methods for the sequencing of DNA, for which they received the 1980 Nobel Prize along with Berg. Sanger's group in England was the first to completely sequence a genome, identifying all 5,386 bases of the bacterial virus $\varphi\chi174$.

Another technological breakthrough occurred in 1985, when the polymerase chain reaction method was developed by Dr. Kary Mullis and colleagues at Cetus Corp. This team devised a method whereby minute samples of DNA can be multiplied a billion-fold for analysis. This technique, which has many applications in diverse fields of biology, is one of the most important scientific breakthroughs in gene analysis. Mullis received the Nobel Prize for this work in 1993.

At this time, however, DNA sequencing was still done by hand. At best, a researcher could manually sequence only a few hundred bases per day. To be able to sequence the human genome, machines would be needed that could sequence a million or more bases per day. In 1986 Leroy Hood developed the first generation of automated DNA sequencers, thereby dramatically increasing the speed with which bases could be sequenced. Thus, by the mid-1980s the stage was set.

With these new techniques, molecular biologists now felt that it might be feasible to sequence the entire human genome. The first serious discussions came in June 1985, when Robert Sinsheimer, chancellor of the University of California at Santa Cruz, called a meeting of leading scientists to discuss the possibility of sequencing the human genome. Sinsheimer was inspired by the success of the Manhattan Project, which was the concerted effort of many physicists to develop atomic weapons during World War II. That project led to rapid development and a massive influx of funding for physicists. Sinsheimer wanted a "Manhattan Project" for molecular biology, to enhance and expand human genome research.

Meanwhile, the DOE continued to be interested in the problem of identifying mutations caused by radiation exposure. Led by associate director Charles DeLisi, the DOE became a strong supporter of the genome-mapping initiative, for it understood that sequencing the entire genome would provide the best way to analyze such mutations. Thus the DOE became the first federal agency to begin funding the Human Genome Project.

Mapping the human genome came to be called the "Holy Grail of Molecular Biology," and many biologists were interested in the project. Most notable among them was Nobel laureate Gilbert who, through his interest, personality, and academic ties, developed enormous enthusiasm for the project. The initial goals set out for the Human Genome Project were threefold: to develop genetic linkage maps; to create a physical map of ordered clones of DNA sequences; and to develop the capacity for large-scale sequencing, because faster and cheaper machines along with other great leaps in technology would be needed to get the job done.

In 1988 the National Institutes of Health (NIH) set up an Office of the Human Genome, and Watson agreed to head the project. It had an estimated budget of approximately $3 billion, and 3 percent of the funding was devoted to the study of the social and ethical issues that would arise from the endeavor. A target date for completion of the project was set for September 30, 2005. By 1990 the Human Genome Project had received the additional endorsement of the National Academy of Sciences, the National Research Council, the DOE, the National Science Foundation, the U.S. Department of Agriculture, and the Howard Hughes Medical Institute. Sequencing of the human genome had now officially begun.

While sequencing the human genome was a primary goal, other sequencing projects were just as important. Many scientists established projects that sought to sequence several organisms of genetic, biochemical, or medical importance (see Table 1). These so-called model organisms, with their smaller genomes, would be useful in testing sequencing methodologies and for providing invaluable information that could be used to identify corresponding genes in the human genome. Sequence databases were established, and computer programs to search these databases were written.

Competition between the Public and Private Sectors

Dr. Craig Venter, a scientist at the NIH, felt that private companies could sequence genomes faster than publicly funded laboratories. For this reason he founded a biotechnology company called the Institute for Genomic Research (TIGR). In 1995 TIGR published the first completely sequenced genome, that of the bacterium *Haemophilus influenzae.* TIGR was soon joined by other biotechnology companies that competed directly with the publicly funded Human Genome Project.

Among these other biotech firms is Celera Genomics, founded in 1998 by Venter in conjunction with the Perkin-Elmer Corporation, manufacturer of the world's fastest automatic DNA sequencers. Celera's goal was to privately sequence the human genome in direct competition with the public efforts supported by the NIH and DOE and the governments of several foreign countries. Using 300 Perkin-Elmer automatic DNA sequencers along with one of the world's most powerful supercomputers,

Celera sequenced the genomes of several model organisms with remarkable speed and, in April 2000, announced that it had a preliminary sequence of the human genome.

In order eventually to make a profit, these biotech companies were patenting DNA sequences and intended ultimately to charge clients, including researchers, for access to their databases. This issue of patenting had already caused controversy. Watson felt strongly that the sequence data flowing from the Human Genome Project should remain within the public domain, freely available to all. Meeting opposition to this view, he stepped down from his position as director of the NIH-sponsored project in 1992 and was succeeded by Francis Collins.

Other researchers shared Watson's view, and in 1996 the international consortium of publicly funded laboratories agreed at a meeting in Bermuda to release all data to GenBank, a genome database maintained by NIH. The agreement reached by these scientists came to be known as "The Bermuda Principles," and it mandated that sequence data would be posted on the Internet within 24 hours of acquisition. Because the information is freely available to the public, the sequences can not be patented. The dispute between Celera Genomics and the International Human Genome Consortium continues, as scientists now begin the task of searching the genome for valuable information.

Progress in the Human Genome Project

Sequencing the human genome has led to some surprising results. For example, we once thought that highly evolved humans would need a great many genes to account for their complexity, and scientists originally estimated the number of human genes to be about 100,000. The draft of the human genome, however, indicates that humans may have only about 30,000 genes, far fewer than originally expected. Indeed, this is only about one-third more than the number of genes found in the lowly roundworm, *Caenorhabditis elegans* (approximately 20,000 genes), and roughly twice the number of genes in the fruit fly *Drosophila melanogaster* (approximately 14,000 genes). Subsequent estimates have placed the number of human genes closer to 70,000; the true number is unknown as of mid-2002. Scientists have learned that most of the genome does not code for proteins, but rather contains "junk DNA" of no known function. In fact, only a small percentage of human DNA actually encodes a gene.

autosomes chromosomes that are not sex determining (not X or Y)

The complete human genome consists of twenty-two pairs of chromosomes plus the X and Y sex chromosomes. On December 2, 1999, more than 100 scientists working together in laboratories in the United Kingdom, Japan, the United States, Canada, and Sweden announced the first completely sequenced human chromosome, chromosome 22, the smallest of the **autosomes**. To assure the accuracy of the sequence data, each segment must be sequenced at least ten times.

Thousands of scientists, working in more than 100 laboratories and 19 different countries around the world, have contributed to the Human Genome Project since its inception. Thanks to the development of later generations of high-speed automatic sequencers and supercomputers to handle the enormous amount of data generated, work on the project progressed well ahead of schedule and well under budget, a rare phenomenon in

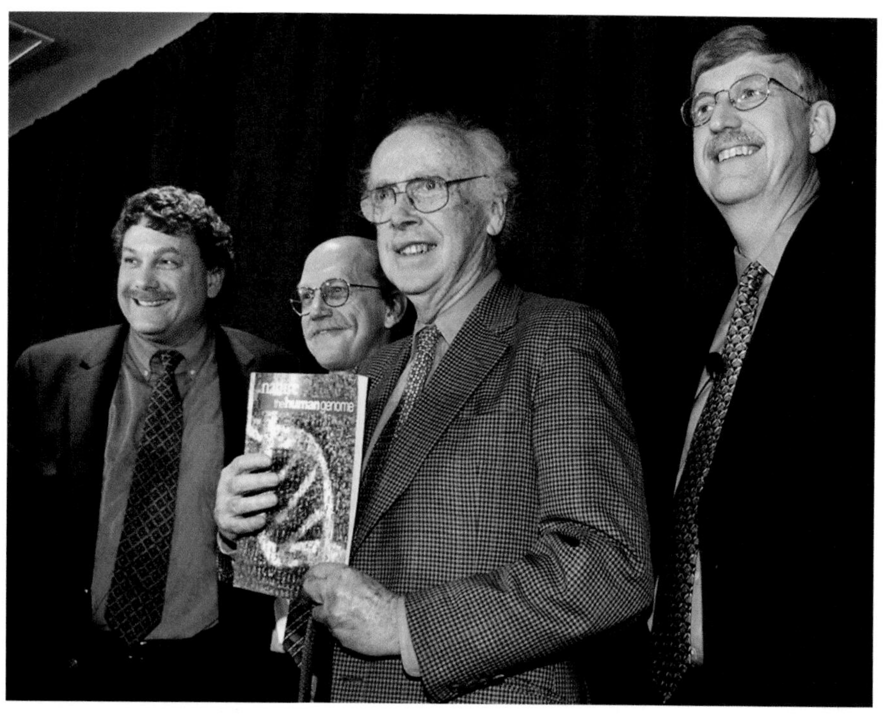

At a 2001 press conference, founder of the Human Genome Project, Dr. James Watson (with a copy of *Nature* in hand) announces the successful completion of human genome sequencing and poses with key supporters for photographers.

government-sponsored projects. In 2001 the first draft of the complete human genome was published. However, considerable work remained to be done, particularly in the sequence of regions of repetitive DNA.

Whose Genome Is It?

Although all humans share more than 99.99 percent of their genome sequences, each human is unique. Geneticists estimate that each person carries many mutations, perhaps hundreds or even thousands of them. Therefore, one of the major questions that has arisen in the Human Genome Project is "whose genome is it?" The final catalog of sequences, whenever it is complete, will have to take into account these individual variations, and ultimately there will be a "consensus sequence," but it will represent no one specific individual.

A related issue arises from the distinct differences that scientists anticipate will occur among different populations. Which sequences should be considered "normal," and which ones should be classed as "mutated"? The Human Genome Diversity Project was proposed in 1997 to catalog and study naturally occurring sequence variations among racial and geographic groups. This project never gained much support, however, because of the social and ethical ramifications to such a catalog. On the other hand, a Human Cancer Genome Anatomy Project was initiated to catalog all the genes that are expressed in cancer cells in order to aid in the detection and treatment of cancers. This project enjoyed much more support.

In 2002 Craig Venter announced that Celera had sequenced his personal genome, not a composite as originally claimed.

Patenting the Genome

From the outset there has been considerable debate among scientists, politicians, and entrepreneurs as to whether the human gene sequences can or

A Celera Genomics technician inputs DNA samples while the DNA sequencer machine searches and catalogs proteins.

should be patented. Indeed, this debate was the reason that Watson resigned as the first director of the NIH Human Genome Project program in 1992. Watson's position was opposed by many biotechnology companies, which hoped to recover the cost of their genome research and began patenting short segments of sequenced DNA without any idea as to their function. As of 2000, the U.S. Patent and Trademark Office (PTO) changed its policy, and began granting patents only to genes that have been identified, rather than just the random sequenced fragments. The data that flow from genome sequencing will be an invaluable scientific resource, particularly in the area of developing new medical treatments, but its use will be restricted if individual organizations can claim exclusive use rights to large segments of it. It is thus clear that debate on the patenting of genes will continue for years to come.

At present much of data from genome research are available to scientists and other interested parties. The data generated by participants in the Bermuda Principles agreement can be accessed on-line at the National Center for Biotechnology Information (NCBI) Web site, at <www.ncbi.nlm.nih.gov/genome/guide/human>. The International Human Genome Consortium Web site provides a current list of genome sites that offer links to most genome databases at <www.ensembl.org/genome/central>. Information about all the genomes that have been sequenced, as well as information on the sequencing of cancer genes, can be found on the Internet at <http://www.ncbi.nlm.nih.gov>.

Genomics and Proteomics

The Human Genome Project has given rise to new fields of research. One of these is **genomics**. This new field combines information science with molecular biology. It is resulting in the "mining of the genome" for valuable sequence data.

genomics the study of gene sequences

An even more recent development is the field of **proteomics**, the study of protein sequences. Research in this field is rapidly expanding, as protein sequences can be predicted from the gene sequence. The folding of the proteins (secondary and tertiary structures) can be predicted by computers, leading to a three-dimensional view of the protein encoded by a particular gene. Proteomics will be the next big challenge for genetics research. Indeed, Celera is already gearing up for massive protein sequencing.

proteomics the study of the full range of proteins expressed by a living cell

Ethical Issues

From the very beginning of the Human Genome Project, many from both the scientific and public sector have been concerned with ethical issues raised by the research. These issues include preserving the confidentiality of an individual's DNA information and avoiding the stigmatization of individuals who carry certain genes. Some fear that insurers will deny coverage for "preexisting" conditions to people carrying a gene that predisposes them to particular diseases, or that employers might start demanding genetic testing of job applicants.

There are also concerns that prenatal genetic testing could lead to genetic manipulation or a decision to abort based on undesirable traits disclosed by the tests. In addition, some raise concerns that a full knowledge of the human genome could raise profound psychological issues. For example, individuals who know that they carry detrimental genes may find the knowledge to be too great a burden to bear. All of these ethical issues will ultimately have to be addressed by society as a whole. SEE ALSO AUTOMATED SEQUENCER; BIOINFORMATICS; CANCER; CLONING GENES; GENOME; GENOMICS; GENOMICS INDUSTRY; MODEL ORGANISMS; MULLER, HERMANN; POLYMERASE CHAIN REACTION; POLYMORPHISMS; REPETITIVE DNA ELEMENTS; SEQUENCING DNA; WATSON, JAMES; YEAST.

Ralph R. Meyer

Bibliography

Collins, Francis S., and Karin G. Jegalian. "Deciphering the Code of Life." *Scientific American* 281, 6 (1999): 86–91.

Cook-Deegan, Robert. *The Gene Wars: Science, Politics and the Human Genome Project.* New York: Norton, 1994.

Davies, Kevin. *Cracking the Genome: Inside the Race to Unlock Human DNA.* New York: Free Press, 2001.

Ezzell, Carol. "Special Report: Beyond the Human Genome Project." *Scientific American* 283, no. 1 (2000): 64–69.

Kevles, Daniel J., and Leroy Hood, eds. *The Code of Codes: Scientific and Social Issues in the Human Genome Project.* Cambridge, MA: Harvard University Press, 1992.

Koshland, Daniel E. Jr. "Sequences and Consequences of the Human Genome." *Science* 246 (1989): 189.

Nature. "The Human Genome." Special Issue 409 (Feb. 15, 2001): 860–921.

Science. "The Human Genome." Special Issue 291 (Feb. 16, 2001): 1145–1434.

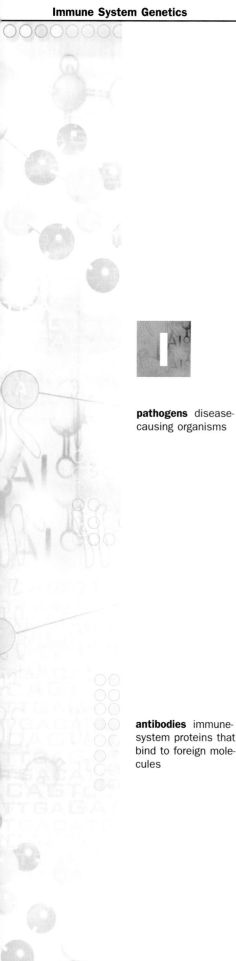

Shostak, Stanley. *Evolution of Sameness and Difference. Perspectives on the Human Genome Project.* Amsterdam: Harwood Academic Publishers, 1999.

Sloan, Phillip R., ed. *Controlling Our Destinies: Historical, Philosophical, Ethical, and Theological Perspectives on the Human Genome Project.* South Bend, IN: Notre Dame Press, 2000.

Watson, James D., and Robert M. Cook-Deegan. "Origins of the Human Genome Project." *FASEB Journal* 5 (1991): 8–11.

Human Immunodeficiency Virus *See HIV*

Huntington's Disease *See Triplet Repeat Disease*

Hybrid Superiority *See Heterozygote Advantage*

Immune System Genetics

The immune system is the set of cells and glands that protects the body from invasion and infection by viruses, bacteria, and other **pathogens**. The immune system must be able to recognize any foreign target, or antigen, of which there are potentially millions. Pathogenic organisms change over time, and new antigens evolve that must also be targeted. At the same time, the immune system must distinguish pathogenic antigens from the body's own tissues, attacking the former and sparing the latter. The key to the scope and specificity of the immune system response is in the genes that give rise to it.

pathogens disease-causing organisms

Overview of the Immune System

The immune system includes several interacting components. Nonspecific immunity (protection against any invasion) is provided by the barriers of the skin and mucous membranes lining the lungs and gut. Additional nonspecific defenses are provided by the inflammatory response and the complement proteins in the bloodstream. We shall not deal further with these defenses.

Specific immunity is the set of defenses mounted against a specific invader. It involves the action of three major types of cells: B cells, T cells, and macrophages. In broad, somewhat oversimplified terms, B cells make proteins called **antibodies** that attach to foreign antigens, serving as warning flags. T cells coordinate the immune attack and destroy virus-infected cells. Macrophages consume flagged antigens and clean up the debris from a T cell attack on infected cells.

antibodies immune-system proteins that bind to foreign molecules

An antibody binds to an invader when its shape fits some shape (the antigen) on the invader's surface. Any particular invader, such as a bacterial cell, may have dozens of such antigens.

The Puzzle of Antibody Diversity

B cells are created in the bone marrow. Many millions of different B cells are made, each containing a unique gene for the specific antibody that it (and all its descendants) will make. A group of B cells with all its descen-

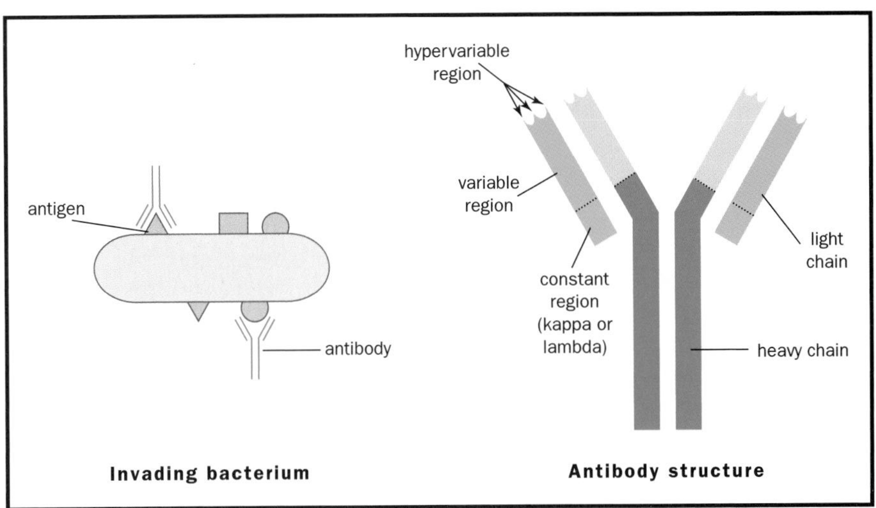

Antibodies bind to antigens on the surface of foreign substances.

dants is called a clone. Thus, the antibody made by one B cell clone differs from that made by any other B cell clone. T cells develop along a slightly different pathway but also contain a unique protein, called the T cell receptor, which is coded for by a gene unique to that T cell clone.

Antibodies are proteins, and like all of the body's proteins, must be encoded by genes. However, the number of distinct antibodies each of us makes (many millions) is vastly greater than the total number of genes in our entire genome (30,000–70,000). How is all this diversity encoded? To understand the answer, it is helpful to look at the structure of an antibody.

Antibody Structure

The antibody is formed from four **polypeptides** that link up in the shape of a Y. There are two identical long heavy (H) chains and two identical short light (L) chains. The tips of each branch of the Y form a pocket, and it is here the antibody binds antigen. Thus, these twin pockets are called the antigen-binding regions of the antibody.

polypeptides chains of amino acids

By comparing the amino acid sequences of antibodies from different B cell clones, several important features can be discovered. Light chains, for instance, have a constant region, with amino acid sequences that differ little from clone to clone, and a variable region, with sequences that differ considerably. The constant region comes in two different forms, termed "kappa" and "lambda." The amino acid sequence of one kappa constant sequence differs little or not at all from clone to clone; similarly, all lambda constant sequences are essentially identical. The variable region does differ considerably between clones. The heavy chain also has a constant region (of which there are five forms) and a variable region.

The constant regions of all the chains are found toward the bottom of the Y, while the variable regions are found toward the tip. Furthermore, within each variable region, there are three hypervariable regions, whose five to ten amino acids differ even more than the other portions of the variable region. These hypervariable regions form the actual points of contact between antibody and antigen.

Gene Segments Combine Randomly to Generate Diversity

The fundamental principle governing antibody generation is combinatorial diversity. A large number of genes are generated by choosing from among a smaller pool of differing gene segments and combining them in different ways. This process, known as **somatic** recombination, is similar in principle to constructing words. The alphabet's twenty-six letters can be combined to make 676 (26^2) two-letter words and almost 12 million five-letter words.

somatic nonreproductive; not an egg or sperm

To understand the molecular details of somatic recombination, let us focus on the creation of a kappa-type light chain. The process is similar for lambda light chains and only marginally more involved for a heavy chain.

We noted that the light chain has both a variable and a constant region. There are forty gene segments that can code for the variable (V) region and a single segment that codes for the constant (C) region. In addition, there are five possible coding segments for the J region, a short region that is also present on light chains. All of these genes and segments are located in sequence on chromosome 2. Each V and J segment is flanked by special noncoding sequences that facilitate the next stage, in which specific segments are joined.

Somatic recombination begins when special recombining proteins randomly bring together the downstream end of one V segment and the upstream end of one J segment. They do this by attaching to the flanking sequences and bending the intervening DNA into a loop. The loop is cut out and degraded, and the remaining DNA is spliced together. The product is the mature antibody gene.

Note in the diagram on the right that the resulting gene may still have some extra upstream V segments. An ingenious mechanism prevents such segments from being transcribed to make messenger RNA, however.

Each V segment contains a **promoter**, the region to which RNA polymerase binds to start transcription. The promoter is inactive, though, until it is brought close to an "enhancer" region between the J and C segments. Thus, transcription will begin at the V segment closest to the enhancer, and only this one V segment is transcribed—the others are too far from the enhancer. The gene may also have extra downstream J segments and **intron** sequences between J and C. These are transcribed, but they are removed by RNA processing.

promoter DNA sequence to which RNA polymerase binds to begin transcription

intron untranslated portion of a gene that interrupts coding regions

Other Sources of Diversity

The random combination of V and J segments alone can produce millions of possible combinations. More diversity arises because the joining of V and J chains is done imprecisely, with the possible loss or gain of several nucleotides, resulting in added or deleted amino acids.

Remember also that each antibody includes both light and heavy chains. Heavy chains are produced by a similar combinatorial process, using a different, larger set of gene segments. The combination of a randomly produced light chain with a randomly produced heavy chain produces even more diversity. Finally, when a B cell multiplies in response to antigens, the

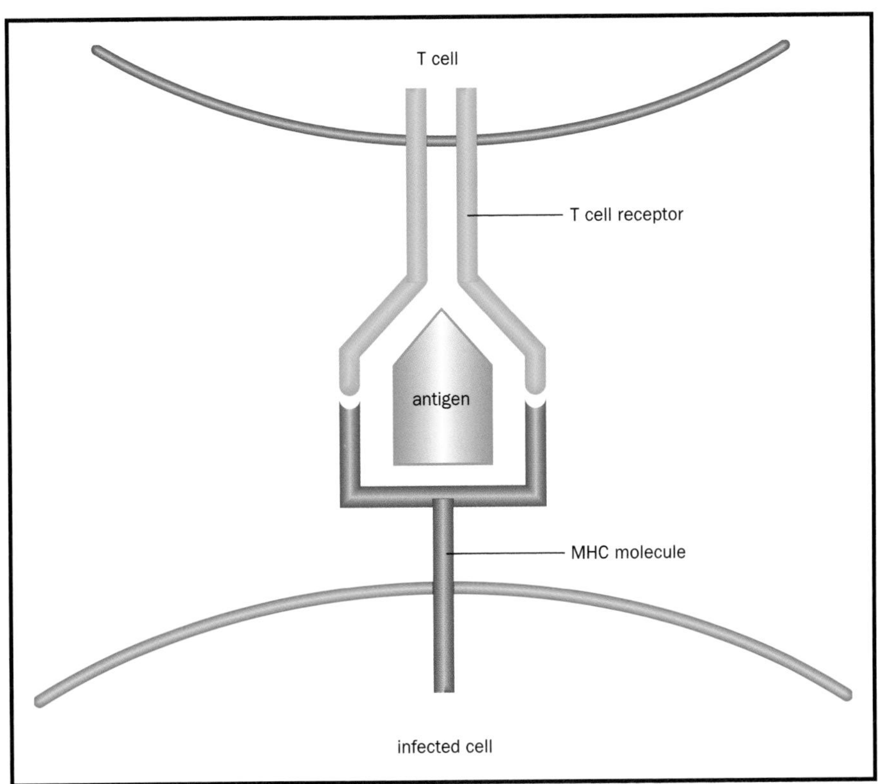

The T cell identifies an infected cell by binding to an antigen held within an MHC molecule on the surface.

rearranged gene can mutate, making some members of the clone different from others. The number of possible antibodies available through all these processes is in the trillions.

T Cell Receptors

As mentioned above, T cells help control the immune response and kill infected cells. Infected cells are recognized because they chop up foreign proteins from the invader and display the bits on their surface. These bits, which are antigens, are held aloft by surface proteins, called MHC (major histocompatibility complex) proteins. The MHC-antigen complex is recognized by the T cell receptor, in cooperation with one or more other T cell surface molecules.

When a T cell discovers a cell whose MHC proteins contain foreign antigens, it marks the cell for destruction. The T cell receptor interacts with antigens in much the same way as an antibody does, although the size of the antigen it recognizes is smaller. T cell receptors come in as many diverse forms as antibodies do, and, while the details differ, their diversity is generated in much the same way, with random recombination of gene segments.

The Major Histocompatibility Complex

The T cell-MHC interaction serves another, related function: It confirms that the cell is part of the self that the immune system should be protecting. Thus, MHCs serve as self-recognition markers. When a T cell recognizes foreign MHCs, as would occur in an organ transplant, it sets in motion an immune attack to reject the foreign tissue. Indeed, "histocompatibility" means compatibility of tissues, and these proteins control that process.

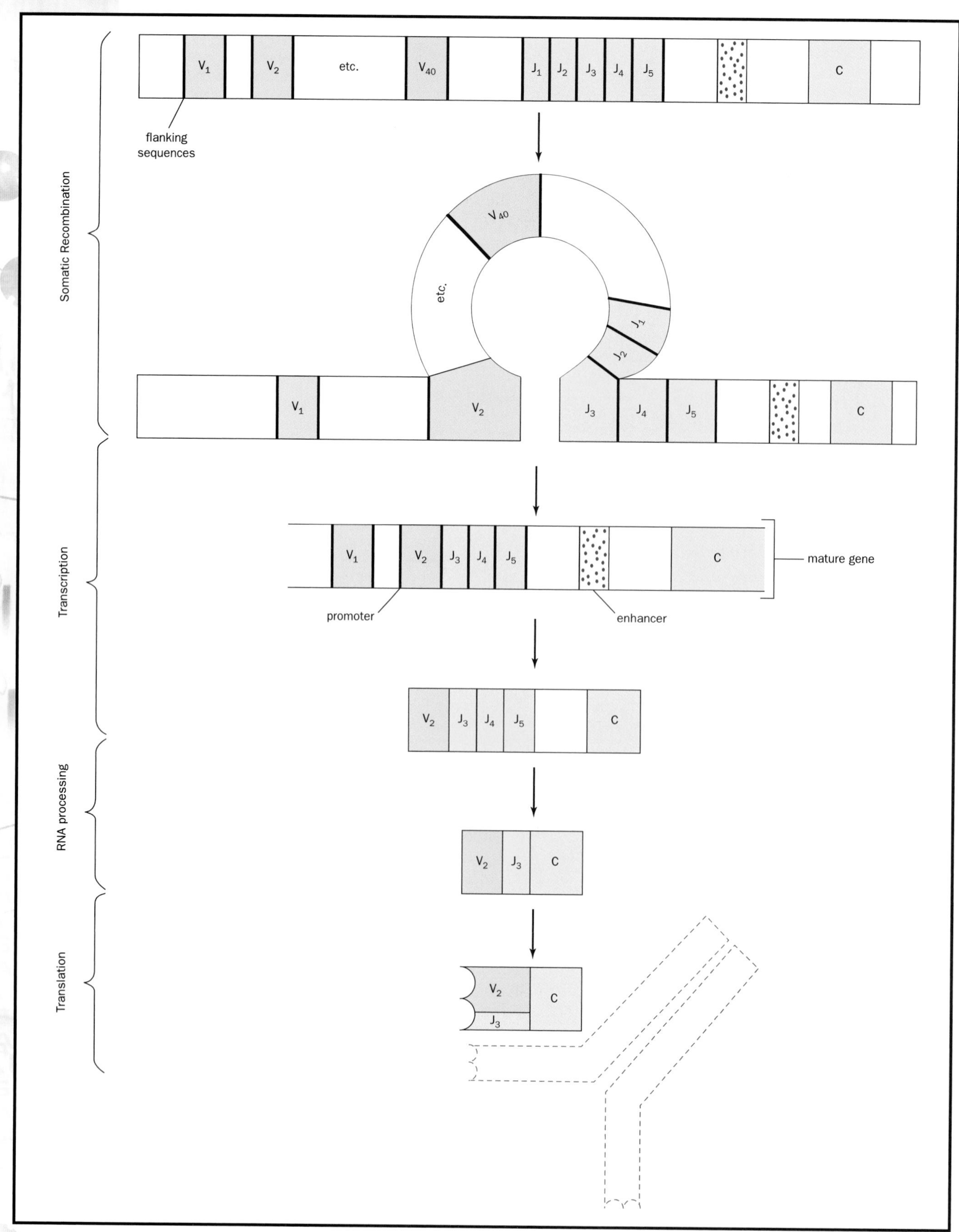

Somatic recombination in a B cell brings together a V segment and a J segment, both chosen at random. Intervening V and J segments are removed and degraded. Upstream V segments are not transcribed because their promoters are not close enough to the enhancer. Downstream J segments are transcribed but removed during RNA processing. The remaining V and J segments each code for part of the variable region of the antibody, while the C segment codes for the constant region.

There are two major classes of MHC proteins, called class I and II, with different functions in antigen presentation. Class I contains three members, each coded for by different genes, and class II contains four members. For almost every gene, there are multiple **alleles**. The number of alleles per gene ranges from a handful to more than 100. Since each person will inherit and express a unique set of MHC alleles, once again we can see the combinatorial possibilities: There are millions of different combinations of MHC alleles, and very few people are likely to have exactly the same set. This is what makes organ transplants so difficult. Matching MHC types is the key to success, but even close relatives may have different allele sets.

The MHC genes are believed to be the most allelically diverse of all human genes.

alleles particular forms of genes

Richard Robinson

Bibliography

Alberts, Bruce, et al. *Molecular Biology of the Cell*, 4th ed. New York: Garland Science, 2002.

Janeway, Charles A., Jr., et al. *Immunobiology: The Immune System in Health and Disease*, 5th ed. New York: Garland Publishing, 2001.

Imprinting

Imprinting refers to the chemical modification of the DNA in some genes that affects how or whether those genes are expressed. One particular kind of DNA imprinting found in mammals is known as parental genomic imprinting, in which the sex of the parent from whom a gene is inherited determines how the gene is modified. While imprinting has been found in only about fifty human genes to date, some estimates suggest it may occur in several hundred more, in perhaps up to 1 percent of all genes. Imprinting defects are responsible for several human diseases, including some forms of cancer. Imprinting also occurs in other organisms, from yeast to plants to fruit flies.

Gene Expression in Imprinted and Nonimprinted Genes

Chromosomes, and the genes they contain, are inherited in pairs, with one copy of each supplied from each parent. For most genes, both members of the pair, called the maternal and paternal alleles, are used equally. Both are expressed (read by the transcription machinery to make protein) in roughly equal amounts.

In contrast, for most imprinted genes, only one allele is expressed, while the other copy is silenced by imprinting. For some genes it is the maternal copy, for others it is the paternal copy. This is an exception to the Mendelian assumption that the two parents contribute equally to the **phenotype** controlled by autosomal genes. For some genes, both alleles are expressed, but one copy is expressed much more than the other. For some genes, the silencing occurs in some tissues but not others.

phenotype observable characteristics of an organism

Imprinted genes should not be confused with sex-linked genes, which are carried on the X or Y chromosome. Most imprinted alleles are located on **autosomes**, but are "stamped" with the sex of the parent that contributed them.

autosomes chromosomes that are not sex determining (not X or Y)

Imprinting should also not be confused with dominant and recessive alleles, in which one allele always controls the phenotype at the expense of

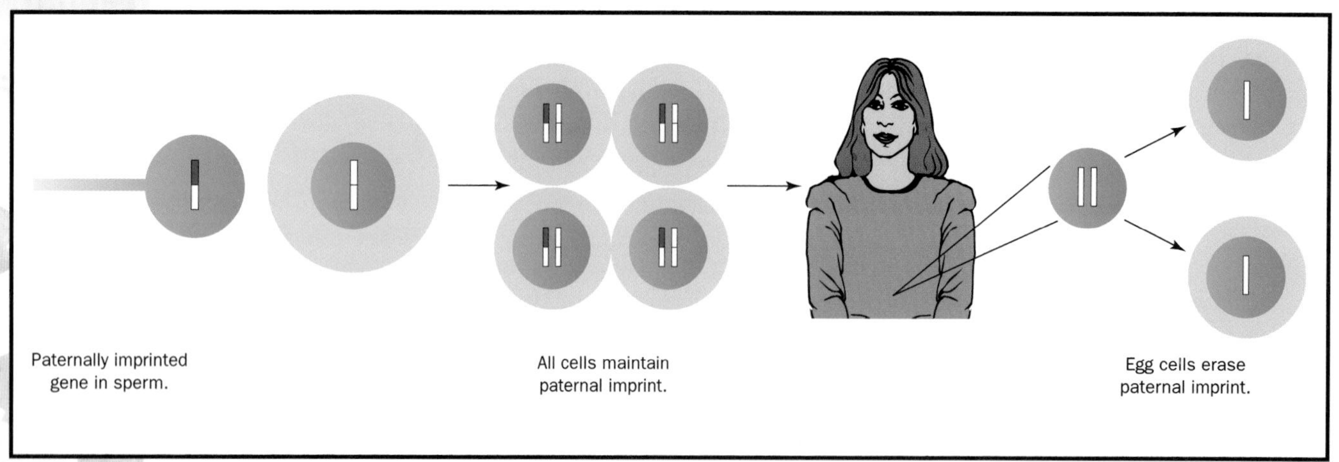
Paternally imprinted gene in sperm.
All cells maintain paternal imprint.
Egg cells erase paternal imprint.

Imprinting silences a gene based on the sex of the parent it came from. The imprint is reset in each new generation.

recessive requiring the presence of two alleles to control the phenotype

dominant controlling the phenotype when one allele is present

the other, because of differences in the alleles themselves. The "dominance" seen in imprinting is determined by the sex of the parent contributing the allele, not any property of the allele itself. Thus, a particular allele will appear to be **recessive** when inherited from one parent, but **dominant** when inherited from the other. Such an effect, in which the expression difference is not due to the alleles but to forces acting on them from outside, is termed an "epigenetic effect."

Imprinting is thought to be responsible for many cases of incomplete penetrance, an inheritance pattern in which a dominant gene (as for a genetic disease) is not expressed in some individuals despite being present. Imprinting offers a mechanism by which a particular allele can be turned on or turned off as it is passed down through successive generations.

Timing and Mechanism of Imprinting

gametes reproductive cells, such as sperm or eggs

Although the details of imprinting are still unknown, it is clear that imprinting must occur either during the formation of the **gametes** or immediately after fertilization, while the two chromosome sets are still distinct. The imprint must be reliably passed on to each new daughter chromosome during DNA replication.

RNA polymerase enzyme complex that creates RNA from DNA template

The exact molecular mechanism of imprinting is also unknown, but it is thought to involve the modification of a gene's promoter. The promoter is the upstream region to which **RNA polymerase** binds to begin transcription. Imprinting prevents or restricts binding of RNA polymerase, thus preventing gene transcription.

nucleotides the building blocks of RNA or DNA

transcription factors proteins that increase the rate of gene transcription

One method by which a gene becomes imprinted is believed to be by the addition of methyl groups ($-CH_3$) to cytosine **nucleotides** in the promoter region. The evidence for methylation is strong. Methylation is a common mechanism for gene silencing, because these bulky side groups interfere with the efficient binding of the various **transcription factors** required to attract the polymerase enzyme. Methylation patterns are known to be altered during gamete formation, and are reliably passed on during replication. Further evidence comes from the observation that altered methylation patterns in some imprinted genes are associated with the aberrant expression of the normally silent allele.

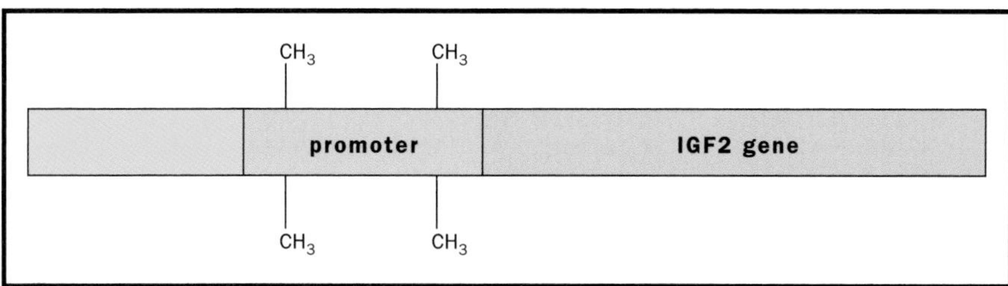

Methylation of the promoter region is believed to be an important silencing region.

Example of Imprinting: The IGF2 Gene

One of the best-studied imprinted genes is the one that encodes an insulin-like growth factor called growth factor 2 (IGF2). In this gene, the paternal copy is active, whereas the maternal copy is inactive. Imagine that two parents have produced a female child. During egg formation in the mother (or shortly after fertilization), the mother's copy of the IGF2 gene is **methylated**, rendering it transcriptionally silent. The child uses only the paternal allele to make the growth factor. However, when this child makes her own eggs, neither copy of the gene will remain active, because the alleles will have been "restamped" as coming from a female. The active allele she used throughout life is passed on in an inactive form to her children.

methylated a methyl group, CH_3, added

The protein encoded by the IGF2 gene is a growth factor, which stimulates the growth of target cells. Failure to properly imprint the maternal allele, or inheritance of two copies of the male allele, can have important consequences. For example, the expression of two copies of the IGF2 gene is associated with Beckwith-Wiedemann syndrome, a growth disorder, accompanied by an increase in a type of cancer called **Wilms tumor**. Other human cancers are also associated with improper imprinting (of other genes), causing either too much or too little gene expression.

Wilms tumor a cancerous cell mass of the kidney

Uniparental Disomy and Human Disease

Inheritance of two copies of one parent's chromosome (or part of it) is called uniparental disomy, a type of chromosome aberration. Detection of uniparental disomy in individuals with genetic disorders was one of the first clues that imprinting had important developmental and medical consequences.

Prader-Willi syndrome and Angelman syndrome can both be caused by uniparental disomy of chromosome 15, which carries a maternally expressed, paternally imprinted gene. Two maternal copies of the gene causes Prader-Willi syndrome, which is marked by mild mental retardation, decreased growth of the **gonads**, and obesity. Two paternal copies of this same gene causes Angelman syndrome, marked by severe mental retardation, small head size, seizures, inappropriate laughter, and distinctive facial features. (The gene itself codes for a protein involved in degrading other proteins.) Imprinting defects can also cause these syndromes in the absence of uniparental disomy, since the result is the same: either zero or two copies of the gene are expressed.

gonads testes or ovaries

Why Imprint?

The evolutionary reason for imprinting is not yet clear, although some scientists propose that, at least in mammals, it arose from an evolutionary tug

of war between males and females. In this scheme, fathers (who contribute only sperm) benefit when the embryo grows as fast as possible. Thus, silencing genes that slow down embryonic growth is in their interest, even if it depletes resources from the mother. Mothers, on the other hand, need to conserve their resources. Silencing genes that promote rapid growth is therefore in their interest. Supporting this hypothesis is the fact that many of the known imprinted genes regulate growth. Paternally expressed (maternally imprinted) genes such as IGF2 tend to promote growth, whereas maternally expressed (paternally imprinted) genes tend to inhibit it. SEE ALSO CHROMOSOMAL ABERRATIONS; FERTILIZATION; INHERITANCE PATTERNS; METHYLATION; RNA POLYMERASES.

Richard Robinson

Bibliography

Everman, David B., and Suzanne B. Cassidy. "Genomic Imprinting: Breaking the Rules." *Journal of the American Academy of Child and Adolescent Psychiatry* 39, no. 3 (March 2000): 386–389.

Greally, John M., and Matthew W. State. "Genomic Imprinting: The Indelible Mark of the Gamete." *Journal of the American Academy of Child and Adolescent Psychiatry* 39, no. 4 (April 2000): 532–535.

Paulsen, Martina, and Anne C. Ferguson-Smith. "DNA Methylation in Genomic Imprinting, Development, and Disease." *Journal of Pathology* 195, no. 1 (2001): 97–110.

Internet Resources

Geneimprint.com. <http://www.geneimprint.com/>.

Yale University School of Medicine and Yale-New Haven Hospital. <http://info.med.yale.edu>.

In Situ Hybridization

In situ hybridization is a technique used to detect specific DNA and RNA sequences in a biological sample. Deoxyribonucleic acid (DNA) and ribonucleic acid (RNA) are **macromolecules** made up of different sequences of four nucleotide bases (adenine, guanine, uracil, cytosine, and thymidine). *In situ* hybridization takes advantage of the fact that each nucleotide base binds with a **complementary** nucleotide base. For instance, adenine binds with thymidine (in DNA) or uracil (in RNA) using **hydrogen bonding**. Similarly, guanine binds with cytosine.

macromolecules large molecules such as proteins, carbohydrates, and nucleic acids

complementary matching opposite, like hand and glove

hydrogen bonding weak bonding between the H of one molecule or group and a nitrogen or oxygen of another

In a specialized molecular biology laboratory, researchers can make a sequence of nucleotide bases that is complementary to a target sequence that occurs naturally in a cell (in a gene, for example). When this complementary sequence is exposed to the cell, it will bind with that naturally occuring target DNA or RNA in that cell, thus forming what is known as a hybrid. The complementary sequence thus can be used as a "probe" for cellular RNA or DNA.

Thus, the term "hybridization" refers to the chemical reaction between the probe and the DNA or RNA to be detected. If hybridization is performed on actual tissue sections, cells, or isolated chromosomes in order to detect the site where the DNA or RNA is located, it is said to be done "*in situ*." By contrast, "in vitro" hybridization takes place in a test tube or other apparatus, and is used to isolate DNA or RNA, or to determine sequence similarity of two nucleotide segments.

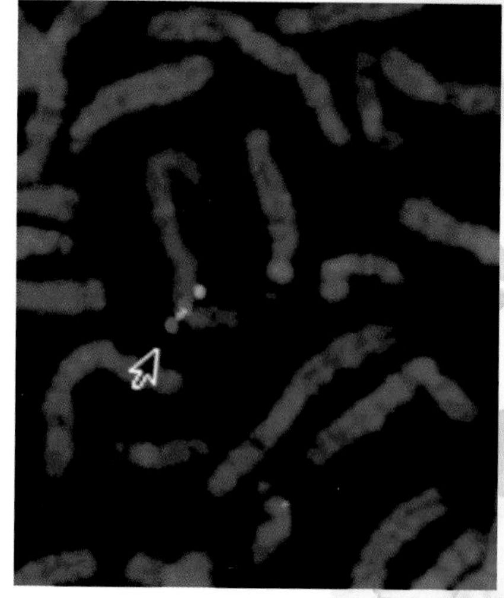

The arrow points to a chromosome section illuminated by the FISH procedure.

Application of the Probe for DNA or RNA to Tissues or Cells

In situ hybridization allows us to learn more about the geographical location of, for example, the messenger RNA (mRNA) in a cell or tissue. It can also tell us where a gene is located on a chromosome. Obviously, a detection system must be built into the technique to allow the **cytochemist** to visualize and map the geography of these molecules in the cells in question.

When *in situ* hybridization was first introduced, it was applied to isolated cell nuclei to detect specific DNA sequences. Early users applied the techniques to isolated chromosomal preparations in order to map the location of genes in those chromosomes. The technique has also been used to detect viral DNA in an infected cell. *In situ* hybridization of RNA has also been used to show that RNA synthesis (transcription) occurs in the nucleus, while protein synthesis (translation) occurs in the **cytoplasm**.

cytochemist chemist specializing in cellular chemistry

cytoplasm the material in a cell, excluding the nucleus

protocol laboratory procedure

Conditions that Promote Optimal *In Situ* Hybridization

Hybrid probes are known as cDNA or cRNA, because they are complementary to the target molecule. In developing an *in situ* hybridization **protocol**, it is vital to learn optimal temperatures and times needed for formation of the hybrid between the cDNA probe or cRNA probe and unique RNA or DNA in the cell. The optimal hybridization temperature depends on several factors, including the types of bases in the target sequence and the concentration of certain ingredients in the media. The concentration of cytosine and guanine in the sequence plays an important role. A cytochemist will use these factors to calculate optimal temperature when planning the experiment. One must be as careful in setting up an *in situ* hybridization experiment as one is when setting up a test tube hybridization assay. The cytochemist and molecular biologist work together to optimize the conditions.

Another vital consideration in developing good *in situ* hybridization techniques is the specificity of the probe itself. If the investigators know the exact nucleotide sequence of the mRNA or DNA in the cell, they can design a complementary probe and have it made in a molecular biology lab. However, if the investigators do not know the exact sequence, they may try a sequence that is as close to exact as possible (such as from a related species). For example, they might try a cDNA probe for a mouse DNA sequence on a tissue preparation from a rat. This may or may not work because, if over 5 percent of the base pairs are not complementary, the probe will bind only loosely to the target. This loose binding may cause it to be dislodged in the washing or detection steps and hence the reaction will not be detected, or only some of the sites may be detected and the labeling will not accurately reflect all of the target sites.

Application of Tools to Detect the Hybridization Sites

Forming the hybrid is not sufficient to allow it to be mapped, because the molecules are too small to be seen in the microscope. Thus, to map the geographical distribution of the gene or gene product, the cytochemist applies

antibodies immune system proteins that bind to foreign molecules

sensitive detection systems that allow the hybrid to be seen. This is done by labeling the cDNA or cRNA probe itself with a molecule that can either be visualized directly (such as fluorescein, which glows when exposed to fluorescent light in a microscope) or indirectly (such as radioactive sulfur, which can be detected by autoradiography; or biotin, which can be detected by avidin or by specific **antibodies** to biotin). Whichever type of molecule is chosen, it should be small, so that it does not interfere with the hybridization process.

The molecule used to visualize the hybrid is called the reporter molecule because it "reports" the site of the hybridization of the probe to the cellular DNA or RNA. Many different detection systems are available to cytochemists, employing a wide variety of reporter molecules. These include fluorescent compounds, colloidal gold compounds, or enzyme reactions or radioactive elements. Cytochemists will choose the most sensitive detection system that is also appropriate for their laboratories. For example, laboratories that do not want to work with radioactive compounds may choose one of the many nonradioactive methods.

One of the best-known nonradioactive *in situ* hybridization methods is FISH or "fluorescence *in situ* hybridization." It allows the detection of many genes in a chromosome or a nucleus, and different combinations of fluorescent reporter molecules are used to produce different colors. Using FISH, a veritable rainbow of colors can be used to map the location of genes on chromosomes. Another method uses enzyme reactions to form a product over the site of the mRNA or DNA in the cell. One can use different enzymes or enzyme substrates in the detection system and thus detect multiple gene products in the same tissue or cells.

Preserving the Tissues or Cells and Preventing Loss of DNA or RNA

Another important consideration in developing *in situ* hybridization technology involves the preservation of the cells or chromosomes. It is important to preserve the morphology (shape) and geographical site in the cell or chromosome where the target DNA or mRNA is located. Investigators may choose to use frozen sections, or they may treat cells or tissues with fixatives that cross-link proteins and stabilize cell structure. This prevents destruction during the hybridization and washing protocols.

Preserving DNA is easy because it is a highly stable molecule. However, preserving RNA is much more difficult because of a very stable enzyme called RNase, which may be found on glassware, in lab solutions, or on the hands of the cytochemist. RNase will quickly destroy any RNA in the cell or the RNA probe itself. Thus, investigators that work with RNA must use sterile techniques, gloves, and solutions to prevent RNase from contaminating and destroying the probe or tissue RNA.

Controlling the Specificity of the Cytochemical Assays

Good cytochemists know that experimental results must be checked and verified. For this reason, in addition to testing for reactions with their target DNA or RNA, they also test for reactions with unrelated nucleotide sequences. Likewise, they test for reactions with other components of the detection systems.

There are a series of controls that are run that detect if the labeling pattern is due to the proper sequence of reactants. For example, if the cytochemist leaves out the probe in the hybridization solution, there should be no reaction. Similarly, if the cytochemist changes the sequence of the probe, or uses a noncomplementary probe, there should be no reaction (unless that new sequences reacts with another sequence in the cell). Tests of the detection system must also be run by leaving out one or more components to learn if the reaction is dependent totally on the complete sequence of reactants. SEE ALSO DNA; NUCLEOTIDE; RNA.

Gwen V. Childs

Bibliography

Bloom, Mark V., Greg A. Freyer, and David A. Micklos. *Laboratory DNA Science: An Introduction to Recombinant DNA Techniques and Methods of Genome Analysis.* Menlo Park, CA: Addison-Wesley, 1996.

Brahic, M., and A.T. Haase. "Detection of Viral Sequences of Low Reiteration Frequency by *In situ* Hybridization." *Proceedings of the National Academy of Science USA* 75 (1978): 6125–6127.

Buongiorno-Nardelli, S., and F. Amaldi. "Autoradiographic Detection of Molecular Hybrids between rRNA and DNA in Tissue Sections." *Nature* 225 (1970): 946–948.

Childs, G. V. "*In situ* Hybridization with Nonradioactive Probes." In *Methods in Molecular Biology,* vol. 123: *In situ* Hybridization Protocols, I. A. Darby, ed. Totowa, NJ: Humana Press, Inc, 1999.

Gall, J. G., and M. Pardue. "Formation and Detection of RNA-DNA Hybrid Molecules in Cytological Preparation." *Proceedings of the National Academy of Science USA* 63 (1969): 378–383.

Haase, A.T., P. Venture, C. Gibbs, and W. Touretellotte. "Measles Virus Nucleotide Sequences: Detection by Hybridization *In situ.*" *Science* 212 (1981): 672–673.

John, H. A., M. L. Birnstiel, and K. W. Jones. "RNA-DNA Hybrids at the Cytological Level." *Nature* 223 (1969): 582–587.

Inbreeding

Inbreeding is defined as mating between related individuals. It is also called consanguinity, meaning "mixing of the blood." Although some plants successfully self-fertilize (the most extreme case of inbreeding), biological mechanisms are in place in many organisms, from fungi to humans, to encourage cross-fertilization. In human populations, customs and laws in many countries have been developed to prevent marriages between closely related individuals (e.g., siblings and first cousins). Despite these proscriptions, genetic counselors are frequently presented with the question "If I marry my cousin, what is the chance that we will have a baby who has a disease?" The answer is that when two partners are related their chance to have a baby with a disease or birth defect is higher than the background risk in the general population.

Increased Disease Risk

Many genetic diseases are recessive, meaning only people who inherit two disease **alleles** develop the disease. All of us carry several single alleles for genetic diseases. Since close relatives have more genes in common than unrelated individuals, there is an increased chance that parents who are closely related will have the same disease alleles and thus have a child who is homozygous for a recessive disease.

alleles particular forms of genes

locus site on a chromosome (plural, loci)

deleterious harmful

For instance, cousins share approximately one-eighth or 12.5 percent of their alleles. So, at any **locus** the chance that cousins share an allele inherited from a common parent is one-eighth. The chance that their offspring will inherit this allele from both parents, if each parent has one copy of the allele, is one-fourth. Thus, the risk the offspring will inherit two copies of the same allele is 1/8 × 1/4, or 1/32, about 3 percent. If this allele is **deleterious**, then the homozygous child will be affected by the disease. Overall, the risk associated with having a child affected with a recessive disease as a result of a first cousin mating is approximately 3 percent, in addition to the background risk of 3 to 4 percent that all couples face.

Inbreeding can be measured by the inbreeding coefficient (often denoted F). This is the probability that two genes at any locus in one individual are identical by descent (have been inherited from a common ancestor). F is larger the more closely related the parents are. For example, the coefficient of inbreeding for an offspring of two siblings is one-fourth (0.25), for an offspring of two half-siblings it is one-eighth (0.125), and for an offspring of two first cousins it is one-sixteenth (0.0625). (This is a different calculation than the calculation of shared alleles between cousins, above.)

In general, inbreeding in human populations is rare. The average inbreeding coefficient is 0.03 for the Dunker population in Pennsylvania and 0.04 for islanders on Tristan da Cunha. Inbreeding occurs in both those populations. Some isolated populations actively avoid inbreeding and have maintained low average inbreeding coefficients even though they are small. For example, polar Eskimos have an average inbreeding coefficient that is less than 0.003.

Beneficial changes can also come from inbreeding, and inbreeding is practiced routinely in animal breeding to enhance specific characteristics, such as milk production or low fat-to-muscle ratios in cows. However, there can often be deleterious effects of such selective breeding when genes controlling unselected traits are influenced too. Generations of inbreeding decrease genetic diversity, and this can be problematic for a species. Some endangered species, which have had their mating groups reduced to very small numbers, are losing important diversity as a result of inbreeding.

Genetic Studies of Inbred Populations

Inbred populations can offer a rich resource for genetic studies. They have the advantage of often being relatively homogeneous in both their genetics and environment. A method that has been used successfully to identify several recessive mutations in inbred groups is homozygosity mapping.

This approach looks for regions of alleles at genetic loci that are linked to one another and are homozygous. With inbreeding, there is an increased chance that, in an affected individual, the two alleles at the disease locus will have descended from a common ancestor. Therefore tightly linked markers (identifiable DNA segments) surrounding the disease locus will also tend to come from the same ancestral chromosome and thus be identical on both homologous chromosomes.

Together with colleagues, Erik Puffenberger, a research scientist and laboratory director at the Clinic for Special Children in Strasburg, Pennsylvania, capitalized on the inbreeding in a large Mennonite kindred to identify the location of a gene for Hirschprung disease on chromosome 13. In this family, parents of an affected child are, on average, related as closely as second or third cousins. The region was located because, true to theory, affected individuals shared alleles that were identical by descent at the region containing the disease gene. SEE ALSO Founder Effect; Inheritance Patterns; Pedigree; Population Genetics.

Eden R. Martin and Marcy C. Speer

Bibliography

Cavalli-Sforza, Luigi L., and Walter F. Bodmer. *The Genetics of Human Populations.* Mineola, NY: Dover Publications, 1999.

Puffenberger Erik G., et al. "Identity-by-Descent and Association Mapping of a Recessive Gene for Hirschprung Disease on Human Chromosome 13q22." *Human Molecular Genetics* 3 (1994): 1217–1225.

Individual Genetic Variation

"Variety is the spice of life," or so the saying goes. In fact, it is probably more precise to say that variety is the key to life. It is genetic variation that contributes to the diversity in **phenotype** that provides for richness in human variation, and it is genetic variation that gives evolution the tool that it needs for selection and for trying out different combinations of alleles and genes.

phenotype observable characteristics of an organism

locus site on a chromosome (plural, loci)

polymorphic occurring in several forms

deleterious harmful

Some variation is directly observable. Some examples of human genetic variants include the widow's-peak hairline, which is dominant to non-widow's peak; free earlobe, which is dominant to attached earlobe; facial dimples, which are dominant to no facial dimples; and tongue-rolling, which is dominant to non-tongue-rolling. Another example is the ability to taste phenylthio-carbamide (PTC), a bitter-tasting substance. Seven out of ten people can taste the bitterness in PTC paper, and the ability to taste it is dominant to nontasting. Much variation at the genetic level, however, is not observable just by looking at someone. Even these "invisible" traits can nonetheless be scored in the laboratory.

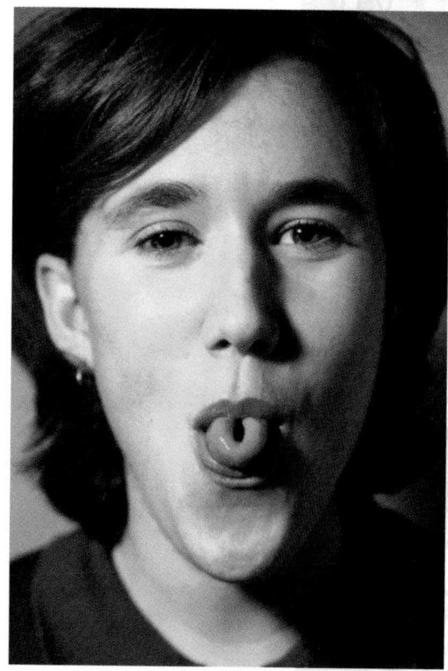

This woman has the inherited trait that allows her to roll her tongue.

Variation and Alleles

Any **locus** having two or more alleles (variant forms of particular genes), each with a frequency of at least 1 percent in the general population, is considered to be **polymorphic**. The difference between two alleles may be as subtle as a single base-pair change, such as the thymine-to-alanine substitution that alters the B chain of hemoglobin A from its wild type to its hemoglobin sickle cell state. These single nucleotide polymorphisms are termed "SNPs" ("snips").

As of March 2002, in the approximately 3.2 billion base pairs of DNA, approximately 3.2 million SNPs have been identified. Some base-pair changes have no **deleterious** effect on the function of the gene; nevertheless, these functionally neutral changes in the DNA still represent different alleles. Alternatively, allelic differences can be as extensive as large, multi-codon deletions, such as those observed in Duchenne muscular dystrophy.

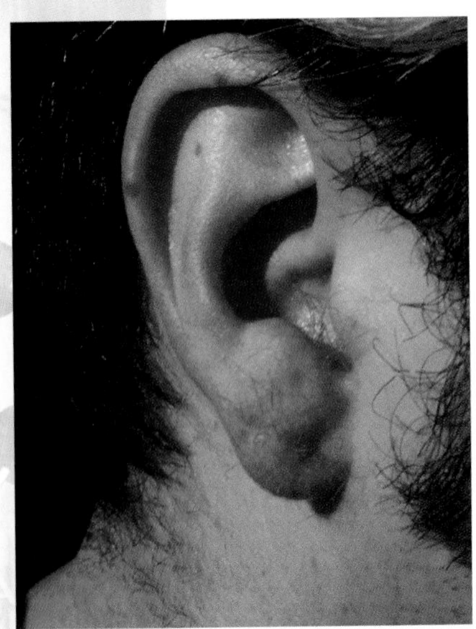

Genetic variations cause differences in appearances such as dimples, widow's peaks, and even hairy ears.

Ear wax consistency is due to single gene with two alleles (wet vs. dry), located near the chromosome 16 centromere.

germ cells cells creating eggs or sperm

somatic nonreproductive; not an egg or sperm

Genetic variation is transmissible from parent to child through **germ cells**, and it can occur anew via mutation in either germ cells or in **somatic** cells. Interestingly, new mutations occur twice as frequently in sperm as in eggs, probably because so many more cell divisions are required to make sperm than eggs.

Scoring Variation in the Lab

Differences in alleles can be scored via laboratory testing. The ability to score allele differences accurately within families, between families, and between laboratories is critically important for linkage analysis in both simple Mendelian and genetically complex common disorders. Linkage analysis traces coinheritance of a disease gene and polymorphic markers such as SNPs to discover where in the chromosomes the disease gene is located.

Allele scoring strategies may be as simple as noting the presence (+) or absence (-) of a deletion or point mutation, or as complicated as assessing the allele size in base pairs of DNA. The latter application is common when highly polymorphic; microsatellite repeat markers are used in linkage analysis.

When genetic counselors talk to patients and families about mutations (a type of genetic variation) that are present in themselves or their children, they are careful to point out that each individual is estimated to carry between five to seven deleterious alleles that, in the right combination with other genes or with specific environmental influences, can lead to disease. Most of us do not know which deleterious genes we carry. Some are recessive and do not influence the genotype unless paired with a second recessive allele. Thus, these alleles will not be noticed without genetic analysis. And, genetic counselors are careful to avoid the term "mutation," because it is potentially stigmatizing. When speaking with patients, they prefer to use the more neutral term "variant." SEE ALSO DISEASE, GENETICS OF; GENOTYPE AND PHENOTYPE.

Marcy C. Speer

Bibliography

Internet Resources

SNP Consortium, Ltd. <http://brie2.cshl.org>.

"Human Genome Project Information." U.S. Department of Energy. <http://www.ornl.gov/hgmis/publicat/primer/intro.html>.

Information Systems Manager

An information systems manager (ISM) is a professional whose skills are needed to handle the large amounts of information generated by and analyzed in the modern genetics laboratory. A successful information systems manager needs to be experienced with the technical aspects of computer hardware and networking systems. The daily work may involve managing a team of information technology workers, so leadership and management skills are usually a necessary qualification. Like most jobs in the scientific

and technical areas, this one involves interacting regularly with staff, other units, upper management, and scientists, so that proficiency in oral and written communication is essential. Although an information systems manager is not usually required to have a deep understanding of the genetic sciences, the ISM must be well acquainted with the information needs of the organization and must possess advanced technical knowledge.

The training needed for a career as an information systems manager begins with formal education and training in computer programming, database fundamentals, systems analysis and design, data communication, and networking, all at the undergraduate level. A major, or at least a concentration, in information systems management at either the bachelor's or master's degree level will also provide a solid foundation for a position as an ISM. Many information systems managers begin their career as systems analysts, programmers, or computer engineers, or were in other computer-related positions. An in-depth understanding of computer technology and applications will make the candidate attractive for many other careers as well, since the projected outlook for all computer-related professions is continued growth through 2008.

The duties of an ISM can be quite diverse. They include setting up networks, including the installation of lines, hardware, and software; administering servers; programming; and setting up intra- and internets. ISMs may even be called upon to do some Web page design. An information systems manager is not an entry-level position, as the ISM has considerable administrative duties. The job requires hiring, training, assigning, and supervising computer specialists so that each employee does his or her particular duty toward meeting the organization's goals. In addition, ISMs need strong budgeting skills and the ability to communicate with contractors, suppliers, and financial groups, for they are involved in making decisions concerning equipment purchases.

The majority of ISMs do most of their work from an office, but may be called upon to go to the client's site to set up networks, hardware, or software. This may include doing their work in research laboratories, where they may be exposed to hazards arising from the investigations being conducted there. ISMs often work under stringent time and budgetary contraints. It is not unusual for ISMs to work more than forty hours per week.

The job can be rewarding when the organization makes progress in meeting their goals. There is usually plenty of room for personal growth, as many ISMs are advanced to higher managerial positions. ISMs typically enjoy a good salary and a substantial benefits package. Salaries usually range from $45,000 to $120,000 per year, with most earning between $50,000 and $95,000. The amount earned depends on education, the amount of experience, and whether the job is in the public, academic, or private sector. Workers in state government usually earn the least, with a median annual income of $63,500, while workers in industrial settings earn the most, with a median annual income of $87,500. SEE ALSO BIOINFORMATICS; COMPUTATIONAL BIOLOGIST; STATISTICAL GENETICIST.

Judith E. Stenger

Bibliography

Ahituv, Niv, and Seev Neumann. *Principles of Information Systems for Management.* Dubuque, IA: Wm. C. Brown Company, 1983.

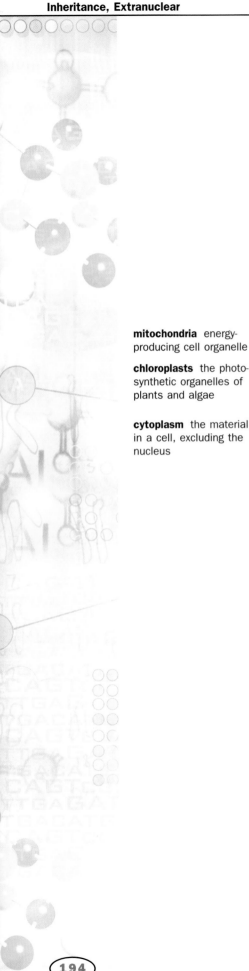

Alter, Steven. *Information Systems: A Management Perspective,* 2nd ed. Menlo Park, CA: Benjamin/Cummings Publishing Company, 1996.

Internet Resource

U.S. Bureau of Labor Statistics. *Occupational Outlook Handbook, 2000–2001 edition.* U.S. Department of Labor. Washington, DC, 2000. <http://stats.bls.gov/ocohome.htm>.

Inheritance, Extranuclear

Less than a decade after the rediscovery of Mendel's laws describing the inheritance of genes in the nucleus, hereditary traits were discovered that obey a different set of laws. The genes involved in this non-Mendelian pattern of inheritance reside outside the nucleus, in the cytoplasm of the cell. Specifically, they were found to reside in **mitochondria**, **chloroplasts**, or intracellular symbiotic bacteria. Those genes play important roles in the cell. Mutations in extranuclear genes are responsible for some hereditary diseases in humans and other organisms, are used in plant breeding, and are used to study population genetics and evolution.

mitochondria energy-producing cell organelle

chloroplasts the photosynthetic organelles of plants and algae

cytoplasm the material in a cell, excluding the nucleus

Genes in Mitochondria and Chloroplasts

The **cytoplasm** of most eukaryotic cells contain organelles called mitochondria, where energy is extracted from food molecules and stored in ATP (adenosine triphosphate) for later use throughout the cell. Virtually all of the oxygen we use is consumed by our mitochondria.

Mitochondria contain their own DNA molecules (mitochondrial DNA, or mtDNA). These molecules carry a few dozen genes that are essential for energy metabolism. For example, the *cob* gene carries the instructions for making a protein, cytochrome *b*, which is an important component of the electron transport system in mitochondria. All the other proteins and RNAs encoded by mtDNA genes are also used in energy metabolism. However, many other key proteins for energy metabolism are encoded by nuclear genes. These are synthesized elsewhere in the cell and imported into the mitochondria. In fact, while the mtDNA genes are absolutely essential for the aerobic production of energy, the majority of all mitochondrial components derive from nuclear genes.

In addition to mitochondria, the cells of plants and algae also contain organelles called chloroplasts, in which photosynthesis takes place. Like the mitochondria, chloroplasts contain DNA molecules (chloroplast DNA, or cpDNA). The cpDNA molecules have genes that encode some of the proteins needed for photosynthesis. Also like the mitochondria, the majority of components needed for photosynthesis are made outside the chloroplast, using information from nuclear genes.

Endosymbiotic Origin of Mitochondria and Chloroplasts

Mitochondria and chloroplasts are self-replicating organelles: They can arise only by division of preexisting mitochondria or chloroplasts. DNA molecules are replicated and divided up among the daughter organelles after division. As a result, organelle genes show hereditary continuity from cell to cell and from parent to offspring, as do the genes in the nucleus. In

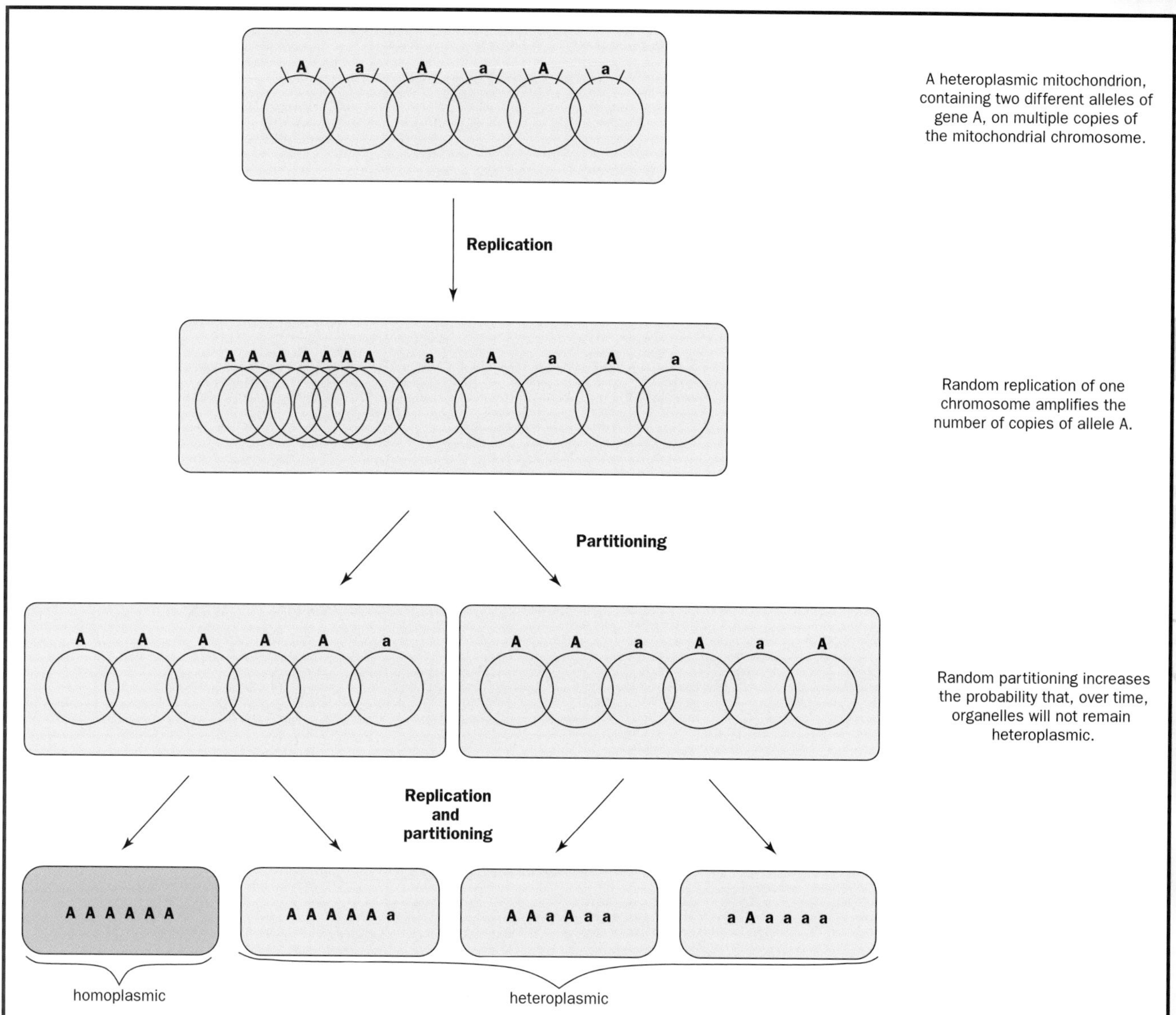

A heteroplasmic mitochondrion, containing two different alleles of gene A, on multiple copies of the mitochondrial chromosome.

Random replication of one chromosome amplifies the number of copies of allele A.

Random partitioning increases the probability that, over time, organelles will not remain heteroplasmic.

fact, the ancestry of mtDNA and cpDNA can be traced back to intracellular **symbionts** (termed "endosymbionts").

Multiple copies of the mitochondrial genome exist in each organelle, which may bear different alleles. Replication and partitioning can create mitochondria with only one allele type.

Early in evolutionary history, an ancestor of the eukaryotes ingested an ancestor of the aerobic α-proteobacteria. This bacterium avoided digestion and became a permanent resident of the host cell, dividing within it and providing it with energy from **aerobic** metabolism. Gradually, over millions of years, the endosymbionts transferred most of their genes to the host nucleus, becoming completely dependent on the host cells. The host cells, in turn, came to depend on the symbiont for aerobic energy production.

Much later, an ancestor of the modern green algae and plants ingested a cyanobacterium capable of photosynthesis. Gradually this endosymbiont also lost most of its genes to the nucleus and became dependent on the host cell, while providing the host with energy from photosynthesis. The resulting **organelles** are self-replicating, like the original symbiont.

Because mitochondria and chloroplasts originated as endosymbiotic bacteria, their genomes differ from the nuclear genome in several important

symbionts organisms living in close association with other organisms

aerobic with oxygen, or requiring it

organelles membrane-bound cell compartments

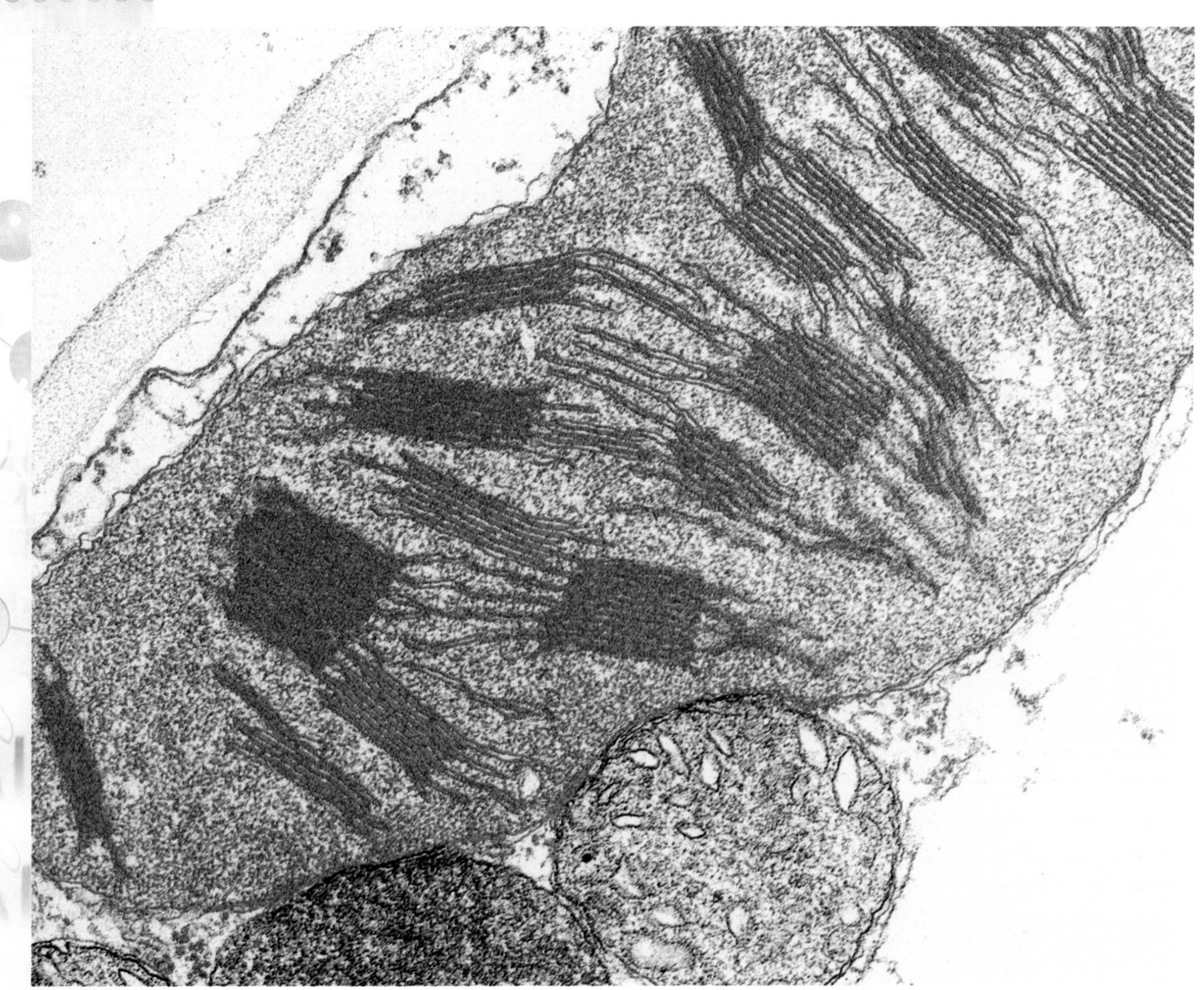

Tracked mutations in extranuclear genes, as in the chloroplasts of the plant cell shown here can be used to study the evolution of a species.

ways. First, all of the organelle genes are located on a single, circular DNA molecule. Second, the genes are virtually contiguous, with little or no intergenic DNA. Third, the gene coding sequences are continuous. In other words, there are no (noncoding) introns separating gene-coding sequences. Also, each organelle has many copies of the DNA molecule, and each cell usually has more than one organelle.

Plant cells commonly have from two to several hundred chloroplasts, while animal cells often have hundreds of mitochondria. As a result, each cell has hundreds or thousands of mtDNA or cpDNA molecules, and hence of each mitochondrial or chloroplast gene. In effect, each cell contains a small population of organelle genes. This is in contrast to the nucleus where, with few exceptions, there are only two copies of each chromosome and gene (or one copy in haploid cells).

Non-Mendelian Inheritance of Organelle Genes

The inheritance of genes in chloroplasts and mitochondria differs from that of nuclear genes in several ways. For instance, organelle genes are characterized by uniparental inheritance. During sexual reproduction, the nuclear genes are inherited from both parents (biparental inheritance). In contrast, the organelle genes are often inherited from only one parent. In animals, this is usually the female parent (maternal inheritance). Mitochondrial genes are inherited maternally because the egg is much larger than the sperm and contains tens of thousands of mtDNA molecules, while the sperm contains only a hundred or so mtDNA molecules. As a result, paternal genes are greatly outnumbered by the maternal genes and can be lost during random replication or other chance events. In addition, in some animals the mitochondrial or mtDNA molecules in the sperm are singled out for degradation in the egg.

Organelle genes are inherited maternally in most plants. In some conifers, mitochondrial genes are inherited maternally, whereas chloroplast genes are inherited paternally. In other conifers, both organelle genomes are inherited paternally. Some other plants (for example, the geranium) and some fungi and protists show a mixed pattern of inheritance. In these organisms, some offspring from a mating inherit organelle genes from only one parent, some only from the other, and some from both parents.

Another way in which extranuclear inheritance differs from nuclear genes is that organelles show vegetative segregation. During the mitotic divisions that produce an adult **eukaryote** from a single cell, each daughter cell receives one copy of each preexisting nuclear chromosome and gene. The result is that if a cell is heterozygous (has two different versions or alleles of a gene in the nucleus), all of the daughter cells are also heterozygous.

eukaryote organism with cells possessing a nucleus

In contrast, different alleles of organelle genes segregate from each other during mitosis. For example, a plant egg with a mixture of normal green and mutant white chloroplasts develops into a plant with a mixture of cells, some of which will contain all green chloroplasts, whereas others will contain all white chloroplasts. This process is called vegetative segregation because it was first discovered in plants, where green and mutant white chloroplasts were observed to segregate during vegetative growth. However, it is now known to occur in all eukaryotes.

Vegetative segregation is the result of two remarkable features of organelle genes. One is random replication. Recall that an organelle contains many copies of its DNA molecule. When mtDNA or cpDNA molecules are replicated before cell division, individual molecules are randomly selected for replication until the total number of molecules has doubled. Consequently, some molecules, and some alleles, are replicated more than others.

The other feature is random partitioning. When an organelle divides, the mtDNA or cpDNA molecules are partitioned (divided up) randomly between the daughter organelles. The result is that heteroplasmic mitochondria or chloroplasts produce homoplasmic daughter organelles with a certain probability in each generation. (An organelle is said to be heteroplasmic if it contains two or more forms of a particular gene. If all the gene copies are identical, it is homoplasmic.) Moreover, when a cell divides, the organelles are partitioned randomly between the two daughter cells, so that a heteroplasmic cell can produce homoplasmic daughters. Over a large number of cell divisions, random replication and random partitioning result in the complete replacement of heteroplasmic cells by homoplasmic cells.

Plasmids

Bacteria often carry small circular DNA molecules called plasmids. These molecules sometimes carry genes that have important properties; for example, some plasmid genes make the host cell resistant to antibiotics, with important consequences for human health. Plasmids are widely used in genetics labs in the process of cloning genes. Like organelle DNA in eukaryotes, plasmid molecules are replicated independently of the cell chromosome, and one cell can contain many copies. Plasmid DNA molecules are replicated randomly and partitioned randomly to daughter cells, just like organelle genes. As a result, a cell with two different **genotypes** of plasmids may produce daughter cells with only one or the other. This process, analogous to vegetative segregation, is called plasmid incompatibility.

genotype set of genes present

Genes in Intracellular Symbionts

Many eukaryotes harbor intracellular symbiotic bacteria as well as organelles. These are usually inherited from only one parent and may have significant effects on their host. Many insects have endosymbionts that are inherited only through the female **germ line**. Well-studied examples are bacteria of the genus *Buchnera* in aphids. Since they took up residence in insect cells, *Buchnera* have lost a number of genes, just like the early ancestors of mitochondria and chloroplasts. Protists also harbor hereditary symbiotic bacteria. An example is a bacterium called *kappa*, found in *Paramecium*. *Kappa* makes a toxin that is secreted by its host and kills other *Paramecium* cells that do not contain *kappa*.

germ line cells giving rise to eggs or sperm

The Practical Importance of Extranuclear Genes

Mutations of mitochondrial genes have been found to cause a number of hereditary diseases in humans. Many of these mutations lead to defects in muscles, including the heart, and the nervous system. Because mitochondrial genes are found in nearly all eukaryotes, they are often used to trace the evolutionary history of organisms, including humans. Chloroplast genes are used for evolutionary studies in plants and algae. When organelle genes are inherited from only one parent, they can be used to trace the ancestry of individuals within a species without the complications caused by recombination between maternal and paternal genes.

Mitochondrial genes can also be used to trace the female ancestor of humans, while Y-chromosome genes can be used to trace male ancestors. In this way, differences between males and females in migration or patterns of reproduction can be detected. In addition, mitochondrial genes are used in studies of animal behavior to identify the parents of animals and birds and determine their social structure. Since the organelle genome is so highly simplified, mtDNA or cpDNA can be retrieved and analyzed from ancient or poorly preserved samples in which there would be no chance of retrieving a nuclear marker. SEE ALSO Cell, Eukaryotic; Mitochondrial Diseases; Mitochondrial Genome; Molecular Anthropology; Plasmid; Prion.

C. William Birky, Jr.

Bibliography

Cann, Rebecca L. "Genetic Clues to Dispersal in Human Populations: Retracing the Past from the Present." *Science* 291 (2001): 1742–1748.

Enserink, Martin. "Evolutionary Biology: Thanks to a Parasite, Asexual Reproduction Catches On." *Science* 275 (1997): 1743–1750.

Gray, Michael W., Gertraud Burger, and B. Franz Lang. "Mitochondrial Evolution." *Science* 283 (1999): 1476–1481.

Ochman, Howard, and Nancy A. Moran. "Genes Lost and Genes Found: Evolution of Bacterial Pathogenesis and Symbiosis." *Science* 292 (2001): 1096–1099.

Palmer, Jeffrey D. "Organelle Genomes: Going, Going, Gone!" *Science* 275 (1997): 790.

Wallace, Douglas C. "Mitochondrial Diseases in Man and Mouse." *Science* 283 (1999): 1482–1488.

Yaffe, Michael P. "The Machinery of Mitochondrial Inheritance and Behavior." *Science* 283 (1999): 1493–1497.

Inheritance Patterns

Inheritance patterns are the predictable patterns seen in the transmission of genes from one generation to the next, and their expression in the organism that possesses them. (A gene is said to be expressed when it is read by cellular mechanisms that result in the production of a protein.) While people have long noted that offspring resemble parents, the formal description of inheritance patterns began with Gregor Mendel, whose discoveries laid the foundation for the modern understanding of genetic inheritance.

Phenotype and Genotype

An organism's observable characteristics, such as height, hair texture, skin color, or ear shape, are known as the **phenotype** of that organism. The phenotype is determined partly by the environment and partly by the set of genes that the organism inherited from its parents. Adult height, for instance, is due partly to nutrition (an environmental influence), and partly to a set of genes governing things such as rates of bone growth, sensitivity to specific hormones, and the like. Phenotype includes not only large-scale characteristics such as height, but every expressed trait, including the types and amounts of all the proteins produced in each cell in the body.

phenotype observable characteristics of an organism

The set of genes an organism inherits is known as its **genotype**. Genes are carried on chromosomes in the cell nucleus. Animals and most other multicellular organisms possess two sets of chromosomes in each cell, one set inherited from the mother, and one from the father. Such an organism is said to be **diploid**. In humans, the maternal and paternal sets each include 23 chromosomes, so humans have 46 chromosomes in each cell. Analysis shows that the maternal and paternal chromosome sets are virtually identical, and they can be matched up to form 23 pairs. One pair, however, may not be a pair at all. These are the sex chromosomes, so called because they determine the sex of the organism. In humans, the female carries two identical sex chromosomes, called X chromosomes, while the male carries two dissimilar chromosomes, one X and one Y. The other 22 pairs of chromosomes are called **autosomes**.

genotype set of genes present

diploid possessing pairs of chromosomes, one member of each pair derived from each parent

autosomes chromosomes that are not sex determining (not X or Y)

Alleles

Members of each chromosome pair (except for X and Y) carry the same set of genes, so that a diploid cell carries two copies of (almost) every gene, one

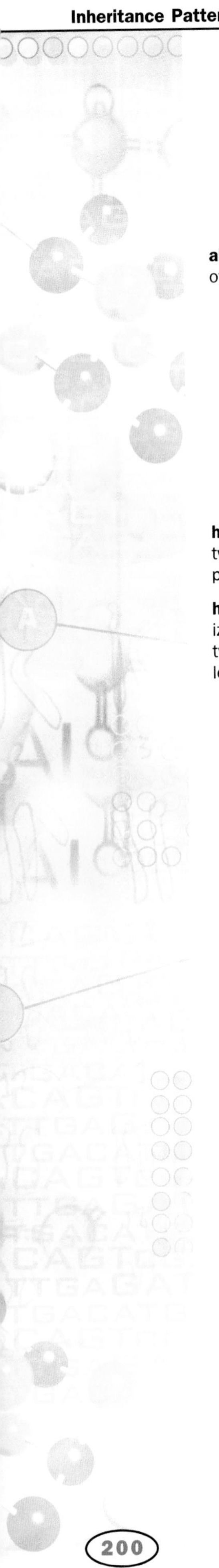

on the maternally derived chromosome, and one on the paternally derived chromosome. These two copies may be precisely identical, meaning the two genes have precisely the same sequence of nucleotides, or their sequences may be slightly different. These sequence differences may have no effect at all on the phenotype, or they may lead to different forms of the same trait, such as brown versus blue eye color, or smooth versus wrinkled pea texture. The two different forms of the gene are called **alleles**, and so we speak of the brown eye color allele or the wrinkled pea texture allele.

alleles particular forms of genes

While a single organism can possess no more than two different alleles for a single gene, many different alleles for a particular gene can exist in a population. For instance, there are three alleles for the ABO blood group gene, namely A, B, and O.

Dominance Relations

When both alleles for a particular trait are identical, the organism is said to be **homozygous** for that trait. When the alleles differ, the organism is **heterozygous**. The presence of two different alleles raises the question of whether one or the other, or both, will determine the phenotype of the organism. For his experiments on peas, Mendel chose traits for which one allele of each pair had a decisive effect, completely determining the phenotype even in the presence of the other allele. He called the determining allele "dominant" and the other allele "recessive." Only when the organism is homozygous for the recessive allele does the phenotype show the recessive trait. For instance, the albino skin pigmentation allele is recessive to other pigmentation alternatives.

homozygous containing two identical copies of a particular gene

heterozygous characterized by possession of two different forms (alleles) of a particular gene

While some sets of alleles do show a complete dominance-recessiveness relationship, most allele sets do not. Instead, each allele contributes to the phenotype. Such a relationship is called codominance. In humans, the A and B blood group alleles are codominant, and a person inheriting both will have blood type AB. Incomplete dominance is another variant in this system. In this case, the phenotype of the heterozygote is intermediate between the two extremes. Skin color in humans often shows this pattern.

Molecular Meaning of Dominance and Recessiveness

The terms "dominant" and "recessive" imply some competitive interaction between alleles over which one will control the phenotype. This is not the case, however. Alleles do not interact in the nucleus. Instead, both alleles are expressed (in most cases), and the phenotype reflects the result. How, then, can one allele determine the phenotype to the exclusion of another?

Genes code for proteins, and their effect on the phenotype is through the proteins they create. By analyzing the quantity and characteristics of protein produced from each allele, it has been shown that, in many cases, recessive alleles code for defective proteins, or for very low levels of protein. If the other allele codes for normal functional protein, and if the organism can make do with half the normal level of protein (or can increase production from the normal allele), the defective allele will have no effect on the phenotype, and will therefore act in a recessive manner. This type of allelic change is called a loss-of-function mutation. This is the case with the albino allele. This allele codes for a nonfunctional enzyme in the pathway that produces the pigment melanin. Even with only one functioning allele, the organism can still make enough melanin to obtain normal skin pigmentation. Thus, the functional allele will appear to be **dominant**.

dominant controlling the phenotype when one allele is present

Alleles coding for defective protein can act in a dominant manner in other situations, however. If the resulting protein is not properly regulated by other cell components, it may perform actions that harm the cell. This is called a toxic-gain-of-function mutation. Huntington's disease is thought to be due to a toxic-gain-of function mutation, although it is not yet clear what the exact toxic mechanism is.

A defective protein can also have a dominant effect if its absence cannot be compensated for by the other functioning allele (this is called a dominant negative effect). This may occur when half the normal protein level is insufficient for normal function (a condition called haploinsufficiency), or when the protein forms part of a multiprotein complex, which is therefore defective in its entirety. Examples include a variety of human collagen-structure disorders. Collagen is the most abundant and important structural protein in the body, and is critical for bone formation and growth. Defects in one of the subunits cause a variety of dominant disorders termed "osteogenesis imperfecta."

Autosomal Dominant Inheritance

Autosomal dominant inheritance is due to a dominant allele carried on one of the autosomes. Autosomal dominant alleles need only be inherited from one parent, either the mother or the father, in order to be expressed in the phenotype. Because of this, any child has a 50 percent chance of inheriting the allele and expressing the trait if one parent has it.

Many normal human traits are due to autosomal dominant alleles, including the presence of dimples, a cleft chin, and a widow's-peak hairline. Note that dominant does not necessarily mean common. Dominant alleles can be rare in a population, and do not spread simply because they are dominant. This phenomenon is explained by the theory known as Hardy-Weinberg equilibrium.

There are hundreds of medical conditions due to autosomal dominant alleles, most of them very rare. They include neurodegenerative disorders such as Huntington's disease, a variety of deafness syndromes, and metabolic disorders such as familial hypercholesterolemia (affecting blood cholesterol levels) and variegate porphyria (affecting the oxygen-carrying porphyrin molecule). Table 1 lists some other examples.

Because inheritance of a harmful dominant allele can be lethal, these alleles tend to be quite rare in the population, and new **mutations** account for many cases of these conditions. Exceptions include late-onset disorders such as Huntington's disease, in which parents may pass on the gene to offspring before developing the symptoms of the disease. Other exceptions arise from incomplete penetrance, in which the allele is present, but (for reasons usually unknown) it is not expressed. Genomic imprinting (see below) may explain some cases of incomplete penetrance. Variable expressivity is also possible, in which different individuals express the trait with different levels of severity.

mutations changes in DNA sequences

Condition	Chromosome Location and Inheritance Pattern	Protein Affected	Symptoms and Comments
Gaucher Disease	1, recessive	glycohydrolase glucocerebrosidase, a lipid metabolism enzyme	Common among European Jews. Lipid accumulation in liver, spleen, and bone marrow. Treat with enzyme replacement
Achondroplasia	4, dominant	fibroblast growth factor receptor 3	Causes dwarfism. Most cases are new mutations, not inherited
Huntington's Disease	4, dominant	huntingtin, function unknown	Expansion of a three-nucleotide portion of the gene causes late-onset neurodegeneration and death
Juvenile Onset Diabetes	6, 11, 7, others	IDDM1, IDDM2, GCK, other genes	Multiple susceptibility alleles are known for this form of diabetes, a disorder of blood sugar regulation. Treated with dietary control and insulin injection
Hemochromatosis	6, recessive	HFE protein, involved in iron absorption from the gut	Defect leads to excess iron accumulation, liver damage. Menstruation reduces iron in women. Bloodletting used as a treatment
Cystic Fibrosis	7, recessive	cystic fibrosis transmembrane regulator, an ion channel	Sticky secretions in the lungs impairs breathing, and in the pancreas impairs digestion. Enzyme supplements help digestive problems
Friedreich's Ataxia	9, recessive	frataxin, mitochondrial protein of unknown function	Loss of function of this protein in mitochondria causes progressive loss of coordination and heart disease
Albinism	11, recessive	tyorsinase	Lack of pigment in skin, hair, eyes; loss of visual acuity
Best Disease	11, dominat	VMD2 gene, protein function unknown	Gradual loss of visual acuity
Sickle Cell Disease	11, recessive	hemoglobin beta subunit, oxygen transport protein in blood cells	Change in hemoglobin shape alters cell shape, decreases oxygen-carrying ability, leads to joint pain, anemia, and infections. Carriers are resistant to malaria. About 8% of US black population are carriers
Phenylketonuria	12, recessive	phenylalanine hydroxylase, an amino acid metabolism enzyme	Inability to breakdown the amino acid phenylalanine causes mental retardation. Dietary avoidance can minimize effects. Postnatal screening is widely done
Marfan Syndrome	15, dominant	fibrillin, a structural protein of connective tissue	Scoliosis, nearsightedness, heart defects, and other symptoms
Tay-Sachs Disease	15, recessive	beta-hexosaminidase A, a lipid metabolism enzyme	Accumulation of the lipid GM2 ganglioside in neurons leads to death in childhood
Breast Cancer	17, 13	BRCA1, BRCA2 genes	Susceptibility alleles for breast cancer are thought to involve reduced ability to repair damaged DNA
Myotonic Dystrophy	19, dominant	dystrophia myotonica protein kinase, a regulatory protein in muscle	Muscle weakness, wasting, impaired intelligence, cataracts
familial hypercholesterolemia	19, incomplete dominance	low-density lipoprotein (LDL) receptor adenosine deaminase, nucleotide metabolism enzyme	Accumulation of cholesterol-carrying LDL in the bloodstream leads to heart disease and heart attack
Severe Combined Immune Deficiency ("Bubble Boy" Disease)	20, recessive	respiratory complex proteins	Immature white blood cells die from accumulation of metabolic products, leading to complete loss of the immune response. Gene therapy has been a limited success
Leber's Hereditary Optic Neuropathy	mitochondria, maternal inheritance	transfer RNA	degeneration of the central portion of the optic nerve, loss of central vision
Mitochondrial Encephalopathy, Lactic Acidosis, and Stroke (MELAS)	mitochondria, maternal inheritance	lignoceroyl-CoA ligase, in peroxisomes	recurring, stroke-like episodes in which sudden headaches are followed by vomiting and seizures; musle weakness
Adrenoleukodystrophy	X	dystrophin, muscle structural protein	Defect causes build-up of long-chain fatty acids. Degeneration of the adrenal gland, loss of myelin insulation in nerves. Featured in the film "Lorenzo's Oil"
Duchenne Muscular Dystrophy	X	Factor VIII, part of the blood clotting cascade	Lack of dystrophin leads to muscle breakdown, weakness, and impaired breathing
Hemophilia A	X		Uncontrolled bleeding, can be treated with injections or replacement protein
Rett Syndrome	X	methyl CpG-binding protein 2, regulates DNA transcription	Most boys die before birth. Girls develop mental retardation, mutism and movment disorder

Table 1.

Autosomal Recessive Inheritance

Autosomal recessive inheritance is due to recessive alleles carried on autosomes. An individual possessing only one recessive allele is known as a carrier. An individual must inherit two recessive alleles, one from each parent, in order to express the recessive trait. When two carrier parents have offspring, each offspring has a 25 percent chance of inheriting two alleles and expressing the trait. The two recessive alleles need not be precisely identical, as long as each is nonfunctional. An individual possessing two different alleles with the same effect is known as a compound heterozygote. Compound heterozygotes account for some cases of the neurologic disorder known as Friedreich's ataxia.

Medical conditions due to autosomal recessive traits also number in the many hundreds. These include cystic fibrosis (affecting ion transport in the lungs and pancreas), Tay-Sachs disease (affecting lipid metabolism and storage, especially in the brain), and hemochromatosis (affecting iron metabolism and storage in a variety of organs).

The number of people with such conditions is actually much higher than that for autosomal dominant conditions. This is because inheritance of one harmful recessive allele does not produce symptoms, and so the individual can reproduce and pass the allele on to children easily. Thus, most harmful recessive alleles are not deleted from a population's gene pool as rapidly as most dominant ones, and the likelihood of inheriting two copies is consequently higher. Most humans harbor a small handful of known harmful alleles; it is only when they mate with another who has the same set that there is a chance of bearing children that express the disorder. Customs that warn against marrying close relations have the effect of minimizing the likelihood of offspring with homozygous recessive conditions.

Sex-Linked Inheritance

The two sex chromosomes differ in the genes they carry. The Y chromosome is very small, and appears to carry very few genes other than the *SRY* gene that determines male sex. Many genes are carried on the X chromosome, however, and these are as essential for males as they are for females. Genes carried on the X chromosome are said to be X-linked.

X-linked dominant alleles affect both males and females, although males may be more severely affected since they inherit only a single X chromosome and thus lack a compensating normal allele. An example of a disorder caused by an X-linked dominant allele is congenital generalized hypertrichosis, which causes dense hair growth on the face and other regions in both sexes. X-linked dominant alleles can be inherited by both males and females, but fathers cannot pass them on to sons.

X-linked recessive alleles affect males more often and more severely than females. A male inherits his single X chromosome from his mother. Because a male has only one X chromosome, he expresses every allele on it, including harmful recessive ones. Examples of conditions due to recessive X-linked alleles include Duchenne muscular dystrophy, one form of hemophilia, and red-green colorblindness. These conditions are much more common in males than in females. Female carriers have a 50 percent chance of giving birth to a male child affected by the recessive allele. The genetic status of the father with respect to X-linked conditions is not relevant in this case, because he donates a Y chromosome to his male children.

Since females have two X chromosomes, they are less likely than males to express harmful recessive X-linked traits. Female children have only a 25 percent chance of inheriting two recessive alleles from a carrier mother and an affected father.

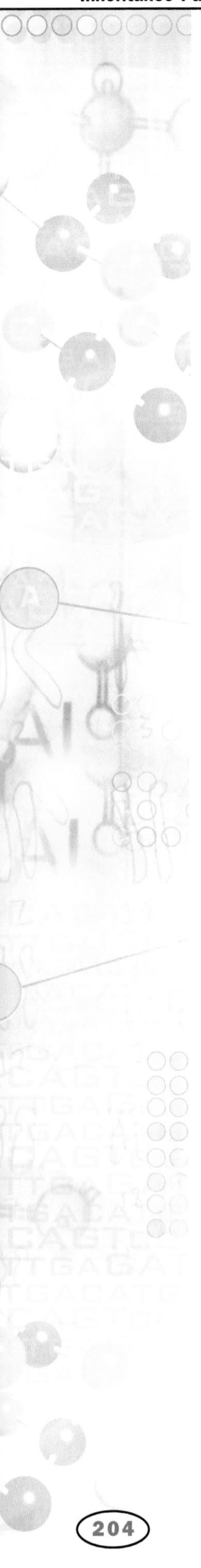

Mosaicism

The reason females are less often affected by recessive alleles is not as simple as it is for autosomes. Since females have twice the number of X chromosomes as males, the question arises as to whether they make twice the amount of each X-encoded protein as males do. In fact they do not, and are prevented from doing so by the random inactivation of one X chromosome in each cell. Therefore, about half of the heterozygote female's cells will express the normal allele, and half will express the harmful recessive allele. This is in contrast to the situation for autosomes, in which each cell expresses both alleles.

Inactivation begins early in development, with some cells shutting down one X and others shutting down the other, followed by faithful inheritance of the inactivated chromosome by each daughter cell following cell division. As a result, many tissues in the adult female are a mosaic of cells with different X chromosomes inactivated. The consequence of this is seen in a woman who is heterozygous for a harmful allele. If the affected tissue is primarily composed of cells expressing only the harmful one, she is likely to express the trait. However, most adult tissues will be a more even mixture, and for many disorders this will prevent her from developing symptoms of the disease.

The trait may also show variable expressivity, with the severity dependent on the proportion of the tissue affected. Duchenne muscular dystrophy in females is an example. Many women with one disease allele will show no symptoms. Others will develop only slightly elevated blood levels of certain enzymes indicating mild muscle damage, while others will develop heart problems and muscle weakness. Such women are referred to as "manifesting carriers."

Mitochondrial Inheritance

Mitochondria are the cell's power plants. They possess their own chromosome, which carries thirty-seven genes. Mitochondria are inherited only from the mother. Mitochondrially inherited disorders include a number of rare muscle diseases (mitochondrial myopathies), as well as some deafness syndromes, optic nerve degeneration, and other neurological disorders.

Penetrance, Expressivity, and Anticipation

The presence of a dominant allele, or two recessive alleles, is not always a guarantee that the trait will be displayed in the phenotype, a phenomenon called incomplete penetrance. Variable expressivity also occurs, with some individuals more affected than others of the same genotype. In most cases the reasons for these differences are unknown, but they are assumed to be due at least in part to other differences between individuals. For instance, if there are differences between individuals in the other genes with which the product of incompletely penetrant or variably expressed allele interacts, this may account for some of these differences in expression.

One phenomenon also associated with changes in both timing and severity of expression is anticipation. Anticipation refers to the successive decrease in the age of symptom onset over several generations, so that a condition might first manifest at age 60 in a grandfather, at age 40 in a father, and at age 20 in a son. This increasingly earlier onset is often accompanied by

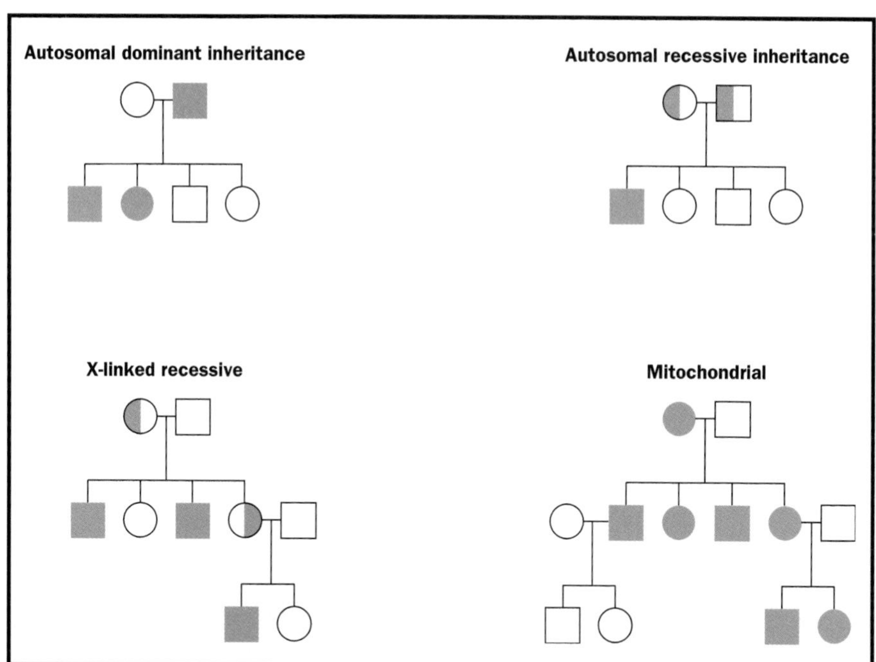

Mitochondrial inheritance passes only through the mother.

increasingly severe symptoms as well. Some cases of anticipation are known to be due to changes in the allele itself over time. Spinocerebellar ataxia, for example, is an autosomal dominant disease that causes balance disorders. The normal allele has a section of its DNA that includes approximately 20 repeats of a nucleotide triplet, CAG. The disease allele has 40 or more CAG repeats. This number can increase between generations, leading to earlier onset and more severe disease over several generations. Other so-called triplet repeat diseases show the same pattern of anticipation.

Imprinting

Some cases of incomplete penetrance appear to be due to imprinting. In this phenomenon, expression of an allele is governed by whether it is derived from the mother or the father. Imprinted alleles are located on autosomes, but are "stamped" with the sex of the parent that contributed it. The chemical basis of the imprint is the addition of methyl (-CH_3) groups to the allele's nucleotides, and the effect is thought to be to silence the allele so it is not expressed. For some genes the maternal allele is silenced, while for others the paternal allele is silenced. When a child inherits the two alleles, they retain these stamps, regardless of the sex of the child. However, when the child makes its own sperm or eggs, the child's own imprinting machinery stamps the alleles so that they correspond to the child's own sex. Therefore, a particular allele can be turned on or turned off as it is passed down through successive generations, from male to female and back again.

Imagine a dominant disease allele that is active when inherited from the mother, but is silenced when inherited from the father. Both the sons and daughters of the mother will develop symptoms of the disease. The daughter's children will also develop symptoms, while the son's children will not, despite having the same genotype. This is an example of incomplete penetrance. Prader-Willi syndrome and Angelman syndrome are examples of disorders arising from imprinted genes.

The pale skin, hair, and eyes typical of an albino presents in this young man, whose genetic makeup determined that he would not have the same dark skin pigment as his father.

Polygenic, Multifactorial, and Complex Traits

Proteins, which are the products of genes, interact with one another in complex ways to determine the phenotype. Almost every trait we observe, such as height, normal metabolic level, or intelligence, is really the product of many genes. Many traits, however, also reflect the influence of the environment. Such traits are called complex traits, to distinguish them from simple traits that are governed by single genes. While any single gene contributing to a complex trait can be described in terms of dominance or recessiveness, autosomal or sex-linked, or other categories, the gene products interact to make a much more subtle phenotypic picture.

A polygenic trait is a complex trait controlled by the alleles of two or more genes, without the influence of the environment. A multifactorial trait is a complex trait controlled by both genes and the environment. Intelligence is multifactorial, with strong influences from both genes (such as those controlling nerve-cell growth and connectivity) and the environment (such as early childhood nutrition and education).

As the number of influences grows, so too does the number of possible phenotypes. Because of this, complex traits show not just a few phenotypes, but a continuum (as can be seen in the wide range of possible human heights). The distribution in a population will usually be described by a bell-shaped curve, with most people displaying the mid-range phenotype.

Pleiotropy and Epistasis

Most single genes affect more than one observable trait, a phenomenon know as **pleiotropy**. For example, the alleles for melanin pigment affect skin color, eye color, and hair color. The ion channel gene affected in cystic fibrosis acts in the lungs, the pancreas, and other passageways, and defects cause symptoms in both these organs, as well as elsewhere in the body.

pleiotropy genetic phenomenon in which alteration of one gene leads to many phenotypic effects

Proteins are also involved in highly ordered metabolic pathways, and a defect "upstream" can mask or prevent expression of other alleles "downstream." This condition is known as **epistasis** ("standing upon"), and the upstream gene is said to be epistatic to the downstream one. For instance, on blood cells, the well-known ABO markers are actually branched sugars attached to proteins embedded in the surface of the cell. In order for the cell to attach these sugars, it must first express a gene (called fucosyltransferase) that attaches one sugar group (fucose) that is common to all blood types. Absence of functional fucosyltransferase prevents the expression of the ABO alleles. SEE ALSO COLOR VISION; CROSSING OVER; DISEASE, GENETICS OF; EPISTASIS; FERTILIZATION; GROWTH DISORDERS; HARDY-WEINBERG EQUILIBRIUM; HEMOPHILIA; IMPRINTING; MEIOSIS; MENDELIAN GENETICS; MITOCHONDRIAL DISEASES; MOSAICISM; MUSCULAR DYSTROPHY; PEDIGREE; PLEIOTROPY; TAY-SACHS DISEASE; TRIPLET REPEAT DISEASE.

epistasis supression of a characteristic of one gene by the action of another gene

Richard Robinson

Bibliography

Hartwell, Leland H., et al. *Genetics: From Genes to Genomes.* New York: McGraw-Hill, 2000.

Lewis, Ricki. *Human Genetics*, 4th ed. New York: McGraw-Hill, 2001.

Internet Resource

Online Mendelian Inheritance in Man. <http://www.ncbi.nlm.nih.gov/omim>.

Intelligence

The roles of genes and environment in the determination of intelligence have been controversial for more than 100 years. Studies of the question have often been marred by untested assumptions, poor design, and even racism, faults that more modern studies have striven to avoid. Nonetheless, examining the biology of intelligence is an enterprise that continues to be fraught with difficulty, and there remains no real consensus even on how to define the term.

IQ Tests

Conventional measures of intelligence are obtained using standard tests, called intelligence quotient tests or, more commonly, IQ tests. These tests have been shown to be reliable and valid. Reliability means that they measure the same thing from person to person, whereas validity means that they measure what they claim to measure. IQ tests measure a person's ability to reason and to solve problems. These abilities are frequently called general cognitive ability, or "g."

Almost all genetic studies of the **heritability** of intelligence (how much is due to genetics and how much is due to the environment) have been obtained from IQ tests. To understand the studies, therefore, it is important to understand what IQ tests measure, and how their use and interpretation have changed over time.

heritability proportion of variability due to genes; ability to be inherited

The standard IQ-measurement approach to intelligence is among the oldest of approaches and probably began in 1876, when Francis Galton investigated how much the similarity between twins changed as they developed

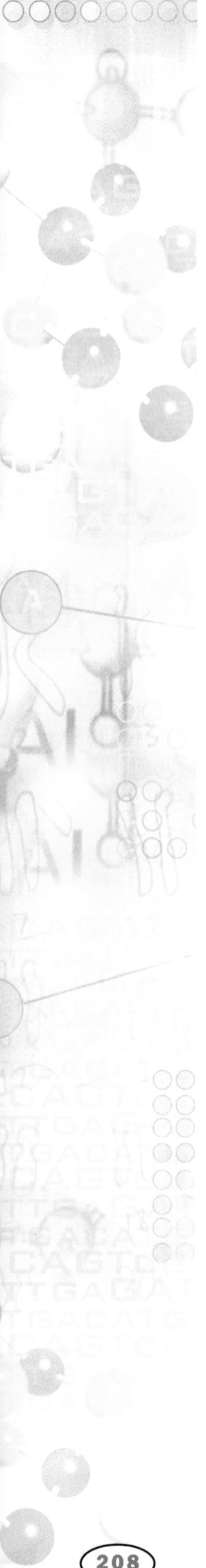

MENSA is an international organization whose only membership criteria is a score in the top 2 percent of the population on a standard IQ test.

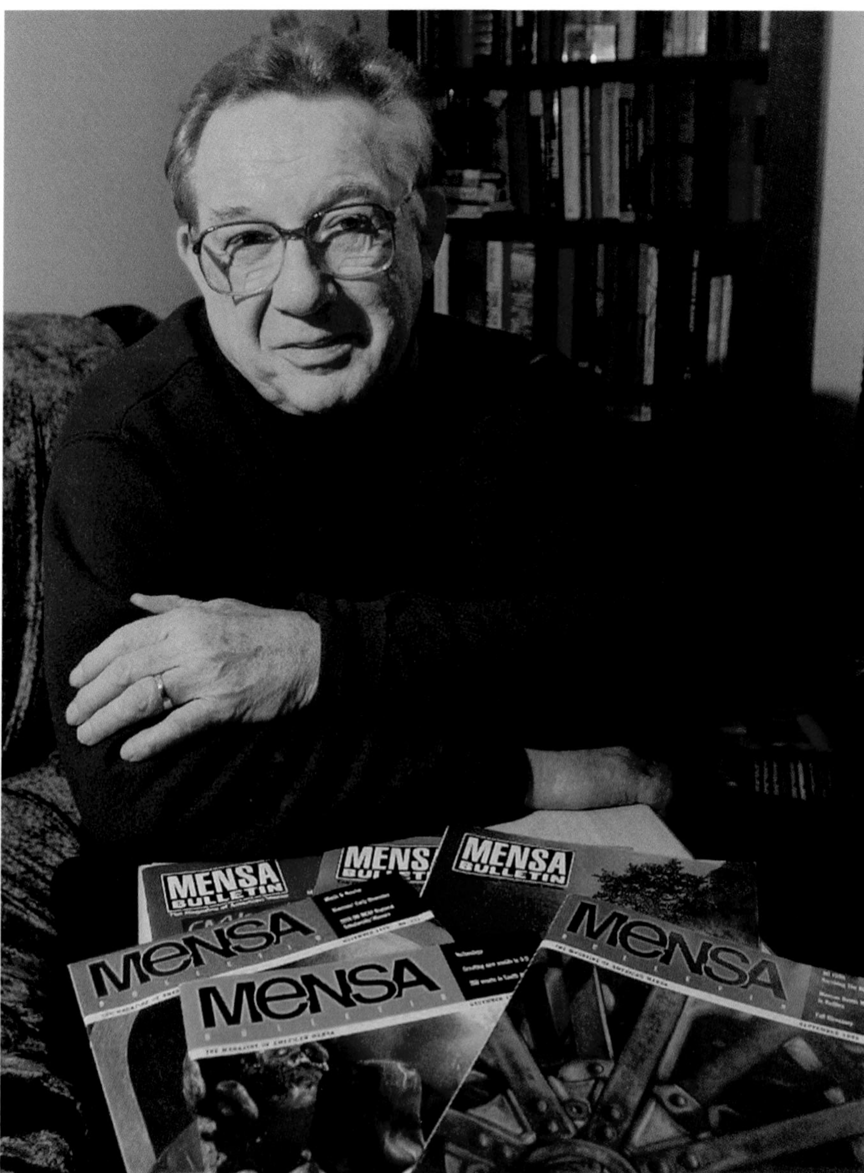

over time. Galton's study was concerned with measuring psychophysical abilities, such as strength of handgrip or visual acuity. The concept of general cognitive ability was first described by Charles Spearman in 1904. Later, Alfred Binet and Theophile Simon (1916) evaluated intelligence based on judgment, involving adaptation to the environment, direction of one's efforts, and self-criticism.

Most standard test results now include three scores: VIQ, PIQ, and FSIQ. The VIQ score measures verbal ability (verbal IQ), PIQ measures performance ability (performance IQ), and FSIQ provides an overall measurement (full scale IQ). Commonly used IQ tests include the Stanford-Binet Intelligence Scale, the Wechsler Intelligence Scale for Children (WISC), and the Wechsler Adult Intelligence Scales. The results achieved by individual testtakers on one of these IQ tests are likely to be similar to the results they achieve on the others, and they all aim to measure general cognitive ability (among other things). Measures of scholastic achievement, such as the SAT and the ACT correlate highly with "g."

Environmental Effects on Intelligence

The study of intelligence must take environmental effects into account. The Flynn effect describes a phenomenon that indicates that IQ has increased about 3 points per decade over the last fifty years, with children scoring higher than parents in each generation. This increase has been linked to multiple environmental factors, including better nutrition, increased schooling, higher educational attainment of parents, less childhood disease, more complex environmental stimulation, lower birth rates, and a variety of other factors.

Males and females have equivalent "g" scores. The question of racial differences and IQ arose when a 10 point IQ difference between African Americans and Americans of European descent was documented. Two adoption studies indicate that the effect may be in part related to environment factors, including culture. Also, environmental differences similar to those identified with the Flynn effect can be postulated. Studies of black Caribbean children and English children raised in an orphanage in England found that the black Caribbean children had higher IQs than the English children, with mixed racial children in between. A study comparing black children adopted by white families and those who were adopted into black families in the United States showed that black children raised by whites had higher IQ scores, again suggesting that the environment played a role.

Expanded Concepts of Intelligence

Many of the standard measures of IQ, such as the WISC and the Stanford-Binet, have changed their content over the years. Although they both still report verbal, performance, and total scores, the Wechsler model now offers scores for four additional factors (verbal comprehension, perceptual organization, processing speed, and freedom from distractibility). The Stanford-Binet also yields additional scores, including abstract-visual reasoning, quantitative reasoning, and short-term memory.

However, the majority of research into genetic and environmental variance in IQ has centered on the assumption that general cognitive ability is the essence of intelligence. Newer tests that measure specific abilities have not been included in genetic studies. These include, for example, tests that measure creativity in a model for intelligence. The addition of new factors in the Wechsler and Stanford-Binet IQ tests represents a trend toward a broader approach to IQ, and away from the notion that IQ can be understood by the single factor, "g."

Family, Twin, and Adoption Studies

Genetic studies have traditionally used models that evaluate how much of the variability in IQ is due to genes and how much is associated with environment. These studies include family studies, twin studies, and adoption studies.

General cognitive ability runs in families. For first-degree relatives (parents, children, brothers, sisters) living together, correlations of "g" for over 8,000 parent-offspring pairs averaged 0.43 (0.0 is no correlation, 1.0 is complete correlation). For more than 25,000 sibling pairs, "g" correlations

averaged 0.47. Heritability estimates range from 40 to 80 percent, meaning that 40 to 80 percent of "g" is due to genes.

In twin studies of over 10,000 pairs of twins, monozygotic (genetically identical) twins averaged an 0.85 correlation of "g," whereas for dizygotic (fraternal, like brothers or sisters) same-sex twins the "g" correlations were 0.60. These twin studies suggest that the heritability (genetic effect) accounts for about half of the variance in "g" scores.

Adoption studies also provide evidence for substantial heritability of "g." The "g" estimate for identical twins raised apart is similar to that of identical twins raised together, proving that for genetically identical individuals, environmental differences did not affect "g." The Colorado Adoption Study (CAP) of first-degree relatives who were adopted also indicated significant heritability of "g." Thus, classical genetic studies indicate that there is a statistically significant and substantial genetic influence on "g."

Newer genetic research on general cognitive ability has focused on developmental changes in IQ, multivariate relations (contributions of multiple factors) among cognitive abilities, and specific genes responsible for the heritability of "g." Developmental changes over time were first studied by Galton in 1876. The CAP study was conducted over twenty-five years and evaluated 245 children who had been separated from their parents at birth and adopted by one month of age. This study, and others, showed that the variance in "g" due to environment for an adopted child in his or her adoptive family is largely unconnected with the shared adoptive family upbringing, that is, a shared parent-sibling environment. For adoptive parents and their adopted children, the parent-offspring correlations for heritability were around zero. For adopted children and their biologic mothers or for children raised with their biologic parents, heritability was the same, increasing with age.

Recent studies indicate that heritability increases over time, with infant measures of about 20 percent, childhood measures at 40 percent, and adult measures reaching 60 percent. Why is there an age effect for the heritability of "g"? Part of this could be due to different genes being expressed over time, as the brain develops. The stability of the heritability measure correlates with changes in brain development, with "maturity" of brain structure achieved after adolescence. Also, it is likely that small gene effects early in life become larger as children and adolescents select or create environments that foster their strengths.

Multivariate relations among cognitive abilities affect more than general cognitive ability as measured by "g." Current models of cognitive abilities include specific components such as spatial and verbal abilities, speed of processing, and memory abilities. Less is known about the heritabilities of these specific cognitive skills. They also show substantial genetic influence, although this influence is less than what has been found for "g." Multivariate genetic analyses indicate that the same genetic factors influence different abilities. In other words, a specific gene found to be associated with verbal ability may also be associated with spatial ability and other specific cognitive abilities. Four studies have shown that genetic effects on measures of school achievement are highly correlated with genetic effects on "g." Also, discrepancies between school achievement and "g," as occurs with underachievers, are predominantly of environmental origin.

Genes for Intelligence

The search for specific genes associated with IQ is proceeding at a rapid pace with the completion of the Human Genome Project. While defects in single genes, such as the fragile X gene, can cause mental retardation, the heritability of general cognitive ability is most likely due to multiple genes of small effect (called quantitative trait loci, or QTLs) rather than a single gene of large effect. QTLs contribute additively and interchangeably to intelligence.

Genetic studies have identified QTLs associated with "g" on chromosomes 4 and 6. These studies involved both children with high "g" and children with average "g." QTLs on chromosome 6 have been identified and shown to be active in the regions of the brain involved in learning and memory. The gene identified is for insulin-like growth factor 2 receptor, or IGF2R, the exact function of which is still unknown. One allele (alternative form) of IGR2R was found to be present 30 percent of the time in two groups of children with high "g." This was twice the frequency of its occurrence in two groups of children with average "g," and these findings have been successfully replicated in other studies. QTLs associated with "g" have also been identified on chromosome 4. Future identification of QTLs will allow geneticists to begin to answer questions about IQ and development and gene-environment interaction directly, rather than relying on less specific family, adoption, and twin studies.

In summary, intelligence measurements ranging from specific cognitive abilities to "g" have a complex relationship. Genetic contributions are large, and heritability increases with age. Heritability remains high for verbal abilities during adulthood. Finally, the identification of QTLs associated with "g" and with specific cognitive abilities is just beginning. SEE ALSO BEHAVIOR; COMPLEX TRAITS; EUGENICS; FRAGILE X SYNDROME; GENETIC DISCRIMINATION; QUANTITATIVE TRAITS; TWINS.

Harry Wright and Ruth Abramson

Bibliography

Casse, D. "IQ since 'The Bell Curve.'" *Commentary Magazine* 106, no. 2 (1998): 33–41.

Chiacchia, K. B. "Race and Intelligence." In *Encyclopedia of Psychology*, 2nd ed., Bonnie Strickland, ed. Farmington Hills, MI: Gale Group, 2001.

Deary, I. J. "Differences in Mental Ability." *British Medical Journal* 317 (1998): 1701–1703.

Fuller, J. L., and W. R. Thompson. "Cognitive and Intellectual Abilities." In *Foundations of Behavior Genetics.* St. Louis, MO: C.V. Mosby Co., 1978.

Plomin, R. "Genetics of Childhood Disorders, III: Genetics and Intelligence." *Journal of the American Academy of Childhood and Adolescent Psychiatry* 38 (1999): 786–788.

Sternberg, R. J., and J. C. Kaufman. "Human Abilities." *Annual Review Psychology* 49: 479–502.

Sternberg, R. J., and E. L. Grigorenko. "Genetics of Childhood Disorders, I: Genetics and Intelligence." *Journal of the American Academy of Childhood and Adolescent Psychiatry* 38 (1999): 486–488.

Internet

Biologists often use two terms to describe alternative approaches for conducting experiments. "In vitro" (Latin for "in glass") refers to experiments

typically carried out in test tubes with purified biochemicals. "In vivo" ("in life") experiments are performed directly on living organisms. In recent years, the indispensable use of computers and the Internet for genetic and molecular biology research has introduced a new term into the language: "in silico" ("in silicon"), referring to the silicon used to manufacture computer chips. In silico genetics experiments are those that are performed with a computer, often involving analysis of DNA or protein sequences over the Internet.

Geneticists and molecular biologists use the Internet much the same way most people do, communicating data and results through e-mail and discussion groups and sharing information on Web sites, for instance. They also make wide use of powerful Internet-based databases and analytical tools. Researchers are determining the DNA sequences of entire **genomes** at an ever accelerating pace, and are devising methods for cataloging entire sets of proteins (termed "proteomes") expressed in organisms. The databases to store all this information are growing at an equal pace, and the computer tools to sort through all the data are becoming increasingly sophisticated.

genomes the total genetic material in cells or organisms

One of the most important Web sites for biological computer analysis (sometimes called **bioinformatics**) is that of the National Center for Biotechnology Information (NCBI), a part of the National Library of Medicine, which, in turn, is part of the National Institutes of Health. The NCBI Web site hosts DNA and protein sequence databases, protein three-dimensional structure databases, scientific literature databases, and search engines for retrieving files of interest. All of these resources are freely accessible to anyone on the Internet.

bioinformatics use of information technology to analyze biological data

Of all the powerful analytical tools available at NCBI, probably the most important and heavily used is a set of computer programs called BLAST, for Basic Local Alignment Search Tool. BLAST can rapidly search many sequence databases to see whether any DNA or protein sequence (a "query sequence," supplied by the user) is similar to other sequences. Since sequence similarity usually suggests that two proteins or DNA molecules are homologous (i.e., that they are evolutionarily related and therefore may have—or encode proteins—with similar functions), discovering a blast match between an unknown protein or nucleic acid sequence and a well-characterized sequence provides an immediate clue about the function of the unknown sequence. An important scientific discovery that, in the past, may have taken many years of in vitro and in vivo analysis to arrive at is now made in a few seconds, with this simple in silico experiment. SEE ALSO BIOINFORMATICS; GENOME; GENOMICS; HOMOLOGY; PROTEOMICS; SEQUENCING DNA.

Paul J. Muhlrad

Bibliography

Internet Resources

Basic Local Alignment Search Tool. National Center for Biotechnology Information. <http://www.ncbi.nlm.nih.gov/BLAST/>.

Baxevanis, Andreas D. "The Molecular Biology Database Collection: 2002 Update." *Nucleic Acids Research.* Oxford University Press. <http://www3.oup.co.uk/nar/database/>.

ExPASy Molecular Biology Server. Swiss Institute of Bioinformatics. <http://ca.expasy.org/>.

Virtual Library of Genetics. U.S. Department of Energy. <http://www.ornl.gov/TechResources/Human_Genome/genetics.html>.

Wellcome Trust Sanger Institute. <http://www.sanger.ac.uk/>.

WWW Virtual Library: Model Organisms. George Manning. <http://ceolas.org/VL/mo/>.

Photo Credits

Unless noted below or within its caption, the illustrations and tables featured in *Genetics* were developed by Richard Robinson, and rendered by GGS Information Services. The photographs appearing in the text were reproduced by permission of the following sources:

Volume 1

Accelerated Aging: Progeria (p. 2), Photo courtesy of The Progeria Research Foundation, Inc. and the Barnett Family; *Aging and Life Span* (p. 8), Fisher, Leonard Everett, Mr.; *Agricultural Biotechnology* (p. 10), © Keren Su/Corbis; *Alzheimer's Disease* (p. 15), AP/Wide World Photos; *Antibiotic Resistance* (p. 27), © Hank Morgan/Science Photo Library, Photo Researchers, Inc.; *Apoptosis* (p. 32), © Microworks/Phototake; *Arabidopsis thaliana* (p. 34), © Steinmark/ Custom Medical Stock Photo; *Archaea* (p. 38), © Eurelios/Phototake; *Behavior* (p. 47), © Norbert Schafer/Corbis; *Bioinformatics* (p. 53), © T. Bannor/Custom Medical Stock Photo; *Bioremediation* (p. 60), Merjenburgh/ Greenpeace; *Bioremediation* (p. 61), AP/Wide World Photos; *Biotechnology and Genetic Engineering, History of* (p. 71), © Gianni Dagl Orti/Corbis; *Biotechnology: Ethical Issues* (p. 67), © AFP/Corbis; *Birth Defects* (p. 78), AP/Wide World Photos; *Birth Defects* (p. 80), © Siebert/ Custom Medical Stock Photo; *Blotting* (p. 88), © Custom Medical Stock Photo; *Breast Cancer* (p. 90), © Custom Medical Stock Photo; *Carcinogens* (p. 98), © Custom Medical Stock Photo; *Cardiovascular Disease* (p. 102), © B&B Photos/Custom Medical Stock Photo; *Cell, Eukaryotic* (p. 111), © Dennis Kunkel/ Phototake; *Chromosomal Aberrations* (p. 122), © Pergement, Ph.D./Custom Medical Stock Photo; *Chromosomal Banding* (p. 126), Courtesy of the Cytogenetics Laboratory, Indiana University School of Medicine; *Chromosomal Banding* (p. 127), Courtesy of the Cytogenetics Laboratory, Indiana University School of Medicine; *Chromosomal Banding* (p.128), Courtesy of the Cytogenetics Laboratory, Indiana University School of Medicine; *Chromosome, Eukaryotic* (p. 137), Photo Researchers, Inc.; *Chromosome, Eukaryotic* (p. 136), © Becker/Custom Medical Stock Photo; *Chromosome, Eukaryotic* (p. 133), Courtesy of Dr. Jeffrey Nickerson/University of Massachusetts Medical School; *Chromosome, Prokaryotic* (p. 141), © Mike Fisher/Custom Medical Stock Photo; *Chromosomes, Artificial* (p. 145), Courtesy of Dr. Huntington F. Williard/University Hospitals of Cleveland; *Cloning Organisms* (p. 163), © Dr.Yorgos Nikas/Phototake; *Cloning: Ethical Issues* (p. 159), AP/Wide World Photos; *College Professor* (p. 166), © Bob Krist/Corbis; *Colon Cancer* (p. 169), © Albert Tousson/Phototake; *Colon Cancer* (p. 167), © G-I Associates/Custom Medical Stock Photo; *Conjugation* (p. 183), © Dennis Kunkel/Phototake; *Conservation Geneticist* (p. 191), © Annie Griffiths Belt/ Corbis; *Delbrück, Max* (p. 204), Library of Congress; *Development, Genetic Control of* (p. 208), © JL Carson/Custom Medical Stock Photo; *DNA Microarrays* (p. 226), Courtesy of James Lund and Stuart Kim, Standford University; *DNA Profiling* (p. 234), AP/Wide World Photos; *DNA Vaccines* (p. 254), Penny Tweedie/Corbis-Bettmann; *Down Syndrome* (p. 257), © Custom Medical Stock Photo.

Volume 2

Embryonic Stem Cells (p. 4), Courtesy of Dr. Douglas Strathdee/University of Edinburgh, Department of Neuroscience; *Embryonic Stem Cells* (p. 5), Courtesy of Dr. Douglas Strathdee/University of Edinburgh, Department of Neuroscience; *Escherichia coli* (*E. coli* bacterium) (p. 10), © Custom Medical Stock Photo; *Eubacteria* (p. 14), © Scimat/Photo Researchers; *Eubacteria* (p. 12), © Dennis Kunkel/Phototake; *Eugenics* (p. 19), American Philosophical Society; *Evolution, Molecular* (p. 22), OAR/National Undersea Research Program (NURP)/National Oceanic and Atmospheric Administration; *Fertilization* (p. 34), © David M. Phillips/Photo Researchers, Inc.; *Founder Effect* (p. 37), © Michael S. Yamashita/Corbis; *Fragile X Syndrome* (p. 41), © Siebert/Custom Medical Stock Photo; *Fruit Fly:* Drosophila (p. 44), © David M. Phillips, Science Source/Photo Researchers, Inc.; *Gel Electrophoresis* (p. 46), © Custom Medical Stock Photo; *Gene Therapy* (p. 75), AP/Wide World; *Genetic Counseling* (p. 88), © Amethyst/Custom Medical Stock Photo; *Genetic Testing* (p. 98), © Department of Clinical Cytogenetics, Addenbrookes Hospital/Science Photo Library/Photo Researchers, Inc.; *Genetically Modified Foods* (p. 109), AP/Wide World; *Genome* (p. 113), Raphael Gaillarde/Getty Images; *Genomic Medicine* (p. 119), © AFP/Corbis; *Growth Disorders* (p. 131), Courtesy Dr. Richard Pauli/U. of Wisconsin, Madison, Clinical Genetics Center; *Hemoglobinopathies* (p. 137), © Roseman/Custom Medical Stock Photo; *Heterozygote Advantage* (p. 147), © Tania Midgley/Corbis; *HPLC: High-Performance Liquid Chromatography* (p. 167), © T. Bannor/Custom Medical Stock Photo; *Human Genome Project* (p. 175), © AFP/Corbis; *Human Genome Project* (p. 176), AP/Wide World; *Individual Genetic Variation* (p. 192), © A. Wilson/Custom Medical Stock Photo; *Individual Genetic Variation* (p. 191), © A. Lowrey/Custom Medical Stock Photo; *Inheritance, Extranuclear* (p. 196), © ISM/Phototake; *Inheritance Patterns* (p. 206), photograph by Norman Lightfoot/National Audubon Society Collection/Photo Researchers, Inc.; *Intelligence* (p. 208), AP/Wide World Photos.

Volume 3

Laboratory Technician (p. 2), Mark Tade/Getty Images; *Maize* (p. 9), Courtesy of Agricultural Research Service/USDA; *Marker Systems* (p. 16), Custom Medical Stock Photo; *Mass Spectrometry* (p. 19), Ian Hodgson/© Rueters New Media; *McClintock, Barbara* (p. 21), AP/Wide World Photos; *McKusick, Victor* (p. 23), The Alan Mason Chesney Medical Archives of The Johns Hopkins Medical Institutions; *Mendel, Gregor* (p. 30), Archive Photos, Inc.; *Metabolic Disease* (p. 38), AP/Wide World Photos; *Mitochondrial Diseases* (p. 53), Courtesy of Dr. Richard Haas/University of California, San Diego, Department of Neurosciences; *Mitosis* (p. 58), J. L. Carson/Custom Medical Stock Photo; *Model Organisms* (p. 61), © Frank Lane Picture Agency/Corbis; *Molecular Anthropology* (p. 66), © John Reader, Science Photo Library/PhotoResearchers, Inc.; *Molecular Biologist* (p. 71), AP/Wide World Photos; *Morgan, Thomas Hunt* (p. 73), © Bettmann/Corbis; *Mosaicism* (p. 78), Courtesy of Carolyn Brown/Department of Medical Genetics of University of British Columbia; *Muller, Hermann* (p. 80), Library of Congress; *Muscular Dystrophy* (p. 85), © Siebert/Custom Medical Stock Photo; *Muscular Dystrophy* (p. 84), © Custom Medical Stock Photo; *Nature of the Gene, History* (p. 103), Library of Congress; *Nature of the Gene, History* (p. 102), Archive Photos, Inc.; *Nomenclature* (p. 108), Courtesy of Center for Human Genetics/Duke University Medical Center; *Nondisjunction* (p. 110), © Gale Group; *Nucleotide* (p. 115), © Lagowski/Custom Medical Stock Photo; *Nucleus* (p. 120), © John T. Hansen, Ph.D./Phototake; *Oncogenes* (p. 129), Courtesy of National Cancer Institute; *Pharmacogenetics and Pharmacogenomics* (p. 145), AP/Wide World Photos; *Plant Genetic Engineer* (p. 150), © Lowell Georgia/Corbis; *Pleiotropy* (p. 154), © Custom Medical Stock Photo; *Polyploidy* (p. 165), AP/Wide World Photos; *Population Genetics* (p. 173), AP/Wide World Photos; *Population Genetics* (p. 172), © JLM Visuals; *Prenatal Diagnosis* (p. 184), © Richard T. Nowitz/Corbis; *Prenatal Diagnosis* (p. 186), © Brigham Narins; *Prenatal Diagnosis* (p. 183), © Dr. Yorgos Nikas/Phototake; *Prion* (p. 188), AP/Wide World Photos; *Prion* (p. 189), AP/

Wide World Photos; *Privacy* (p. 191), © K. Beebe/Custom Medical Stock Photo; *Proteins* (p. 199), NASA/Marshall Space Flight Center; *Proteomics* (p. 207), © Geoff Tompkinson/ Science Photo Library, National Aubodon Society Collection/Photo Researchers, Inc.; *Proteomics* (p. 206), Lagowski/Custom Medical Stock Photo.

Volume 4

Quantitative Traits (p. 2), Photo by Peter Morenus/Courtesy of University of Connecticut; *Reproductive Technology* (p. 20), © R. Rawlins/Custom Medical Stock Photo; *Reproductive Technology: Ethical Issues* (p. 27), AP/Wide World Photos; *Restriction Enzymes* (p. 32), © Gabridge/Custom Medical Stock Photo; *Ribosome* (p. 43), © Dennis Kunkel/ Phototake; *Rodent Models* (p. 61), © Reuters NewMedia Inc./Corbis; *Roundworm:* Caenorhabditis elegans (p. 63), © J. L. Carson/Custom Medical Stock Photo; *Sanger, Fred* (p. 65), © Bettmann/Corbis; *Sequencing DNA* (p. 74), © T. Bannor/Custom Medical Stock Photo; *Severe Combined Immune Deficiency* (p. 76), © Bettmann/Corbis; *Tay-Sachs Disease* (p. 100), © Dr. Charles J. Ball/ Corbis; *Transgenic Animals* (p. 125), Courtesy Cindy McKinney, Ph.D./Penn State University's Transgenic Mouse Facility; *Transgenic Organisms: Ethical Issues* (p. 131), © Daymon Hartley/Greenpeace; *Transgenic Plants* (p. 133), © Eurelios/Phototake; *Transplantation* (p. 140), © Reuters New Media/Corbis; *Twins* (p. 156), © Dennis Degnan/Corbis; *X Chromosome* (p. 175), © Gale Group; *Yeast* (p. 180), © Dennis Kunkel; *Zebrafish* (p. 182), Courtesy of Dr. Jordan Shin, Cardiovascular Research Center, Massachusetts General Hospital.

Glossary

α the Greek letter alpha

β the Greek letter beta

γ the Greek letter gamma

λ the Greek letter lambda

σ the Greek letter sigma

E. coli the bacterium *Escherichia coli*

"-ase" suffix indicating an enzyme

acidic having the properties of an acid; the opposite of basic

acrosomal cap tip of sperm cell that contains digestive enzymes for penetrating the egg

adenoma a tumor (cell mass) of gland cells

aerobic with oxygen, or requiring it

agar gel derived from algae

agglutinate clump together

aggregate stick together

algorithm procedure or set of steps

allele a particular form of a gene

allelic variation presence of different gene forms (alleles) in a population

allergen substance that triggers an allergic reaction

allolactose "other lactose"; a modified form of lactose

amino acid a building block of protein

amino termini the ends of a protein chain with a free NH_2 group

amniocentesis removal of fluid from the amniotic sac surrounding a fetus, for diagnosis

amplify produce many copies of, multiply

anabolic steroids hormones used to build muscle mass

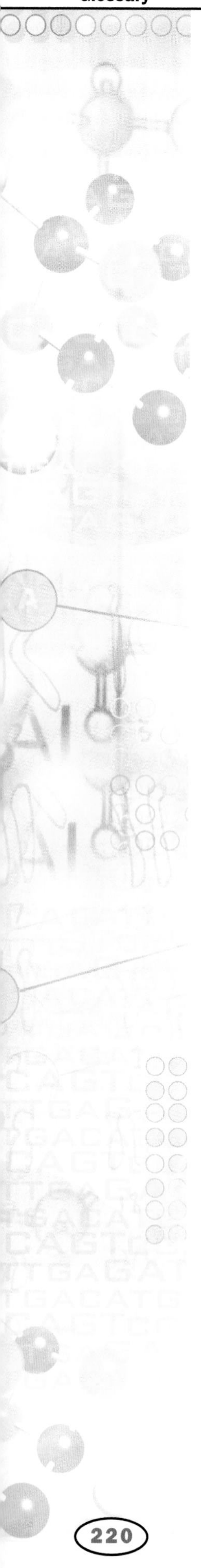

anaerobic without oxygen or not requiring oxygen

androgen testosterone or other masculinizing hormone

anemia lack of oxygen-carrying capacity in the blood

aneuploidy abnormal chromosome numbers

angiogenesis growth of new blood vessels

anion negatively charged ion

anneal join together

anode positive pole

anterior front

antibody immune-system protein that binds to foreign molecules

antidiuretic a substance that prevents water loss

antigen a foreign substance that provokes an immune response

antigenicity ability to provoke an immune response

apoptosis programmed cell death

Archaea one of three domains of life, a type of cell without a nucleus

archaeans members of one of three domains of life, have types of cells without a nucleus

aspirated removed with a needle and syringe

aspiration inhalation of fluid or solids into the lungs

association analysis estimation of the relationship between alleles or genotypes and disease

asymptomatic without symptoms

ATP adenosine triphosphate, a high-energy compound used to power cell processes

ATPase an enzyme that breaks down ATP, releasing energy

attenuation weaken or dilute

atypical irregular

autoimmune reaction of the immune system to the body's own tissues

autoimmunity immune reaction to the body's own tissues

autosomal describes a chromosome other than the X and Y sex-determining chromosomes

autosome a chromosome that is not sex-determining (not X or Y)

axon the long extension of a nerve cell down which information flows

bacteriophage virus that infects bacteria

basal lowest level

base pair two nucleotides (either DNA or RNA) linked by weak bonds

basic having the properties of a base; opposite of acidic

benign type of tumor that does not invade surrounding tissue

binding protein protein that binds to another molecule, usually either DNA or protein

biodiversity degree of variety of life

bioinformatics use of information technology to analyze biological data

biolistic firing a microscopic pellel into a biological sample (from biological/ ballistic)

biopolymers biological molecules formed from similar smaller molecules, such as DNA or protein

biopsy removal of tissue sample for diagnosis

biotechnology production of useful products

bipolar disorder psychiatric disease characterized by alternating mania and depression

blastocyst early stage of embryonic development

brackish a mix of salt water and fresh water

breeding analysis analysis of the offspring ratios in breeding experiments

buffers substances that counteract rapid or wide pH changes in a solution

Cajal Ramon y Cajal, Spanish neuroanatomist

carcinogens substances that cause cancer

carrier a person with one copy of a gene for a recessive trait, who therefore does not express the trait

catalyst substance that speeds a reaction without being consumed (e.g., enzyme)

catalytic describes a substance that speeds a reaction without being consumed

catalyze aid in the reaction of

cathode negative pole

cDNA complementary DNA

cell cycle sequence of growth, replication and division that produces new cells

centenarian person who lives to age 100

centromere the region of the chromosome linking chromatids

cerebrovascular related to the blood vessels in the brain

cerebrovascular disease stroke, aneurysm, or other circulatory disorder affecting the brain

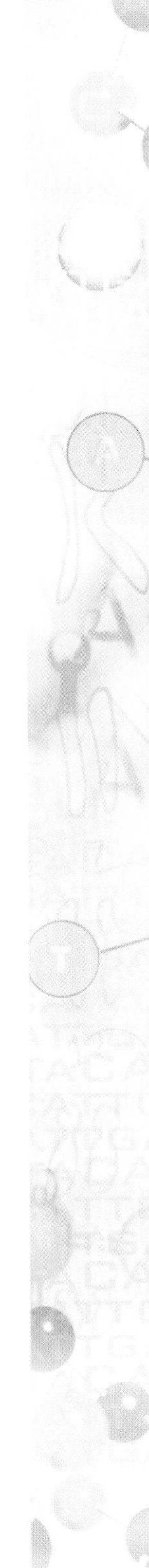

charge density ratio of net charge on the protein to its molecular mass

chemotaxis movement of a cell stimulated by a chemical attractant or repellent

chemotherapeutic use of chemicals to kill cancer cells

chloroplast the photosynthetic organelle of plants and algae

chondrocyte a cell that forms cartilage

chromatid a replicated chromosome before separation from its copy

chromatin complex of DNA, histones, and other proteins, making up chromosomes

ciliated protozoa single-celled organism possessing cilia, short hair-like extensions of the cell membrane

circadian relating to day or day length

cleavage hydrolysis

cleave split

clinical trials tests performed on human subjects

codon a sequence of three mRNA nucleotides coding for one amino acid

Cold War prolonged U.S.-Soviet rivalry following World War II

colectomy colon removal

colon crypts part of the large intestine

complementary matching opposite, like hand and glove

conformation three-dimensional shape

congenital from birth

conjugation a type of DNA exchange between bacteria

cryo-electron microscope electron microscope that integrates multiple images to form a three-dimensional model of the sample

cryopreservation use of very cold temperatures to preserve a sample

cultivars plant varieties resulting from selective breeding

cytochemist chemist specializing in cellular chemistry

cytochemistry cellular chemistry

cytogenetics study of chromosome structure and behavior

cytologist a scientist who studies cells

cytokine immune system signaling molecule

cytokinesis division of the cell's cytoplasm

cytology the study of cells

cytoplasm the material in a cell, excluding the nucleus

cytosol fluid portion of a cell, not including the organelles

de novo entirely new

deleterious harmful

dementia neurological illness characterized by impaired thought or awareness

demography aspects of population structure, including size, age distribution, growth, and other factors

denature destroy the structure of

deoxynucleotide building block of DNA

dimerize linkage of two subunits

dimorphism two forms

diploid possessing pairs of chromosomes, one member of each pair derived from each parent

disaccharide two sugar molecules linked together

dizygotic fraternal or nonidentical

DNA deoxyribonucleic acid

domains regions

dominant controlling the phenotype when one allele is present

dopamine brain signaling chemical

dosage compensation equalizing of expression level of X-chromosome genes between males and females, by silencing one X chromosome in females or amplifying expression in males

ecosystem an ecological community and its environment

ectopic expression expression of a gene in the wrong cells or tissues

electrical gradient chemiosmotic gradient

electrophoresis technique for separation of molecules based on size and charge

eluting exiting

embryogenesis development of the embryo from a fertilized egg

endangered in danger of extinction throughout all or a significant portion of a species' range

endogenous derived from inside the organism

endometriosis disorder of the endometrium, the lining of the uterus

endometrium uterine lining

endonuclease enzyme that cuts DNA or RNA within the chain

endoplasmic reticulum network of membranes within the cell

endoscope tool used to see within the body

endoscopic describes procedure wherein a tool is used to see within the body

endosymbiosis symbiosis in which one partner lives within the other

enzyme a protein that controls a reaction in a cell

epidemiologic the spread of diseases in a population

epidemiologists people who study the incidence and spread of diseases in a population

epidemiology study of incidence and spread of diseases in a population

epididymis tube above the testes for storage and maturation of sperm

epigenetic not involving DNA sequence change

epistasis suppression of a characteristic of one gene by the action of another gene

epithelial cells one of four tissue types found in the body, characterized by thin sheets and usually serving a protective or secretory function

Escherichia coli common bacterium of the human gut, used in research as a model organism

estrogen female horomone

et al. "and others"

ethicists a person who writes and speaks about ethical issues

etiology causation of disease, or the study of causation

eubacteria one of three domains of life, comprising most groups previously classified as bacteria

eugenics movement to "improve" the gene pool by selective breeding

eukaryote organism with cells possessing a nucleus

eukaryotic describing an organism that has cells containing nuclei

ex vivo outside a living organism

excise remove; cut out

excision removal

exogenous from outside

exon coding region of genes

exonuclease enzyme that cuts DNA or RNA at the end of a strand

expression analysis whole-cell analysis of gene expression (use of a gene to create its RNA or protein product)

fallopian tubes tubes through which eggs pass to the uterus

fermentation biochemical process of sugar breakdown without oxygen

fibroblast undifferentiated cell normally giving rise to connective tissue cells

fluorophore fluorescent molecule

forensic related to legal proceedings

founder population

fractionated purified by separation based on chemical or physical properties

fraternal twins dizygotic twins who share 50 percent of their genetic material

frontal lobe one part of the forward section of the brain, responsible for planning, abstraction, and aspects of personality

gamete reproductive cell, such as sperm or egg

gastrulation embryonic stage at which primitive gut is formed

gel electrophoresis technique for separation of molecules based on size and charge

gene expression use of a gene to create the corresponding protein

genetic code the relationship between RNA nucleotide triplets and the amino acids they cause to be added to a growing protein chain

genetic drift evolutionary mechanism, involving random change in gene frequencies

genetic predisposition increased risk of developing diseases

genome the total genetic material in a cell or organism

genomics the study of gene sequences

genotype set of genes present

geothermal related to heat sources within Earth

germ cell cell creating eggs or sperm

germ-line cells giving rise to eggs or sperm

gigabase one billion bases (of DNA)

glucose sugar

glycolipid molecule composed of sugar and fatty acid

glycolysis the breakdown of the six-carbon carbohydrates glucose and fructose

glycoprotein protein to which sugars are attached

Golgi network system in the cell for modifying, sorting, and delivering proteins

gonads testes or ovaries

gradient a difference in concentration between two regions

Gram negative bacteria bacteria that do not take up Gram stain, due to membrane structure

Gram positive able to take up Gram stain, used to classify bacteria

gynecomastia excessive breast development in males

haploid possessing only one copy of each chromosome

haplotype set of alleles or markers on a short chromosome segment

hematopoiesis formation of the blood

hematopoietic blood-forming

heme iron-containing nitrogenous compound found in hemoglobin

hemolysis breakdown of the blood cells

hemolytic anemia blood disorder characterized by destruction of red blood cells

hemophiliacs a person with hemophilia, a disorder of blood clotting

herbivore plant eater

heritability proportion of variability due to genes; ability to be inherited

heritability estimates how much of what is observed is due to genetic factors

heritable genetic

heterochromatin condensed portion of chromosomes

heterozygote an individual whose genetic information contains two different forms (alleles) of a particular gene

heterozygous characterized by possession of two different forms (alleles) of a particular gene

high-throughput rapid, with the capacity to analyze many samples in a short time

histological related to tissues

histology study of tissues

histone protein around which DNA winds in the chromosome

homeostasis maintenance of steady state within a living organism

homologous carrying similar genes

homologues chromosomes with corresponding genes that pair and exchange segments in meiosis

homozygote an individual whose genetic information contains two identical copies of a particular gene

homozygous containing two identical copies of a particular gene

hormones molecules released by one cell to influence another

hybrid combination of two different types

hybridization (molecular) base-pairing among DNAs or RNAs of different origins

hybridize to combine two different species

hydrogen bond weak bond between the H of one molecule or group and a nitrogen or oxygen of another

hydrolysis splitting with water

hydrophilic "water-loving"

hydrophobic "water hating," such as oils

hydrophobic interaction attraction between portions of a molecule (especially a protein) based on mutual repulsion of water

hydroxyl group chemical group consisting of -OH

hyperplastic cell cell that is growing at an increased rate compared to normal cells, but is not yet cancerous

hypogonadism underdeveloped testes or ovaries

hypothalamus brain region that coordinates hormone and nervous systems

hypothesis testable statement

identical twins monozygotic twins who share 100 percent of their genetic material

immunogenicity likelihood of triggering an immune system defense

immunosuppression suppression of immune system function

immunosuppressive describes an agent able to suppress immune system function

in vitro "in glass"; in lab apparatus, rather than within a living organism

in vivo "in life"; in a living organism, rather than in a laboratory apparatus

incubating heating to optimal temperature for growth

informed consent knowledge of risks involved

insecticide substance that kills insects

interphase the time period between cell divisions

intra-strand within a strand

intravenous into a vein

intron untranslated portion of a gene that interrupts coding regions

karyotype the set of chromosomes in a cell, or a standard picture of the chromosomes

kilobases units of measure of the length of a nucleicacid chain; one kilobase is equal to 1,000 base pairs

kilodalton a unit of molecular weight, equal to the weight of 1000 hydrogen atoms

kinase an enzyme that adds a phosphate group to another molecule, usually a protein

knocking out deleting of a gene or obstructing gene expression

laparoscope surgical instrument that is inserted through a very small incision, usually guided by some type of imaging technique

latent present or potential, but not apparent

lesion damage

ligand a molecule that binds to a receptor or other molecule

ligase enzyme that repairs breaks in DNA

ligate join together

linkage analysis examination of co-inheritance of disease and DNA markers, used to locate disease genes

lipid fat or wax-like molecule, insoluble in water

loci/locus site(s) on a chromosome

longitudinally lengthwise

lumen the space within the tubes of the endoplasmic reticulum

lymphocytes white blood cells

lyse break apart

lysis breakage

macromolecular describes a large molecule, one composed of many similar parts

macromolecule large molecule such as a protein, a carbohydrate, or a nucleic acid

macrophage immune system cell that consumes foreign material and cellular debris

malignancy cancerous tissue

malignant cancerous; invasive tumor

media (bacteria) nutrient source

meiosis cell division that forms eggs or sperm

melanocytes pigmented cells

meta-analysis analysis of combined results from multiple clinical trials

metabolism chemical reactions within a cell

metabolite molecule involved in a metabolic pathway

metaphase stage in mitosis at which chromosomes are aligned along the cell equator

metastasis breaking away of cancerous cells from the initial tumor

metastatic cancerous cells broken away from the initial tumor

methylate add a methyl group to

methylated a methyl group, CH_3, added

methylation addition of a methyl group, CH_3

microcephaly reduced head size

microliters one thousandth of a milliliter

micrometer 1/1000 meter

microsatellites small repetitive DNA elements dispersed throughout the genome

microtubule protein strands within the cell, part of the cytoskeleton

miscegenation racial mixing

mitochondria energy-producing cell organelle

mitogen a substance that stimulates mitosis

mitosis separation of replicated chromosomes

molecular hybridization base-pairing among DNAs or RNAs of different origins

molecular systematics the analysis of DNA and other molecules to determine evolutionary relationships

monoclonal antibodies immune system proteins derived from a single B cell

monomer "single part"; monomers are joined to form a polymer

monosomy gamete that is missing a chromosome

monozygotic genetically identical

morphologically related to shape and form

morphology related to shape and form

mRNA messenger RNA

mucoid having the properties of mucous

mucosa outer covering designed to secrete mucus, often found lining cavities and internal surfaces

mucous membranes nasal passages, gut lining, and other moist surfaces lining the body

multimer composed of many similar parts

multinucleate having many nuclei within a single cell membrane

mutagen any substance or agent capable of causing a change in the structure of DNA

mutagenesis creation of mutations

mutation change in DNA sequence

nanometer 10^{-9}(exp) meters; one billionth of a meter

nascent early-stage

necrosis cell death from injury or disease

nematode worm of the Nematoda phylum, many of which are parasitic

neonatal newborn

neoplasms new growths

neuroimaging techniques for making images of the brain

neurological related to brain function or disease

neuron nerve cell

neurotransmitter molecule released by one neuron to stimulate or inhibit a neuron or other cell

non-polar without charge separation; not soluble in water

normal distribution distribution of data that graphs as a bell-shaped curve

Northern blot a technique for separating RNA molecules by electrophoresis and then identifying a target fragment with a DNA probe

Northern blotting separating RNA molecules by electrophoresis and then identifying a target fragment with a DNA probe

nuclear DNA DNA contained in the cell nucleus on one of the 46 human chromosomes; distinct from DNA in the mitochondria

nuclear membrane membrane surrounding the nucleus

nuclease enzyme that cuts DNA or RNA

nucleic acid DNA or RNA

nucleoid region of the bacterial cell in which DNA is located

nucleolus portion of the nucleus in which ribosomes are made

nucleoplasm material in the nucleus

nucleoside building block of DNA or RNA, composed of a base and a sugar

nucleoside triphosphate building block of DNA or RNA, composed of a base and a sugar linked to three phosphates

nucleosome chromosome structural unit, consisting of DNA wrapped around histone proteins

nucleotide a building block of RNA or DNA

ocular related to the eye

oncogene gene that causes cancer

oncogenesis the formation of cancerous tumors

oocyte egg cell

open reading frame DNA sequence that can be translated into mRNA; from start sequence to stop sequence

opiate opium, morphine, and related compounds

organelle membrane-bound cell compartment

organic composed of carbon, or derived from living organisms; also, a type of agriculture stressing soil fertility and avoidance of synthetic pesticides and fertilizers

osmotic related to differences in concentrations of dissolved substances across a permeable membrane

ossification bone formation

osteoarthritis a degenerative disease causing inflammation of the joints

osteoporosis thinning of the bone structure

outcrossing fertilizing between two different plants

oviduct a tube that carries the eggs

ovulation release of eggs from the ovaries

ovules eggs

ovum egg

oxidation chemical process involving reaction with oxygen, or loss of electrons

oxidized reacted with oxygen

pandemic disease spread throughout an entire population

parasites organisms that live in, with, or on another organism

pathogen disease-causing organism

pathogenesis pathway leading to disease

pathogenic disease-causing

pathogenicity ability to cause disease

pathological altered or changed by disease

pathology disease process

pathophysiology disease process

patient advocate a person who safeguards patient rights or advances patient interests

PCR polymerase chain reaction, used to amplify DNA

pedigrees sets of related individuals, or the graphic representation of their relationships

peptide amino acid chain

peptide bond bond between two amino acids

percutaneous through the skin

phagocytic cell-eating

phenotype observable characteristics of an organism

phenotypic related to the observable characteristics of an organism

pheromone molecule released by one organism to influence another organism's behavior

phosphate group PO_4 group, whose presence or absence often regulates protein action

phosphodiester bond the link between two nucleotides in DNA or RNA

phosphorylating addition of phosphate group (PO_4)

phosphorylation addition of the phosphate group PO_4^{3-}

phylogenetic related to the evolutionary development of a species

phylogeneticists scientists who study the evolutionary development of a species

phylogeny the evolutionary development of a species

plasma membrane outer membrane of the cell

plasmid a small ring of DNA found in many bacteria

plastid plant cell organelle, including the chloroplast

pleiotropy genetic phenomenon in which alteration of one gene leads to many phenotypic effects

point mutation gain, loss, or change of one to several nucleotides in DNA

polar partially charged, and usually soluble in water

pollen male plant sexual organ

polymer molecule composed of many similar parts

polymerase enzyme complex that synthesizes DNA or RNA from individual nucleotides

polymerization linking together of similar parts to form a polymer

polymerize to link together similar parts to form a polymer

polymers molecules composed of many similar parts

polymorphic occurring in several forms

polymorphism DNA sequence variant

polypeptide chain of amino acids

polyploidy presence of multiple copies of the normal chromosome set

population studies collection and analysis of data from large numbers of people in a population, possibly including related individuals

positional cloning the use of polymorphic genetic markers ever closer to the unknown gene to track its inheritance in CF families

posterior rear

prebiotic before the origin of life

precursor a substance from which another is made

prevalence frequency of a disease or condition in a population

primary sequence the sequence of amino acids in a protein; also called primary structure

primate the animal order including humans, apes, and monkeys

primer short nucleotide sequence that helps begin DNA replication

primordial soup hypothesized prebiotic environment rich in life's building blocks

probe molecule used to locate another molecule

procarcinogen substance that can be converted into a carcinogen, or cancer-causing substance

procreation reproduction

progeny offspring

prokaryote a single-celled organism without a nucleus

promoter DNA sequence to which RNA polymerase binds to begin transcription

promutagen substance that, when altered, can cause mutations

pronuclei egg and sperm nuclei before they fuse during fertilization

proprietary exclusively owned; private

proteomic derived from the study of the full range of proteins expressed by a living cell

proteomics the study of the full range of proteins expressed by a living cell

protists single-celled organisms with cell nuclei

protocol laboratory procedure

protonated possessing excess H^+ ions; acidic

pyrophosphate free phosphate group in solution

quiescent non-dividing

radiation high energy particles or waves capable of damaging DNA, including X rays and gamma rays

recessive requiring the presence of two alleles to control the phenotype

recombinant DNA DNA formed by combining segments of DNA, usually from different types of organisms

recombining exchanging genetic material

replication duplication of DNA

restriction enzyme an enzyme that cuts DNA at a particular sequence

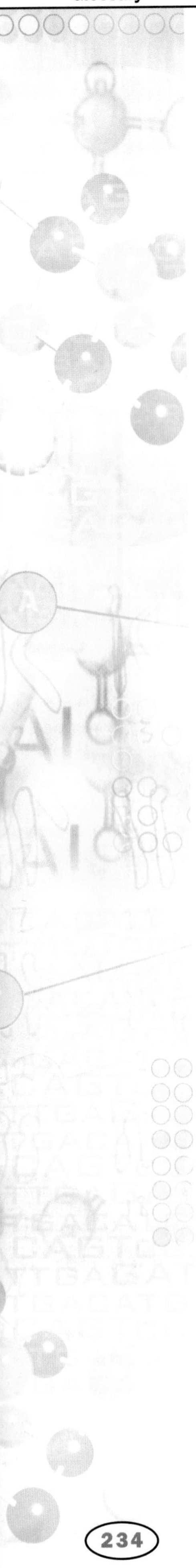

retina light-sensitive layer at the rear of the eye

retroviruses RNA-containing viruses whose genomes are copied into DNA by the enzyme reverse transcriptase

reverse transcriptase enzyme that copies RNA into DNA

ribonuclease enzyme that cuts RNA

ribosome protein-RNA complex at which protein synthesis occurs

ribozyme RNA-based catalyst

RNA ribonucleic acid

RNA polymerase enzyme complex that creates RNA from DNA template

RNA triplets sets of three nucleotides

salinity of, or relating to, salt

sarcoma a type of malignant (cancerous) tumor

scanning electron microscope microscope that produces images with depth by bouncing electrons off the surface of the sample

sclerae the "whites" of the eye

scrapie prion disease of sheep and goats

segregation analysis statistical test to determine pattern of inheritance for a trait

senescence a state in a cell in which it will not divide again, even in the presence of growth factors

senile plaques disease

serum (pl. sera) fluid portion of the blood

sexual orientation attraction to one sex or the other

somatic nonreproductive; not an egg or sperm

Southern blot a technique for separating DNA fragments by electrophoresis and then identifying a target fragment with a DNA probe

Southern blotting separating DNA fragments by electrophoresis and then identifying a target fragment with a DNA probe

speciation the creation of new species

spindle football-shaped structure that separates chromosomes in mitosis

spindle fiber protein chains that separate chromosomes during mitosis

spliceosome RNA-protein complex that removes introns from RNA transcripts

spontaneous non-inherited

sporadic caused by new mutations

stem cell cell capable of differentiating into multiple other cell types

stigma female plant sexual organ

stop codon RNA triplet that halts protein synthesis

striatum part of the midbrain

subcutaneous under the skin

sugar glucose

supercoiling coiling of the helix

symbiont organism that has a close relationship (symbiosis) with another

symbiosis a close relationship between two species in which at least one benefits

symbiotic describes a close relationship between two species in which at least one benefits

synthesis creation

taxon/taxa level(s) of classification, such as kingdom or phylum

taxonomical derived from the science that identifies and classifies plants and animals

taxonomist a scientist who identifies and classifies organisms

telomere chromosome tip

template a master copy

tenets generally accepted beliefs

terabyte a trillion bytes of data

teratogenic causing birth defects

teratogens substances that cause birth defects

thermodynamics process of energy transfers during reactions, or the study of these processes

threatened likely to become an endangered species

topological describes spatial relations, or the study of these relations

topology spatial relations, or the study of these relations

toxicological related to poisons and their effects

transcript RNA copy of a gene

transcription messenger RNA formation from a DNA sequence

transcription factor protein that increases the rate of transcription of a gene

transduction conversion of a signal of one type into another type

transgene gene introduced into an organism

transgenics transfer of genes from one organism into another

translation synthesis of protein using mRNA code

translocation movement of chromosome segment from one chromosome to another

transposable genetic element DNA sequence that can be copied and moved in the genome

transposon genetic element that moves within the genome

trilaminar three-layer

triploid possessing three sets of chromosomes

trisomics mutants with one extra chromosome

trisomy presence of three, instead of two, copies of a particular chromosome

tumor mass of undifferentiated cells; may become cancerous

tumor suppressor genes cell growths

tumors masses of undifferentiated cells; may become cancerous

vaccine protective antibodies

vacuole cell structure used for storage or related functions

van der Waal's forces weak attraction between two different molecules

vector carrier

vesicle membrane-bound sac

virion virus particle

wet lab laboratory devoted to experiments using solutions, cell cultures, and other "wet" substances

wild-type most common form of a trait in a population

Wilm's tumor a cancerous cell mass of the kidney

X ray crystallography use of X rays to determine the structure of a molecule

xenobiotic foreign biological molecule, especially a harmful one

zygote fertilized egg

Topic Outline

APPLICATIONS TO OTHER FIELDS

Agricultural Biotechnology
Biopesticides
Bioremediation
Biotechnology
Conservation Biology: Genetic Approaches
DNA Profiling
Genetically Modified Foods
Molecular Anthropology
Pharmacogenetics and Pharmacogenomics
Plant Genetic Engineer
Public Health, Genetic Techniques in
Transgenic Animals
Transgenic Microorganisms
Transgenic Plants

BACTERIAL GENETICS

Escherichia coli (*E. coli* bacterium)
Ames Test
Antibiotic Resistance
Chromosome, Prokaryotic
Cloning Genes
Conjugation
Eubacteria
Microbiologist
Overlapping Genes
Plasmid
Transduction
Transformation
Transgenic Microorganisms
Transgenic Organisms: Ethical Issues
Viroids and Virusoids
Virus

BASIC CONCEPTS

Biotechnology
Crossing Over
Disease, Genetics of
DNA
DNA Structure and Function, History
Fertilization
Gene
Genetic Code
Genetics
Genome
Genotype and Phenotype
Homology
Human Genome Project
Inheritance Patterns
Meiosis
Mendelian Genetics
Mitosis
Mutation
Nucleotide
Plasmid
Population Genetics
Proteins
Recombinant DNA
Replication
RNA
Transcription
Translation

BIOTECHNOLOGY

Agricultural Biotechnology
Biopesticides
Bioremediation
Biotechnology
Biotechnology and Genetic Engineering, History
Biotechnology: Ethical Issues
Cloning Genes
Cloning Organisms
DNA Vaccines

Genetically Modified Foods
HPLC: High-Performance Liquid Chromatography
Pharmaceutical Scientist
Plant Genetic Engineer
Polymerase Chain Reaction
Recombinant DNA
Restriction Enzymes
Reverse Transcriptase
Transgenic Animals
Transgenic Microorganisms
Transgenic Organisms: Ethical Issues
Transgenic Plants

CAREERS

Attorney
Bioinformatics
Clinical Geneticist
College Professor
Computational Biologist
Conservation Geneticist
Educator
Epidemiologist
Genetic Counselor
Geneticist
Genomics Industry
Information Systems Manager
Laboratory Technician
Microbiologist
Molecular Biologist
Pharmaceutical Scientist
Physician Scientist
Plant Genetic Engineer
Science Writer
Statistical Geneticist
Technical Writer

CELL CYCLE

Apoptosis
Balanced Polymorphism
Cell Cycle
Cell, Eukaryotic
Centromere
Chromosome, Eukaryotic
Chromosome, Prokaryotic
Crossing Over
DNA Polymerases
DNA Repair
Embryonic Stem Cells
Eubacteria
Inheritance, Extranuclear
Linkage and Recombination
Meiosis
Mitosis
Oncogenes
Operon
Polyploidy
Replication
Signal Transduction
Telomere
Tumor Suppressor Genes

CLONED OR TRANSGENIC ORGANISMS

Agricultural Biotechnology
Biopesticides
Biotechnology
Biotechnology: Ethical Issues
Cloning Organisms
Cloning: Ethical Issues
Gene Targeting
Model Organisms
Patenting Genes
Reproductive Technology
Reproductive Technology: Ethical Issues
Rodent Models
Transgenic Animals
Transgenic Microorganisms
Transgenic Organisms: Ethical Issues
Transgenic Plants

DEVELOPMENT, LIFE CYCLE, AND NORMAL HUMAN VARIATION

Aging and Life Span
Behavior
Blood Type
Color Vision
Development, Genetic Control of
Eye Color
Fertilization
Genotype and Phenotype
Hormonal Regulation
Immune System Genetics
Individual Genetic Variation
Intelligence
Mosaicism
Sex Determination
Sexual Orientation
Twins
X Chromosome
Y Chromosome

DNA, GENE AND CHROMOSOME STRUCTURE

Antisense Nucleotides
Centromere
Chromosomal Banding
Chromosome, Eukaryotic
Chromosome, Prokaryotic
Chromosomes, Artificial
DNA
DNA Repair
DNA Structure and Function, History
Evolution of Genes
Gene
Genome
Homology
Methylation
Multiple Alleles
Mutation
Nature of the Gene, History
Nomenclature
Nucleotide
Overlapping Genes
Plasmid
Polymorphisms
Pseudogenes
Repetitive DNA Elements
Telomere
Transposable Genetic Elements
X Chromosome
Y Chromosome

DNA TECHNOLOGY

In situ Hybridization
Antisense Nucleotides
Automated Sequencer
Blotting
Chromosomal Banding
Chromosomes, Artificial
Cloning Genes
Cycle Sequencing
DNA Footprinting
DNA Libraries
DNA Microarrays
DNA Profiling
Gel Electrophoresis
Gene Targeting
HPLC: High-Performance Liquid Chromatography
Marker Systems
Mass Spectrometry
Mutagenesis
Nucleases
Polymerase Chain Reaction
Protein Sequencing
Purification of DNA
Restriction Enzymes
Ribozyme
Sequencing DNA

ETHICAL, LEGAL, AND SOCIAL ISSUES

Attorney
Biotechnology and Genetic Engineering, History
Biotechnology: Ethical Issues
Cloning: Ethical Issues
DNA Profiling
Eugenics
Gene Therapy: Ethical Issues
Genetic Discrimination
Genetic Testing: Ethical Issues
Legal Issues
Patenting Genes
Privacy
Reproductive Technology: Ethical Issues
Transgenic Organisms: Ethical Issues

GENE DISCOVERY

Ames Test
Bioinformatics
Complex Traits
Gene and Environment
Gene Discovery
Gene Families
Genomics
Human Disease Genes, Identification of
Human Genome Project
Mapping

GENE EXPRESSION AND REGULATION

Alternative Splicing
Antisense Nucleotides
Chaperones
DNA Footprinting
Gene
Gene Expression: Overview of Control
Genetic Code
Hormonal Regulation
Imprinting
Methylation
Mosaicism
Nucleus

GENETIC DISORDERS

GENETIC MEDICINE: DIAGNOSIS, TESTING, AND TREATMENT

GENOMES

GENOMICS, PROTEOMICS, AND BIOINFORMATICS

HISTORY

INHERITANCE

MODEL ORGANISMS

MUTATION

ORGANISMS, CELL TYPES, VIRUSES

Archaea
Cell, Eukaryotic
Eubacteria
Evolution, Molecular
Fruit Fly: *Drosophila*
HIV
Maize
Model Organisms
Nucleus
Prion
Retrovirus
Rodent Models
Roundworm: *Caenorhabditis elegans*
Signal Transduction
Viroids and Virusoids
Virus
Yeast
Zebrafish

POPULATION GENETICS AND EVOLUTION

Antibiotic Resistance
Balanced Polymorphism
Conservation Biologist
Conservation Biology: Genetic Approaches
Evolution of Genes
Evolution, Molecular
Founder Effect
Gene Flow
Genetic Drift
Hardy-Weinberg Equilibrium
Heterozygote Advantage
Inbreeding
Individual Genetic Variation
Molecular Anthropology
Population Bottleneck
Population Genetics
Population Screening
Selection
Speciation

RNA

Antisense Nucleotides
Blotting
DNA Libraries
Genetic Code
HIV
Nucleases
Nucleotide
Reading Frame
Retrovirus
Reverse Transcriptase
Ribosome
Ribozyme
RNA
RNA Interference
RNA Polymerases
RNA Processing
Transcription
Translation

Volume 2 Index

Page numbers in **boldface type** indicate article titles; those in *italic type* indicate illustrations. A cumulative index, which combines the terms in all volumes of *Genetics*, can be found in volume 4 of this series.

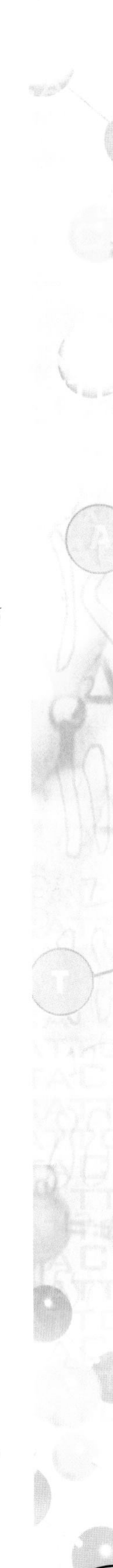

B

C

D

E

H

I

J

K

L

N

Q

R

S

U

V